AF796973
WITHDRAWN

Temporal and spatial patterns of vegetation dynamics

Advances in vegetation science 9

Edited by
EDDY VAN DER MAAREL

Temporal and spatial patterns of vegetation dynamics

Edited by
J. MILES
W. SCHMIDT and
E. VAN DER MAAREL

Reprinted from Vegetatio, volume 77

Kluwer Academic Publishers
DORDRECHT/BOSTON/LONDON

QK
910
.T45
1989

ISBN 0–7923–0103–X

Published by Kluwer Academic Publishers,
P.O. Box 17, 3300 AA Dordrecht, The Netherlands.

Kluwer Academic Publishers incorporates
the publishing programmes of
D. Reidel, Martinus Nijhoff, Dr W. Junk and MTP Press.

Sold and distributed in the U.S.A. and Canada
by Kluwer Academic Publishers,
101 Philip Drive, Norwell, MA 02061, U.S.A.

In all other countries, sold and distributed
by Kluwer Academic Publishers Group,
P.O. Box 322, 3300 AH Dordrecht, The Netherlands.

All Rights Reserved
© 1989 by Kluwer Academic Publishers
No part of the material protected by this copyright notice may be reproduced or utilized in any form or by any means, electronic or mechanical including photocopying, recording or by any information storage and retrieval system, without written permission from the copyright owner.

Printed in the Netherlands

Contents

Vegetatio 77: 7–19, 1988
© Kluwer Academic Publishers, Dordrecht – Printed in the Netherlands

Vegetation dynamics: patterns in time and space

Eddy van der Maarel
Institute of Ecological Botany, Uppsala University, Box 559, 75122 Uppsala, Sweden

Accepted 6.9.1988

Keywords: Fluctuation, Gap dynamics, Island, Isolation, Patch dynamics, Primary succession, Regeneration succession, Secondary succession, Secular succession

Abstract

This paper introduces the collection of contributions in this special volume on temporal and spatial patterns of vegetation dynamics. First, it is pointed out that the dynamics of any piece of vegetation, large or small, is always dependent on the degree of isolation of that piece towards its environment. Then ten types of island situation are treated ranging from very much to very little isolated: remote species-rich oceanic islands, remote species-poor islands, young big islands near a continent, small off-shore islands, emerging islands, isolated hills, landscape islands, isolated patches of vegetation, and gaps in stands of vegetation.

Also, eight forms of vegetation dynamics are treated, ranging from short-term to long-term changes and involving larger and larger units: individuals, patches, communities, landscapes and vegetation regions. The forms of dynamics are fluctuation, gap dynamics, patch dynamics, cyclic succession, regeneration succession, secondary succession, primary succession, and secular succession. Each form of dynamics may occur under varying degrees of isolation.

The general conclusion is that processes and patterns of vegetation dynamics cannot be generalized in any simple manner. The 20 papers collected in this volume, divergent as they are, express the complexity of vegetation dynamics.

Introduction

This volume is a selection of contributions presented at two symposia during the 14th International Botanical Congress in Berlin, under the titles 'Vegetation dynamics' and 'Vegetation dynamics under isolation'. These contributions cover the major part of this field of ecology, though without forming any coherent and comprehensive collection. On the other hand, there is some attraction in the variation in approach, type of object, phytogeographical region, discipline represented and scale in time and space of the various contributions. They may be considered an indication of the state of the art in vegetation dynamics.

If there is any continuous thread to put through this collection of contributions it would be the dependence of dynamical phenomena on scale, both in time and in space, and both within the system under study and in its wider surroundings. Ideally, dynamics of island ecosystems should be seen as more extreme cases of what can be considered the standard situation: dynamics in isolation. Every patch of vegetation develops to some extent in isolation from its surroundings: a gap in a forest, a forest patch in a cultural landscape, a low-lying bog or a lake in an otherwise dry forest landscape, etc. In addition, the size of the patch and the size of the surrounding included in the study will vary. A similar variety exists regarding the time scale: changes in vegetation are always preceded by other changes, either in the environment or in existing vegetation, or in both. The scale of the latter may range from short term to long term and likewise the speed and duration of the vegetation response will vary.

This introduction aims at elucidating some aspects of scale variation, while reviewing some recent literature not covered by the various contributions. It will be restricted to terrestrial situations. There is an equally interesting island situation in the occurrence of lakes, notably in boreal countries, where regional 'archipelagos' of small lakes occur with in some cases much larger lakes as 'continents'. For vegetation dynamics we may take as previous landmarks two symposia organized in the summer of 1985 by the Working Group on Theoretical Vegetation Science in Uppsala, on Theory and models in vegetation science (Prentice & van der Maarel 1987), and the British Ecological Society in Southampton on Colonization, succession and stability (Gray *et al.* 1987) respectively. No such landmarks could be used regarding island biogeography.

First the various types of isolation will be discussed as a general framework for the major issue in this volume: vegetation dynamics.

Pluriformity of isolation

Starting from modern island biogeography, which is mainly built on experiences with a few animal groups (MacArthur & Wilson 1967; Gorman 1979), notably mobile animals, we find that the size of an island and its distance to a larger body of land are the most essential features. This is related to the following requirements and restrictions the average mobile animal has: a fairly large minimum viable population size, restriction to sexual reproduction, a limited reproduction and a relatively short life span. If we simplify island theory to the expected or required equilibrium between immigration and extinction, the first process is related to island isolation and the second to island size.

Plants are different and vegetation, communities of interacting plants, is still more different. Plants are less mobile on average, although some groups have large distance dispersal. Many plants are long-lived, but many others are short-lived. More importantly, plants are much more diverse regarding mobility and life span and it is almost impossible to generalize. Anyway, island size and isolation are apparently less important for plants. (Maybe the focus on size and distance is found with the animal ecologist rather than with the animals themselves).

Another difference between plants and many animal groups so far included in island biogeography is that plants form communities on relatively small areas, which even on small islands only occupy a small part of the island.

This all may not make too much difference for a typology of islands, but it certainly does so for the ecological significance of each type. In the following tentative survey of 10 types of island situation the presumed essential features: 1) distance to larger similar areas, 2) size, and 3) local community size, vary much and therewith also conditions for succession and maintenance of communities. The general conclusion will be that simple island theory based only on immigration and extinction will not be of great significance.

Types of island situation

1. *Old species-rich islands.* These can be remote oceanic islands, such as the Hawaii and New Zealand islands, or continent-near so-called continental islands such as the Canary Islands. Their common characteristic is the long period of isolation, usually since the Tertiary. Most of the old islands have a high percentage of endemic species, so-called palaeoendemics (See examples in Bramwell 1979). For such islands the usual immigration - extinction considerations are hardly relevant. Maintenance of community and species diversity is largely a matter of within-island regeneration processes. As far as there is immigration, it concerns alien species, both animals and plants which have usually been brought in by man. They often spread almost unlimitedly with dramatic effects on the native flora (and fauna). Examples are found in Bramwell (1979) and Holzner *et al.* (1983).

Most of the agressive introduced plant species belong to the category of neophytes, species which arrived on one or more places outside their original distribution area since the time of Columbus (e.g., Sukopp 1962, 1969; see also Kornas 1968, and in Holzner *et al.* 1983, who speaks of kenophytes). In so far the original, pre-Columbian plant communities on old islands are preserved, their community composition and dynamics are largely subject to the same forces as operating on continents.

2. *Remote species-poor islands.* Such islands may be relatively young or/and very remote. They have few endemic species, and usually not many species at all. Many of them may exhibit relatively broad ecological amplitudes, broader than in other parts of their distribution area. Examples are presented for the subantarctic island Marion by Gremmen (1981). The poverty of tree species on the Galapagos islands (see Hamann 1981) is another example. The cyclic succession in the open *Scalesia* woods on these islands described by Lawesson (1988, this volume), where only one species is involved, may even be related to species poverty in the canopy and this would be similar in the classical *Calluna* cycle (Gimingham 1988, this volume, and references therein)!

If such remote islands occur in archipelagos the smaller islands within them may to some extent act as real islands towards any bigger islands acting as the local continent. Itow (1988, this volume) shows how island forest communities (of this and other types) are less saturated with species than comparable forests on continents.

3. *Young big islands near a continent.* The British isles are an example here. The flora of such an island is poorer than nearby continental areas of the same size, because there are few (neo-) endemics to compensate for the number of species which never reached the island.

4. *Small off-shore islands.* They are usually of a different substrate as compared with the mainland, e.g. sandy barrier islands or coral keys. In at least some documented cases such as the Dutch Wadden islands the flora is richer than expected, simply because the habitat diversity is greater than on the nearby continent and because some of the island habitat types have a high alpha diversity (van der Maarel 1982; Mennema & Weeda 1983). Because the main substrate (sand) deviates from that on the nearby continent (clay) there is only a partial floristic exchange with the nearby continent. This effect of habitat diversity has also been described for the Galapagos islands (van der Werff 1983). Because of their small size they are exposed to wind and salt spray and succession may be hampered, i.e. the forest stage may be missing.

A special case of off-shore islands is found in the Aegean, where in part very small islets occur with a summer-dry rocky environment. They may have special species combinations with poor colonization, but apparently also low extinction rates (Höner & Greuter 1988, this volume).

5. *Emerging islands.* They occur along coasts subjected to land upheaval. The classical example is the Baltic skerry with many archipelagos and thousands of often very small islands, small even for plants. Here the processes of colonization are more pronounced than elsewhere (Vartiainen 1988, this volume). Another type of emerging island is the volcanic island, with the species-poor Surtsey near Iceland as an example, and the species-rich Krakatau (recently described regarding its succession to forest by Tagawa *et al.* 1985). Raus (1988, this volume) describes how immigration on young volcanic islands in the Aegean compensates for species losses either due to extinction or destruction.

6. *Islands in lakes.* These are usually small and of the same substrate as the land surrounding the lake. They will obey the laws of immigration and extinction, as was demonstrated by Nilsson & Nilsson (1978) for islands in a South-Swedish lake. Such islands are particular in the sense that the 'continent' is usually at a short distance and moreover it is everywhere around, which implies relatively high immigration rates. In another case, Lake Hjälmaren, the lake islands were created 100 yr ago by artificial lowering of the lake level and cencused from the beginning. Rydin & Borgegård (1988a) showed how Preston's parameter z of the canonical distribution as measured from the species-area relation, was near to the theoretical 'equilibrium' value of 0.26 already six years after colonization and succession had start-

ed and apparently it has no significance in relation to successional status.

Another special island situation was created in some of the estuaries in the SW Netherlands after closing them and lowering the water table. No island biogeographical studies have been published on this situation, but on the other hand extensive succession studies, which will be discussed later.

7. *Isolated hills.* Such hills or small mountains, also known under the German name of Inselberge have a substrate which is different from that of the surrounding lowland. Two well-known groups in Czechoslovakia I have seen are the limestone hills of Pavlov in southern Moravia (Pavlovske kopce) with a very species-rich steppe vegetation varying from hill to hill (Podpera 1928, see also Domin 1928), and the series of ultrabasic volcano hills near Louny, central Bohemia (Česke středohoři), again with a species-rich steppe vegetation (Slavikova *et al.* 1983). Although their flora and vegetation have been well described (as belonging mainly to the Festucetalia valesiacae, see Kolbek 1975) no island-biogeographical studies of such hill systems are known to me.

8. *Landscape islands.* They are the result of changes in the original vegetation of an area. Usually we are dealing with forest islands in a matrix of agricultural and seminatural landscape. Their study (see e.g. Burgess & Sharpe 1981) is becoming the domain of the emerging science of landscape ecology (Forman & Godron 1981, 1986; Risser 1987). One recent thorough study of Polish woodland islands (Dzwonko & Loster 1988) showed that their species numbers were not only a function of area and degree and time of isolation, but also of habitat diversity and successional status.

Landscape islands have also become subject of discussions in nature conservation. In particular the existence of corridors (both actual and, rather more often, potential) between islands (Forman 1982) has been considered important (e.g. Colaris 1983), although still little evidence is available to support this view.

A special case is formed by so-called remnant trees in forest clearings which during the period of agricultural use (fields or meadows) act as shelter for men and animals, but also as sources of forest regeneration if agriculture is abandoned again. Guevara (1986, Guevara *et al.* 1986) distinguished four features with a positive effect on the establishment of woody species from the surrounding forest: 1). Physical and chemical conditions under the canopy of the remnant tree, promoting or inhibiting germination of immigrant diaspores; 2). Size of the tree (sometimes tree group); 3). Distance to the mature forest; and 4). Nature of the remnant tree, particularly its function as a food source for birds, transporting bird-dispersed species (which may make up over 80% of the total).

9. *Isolated patches of vegetation.* In many cases vegetation stands are to some extent isolated from their surroundings, such as in mosaic patterns in bogs and dunes, where hummocks and hollows, and dune tops and dune slacks may isolate each other. A case which is being investigated from the isolation point-of-view concerns dispersed gravelly deposits on the limestone plateau of southern Öland, where stands of the *Veronica spicata-Avenula pratensis* association (Rosén 1982; Krahulec *et al.* 1986) occur in different degrees of isolation from each other (van der Maarel 1988). Although the distances between patches of the same community type may be as short as a few m, clear discontinuities in the distribution of many plant species occur.

10. *Gaps in stands of vegetation.* The least conspicuous form of isolation is found in gaps. Each gap is an island, but with a minimal degree of isolation due to the possibility of colonizing this island by seeds from the seedbank and by vegetative growth from the immediate surroundings.

Aspects of ecological isolation

From this survey it will become clear that there are so many types of islands that generalization is very difficult indeed. Further the concept of ecological isolation emerges from the survey. At least four different aspects of ecological isolation are recognized, as is summarized in Fig. 1. In addition to the

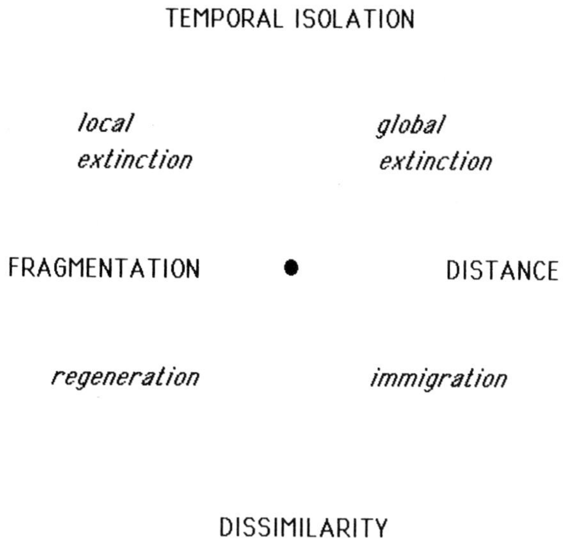

Fig. 1. Four dimensions of ecological isolation and four key processes in vegetation dynamics characterizing them. The position of each key process in between two dimensions suggests a particular significance for the pair of dimensions referred to.

traditional aspects of geographical distance and fragmentation (i.e. small island size) we add temporal isolation and environmental dissimilarity, i.e. between the island and the continent to be compared. The equilibrium status of an island will be dependent on all four aspects.

To each of the four aspects a key process can be linked. The four keywords denote the four major processes operating on any island biota. By putting a keyword to an island aspect it is suggested that the pertaining process is of relatively high significance there. The immigration of species is crucial in relation to geographical distance, extinction is especially important in relation to fragmentation. In fact the term extinction as used in island biogeography is misleading; we better call it local extinction! Global extinction, as different from local extinction is a process of particular significance to old islands, meaning that temporal isolation, usually connected with level of endemism, is related to the vulnerability of the island for disturbance by introduced alien organisms, leading to the general and definitive extinction of species. The fourth process is that of internal regulation which is decisive for the maintenance of biotic diversity in ecologically isolated islands and island situations.

Pluriformity of vegetation dynamics

There are many different ways along which vegetation may change. Although we know this (if not from own studies then from reviews like those by Whittaker & Levin 1977; McIntosh 1980 and Miles 1987) and although we have since long formalized some knowledge in an array of concepts on dynamics, we use to forget it every time we try to find unifying concepts, general mechanisms, etc. (Recent examples are Denslow 1980; Finegan 1984; Shugart 1984; Rosenzweig 1987 and Walker & Chapin 1987.)

Many of such general or generalizing approaches have been introduced or formalized in a more specific context, explicitely or not. To mention just one example: Connell & Slatyer's (1977) three main models (also called mechanisms) of succession, although presented in a more general framework, refer in fact, largely to what I will treat below as 'regeneration succession' following disturbance such as fire (cf McIntosh 1980). The companion system of vital attributes by Noble & Slatyer (1980) even explicitely refers to plant communities subject to recurrent disturbance. And yet, both systems are used for many types of vegetation dynamics. Pickett *et al.* (1987) provide further examples of what they call misapplication of the Connell and Slatyer's models and the originators themselves explain once more that their models were never designed to model any aspect of succession (Connell *et al.* 1987).

On the concept of succession

It may be useful to first define succession, the most central and most appealing form of dynamics. The definition by van der Maarel & Werger (1978), with the formal, phytosociological term phytocoena replaced by the more generally recognized one, plant communities, comes close to what many ecologists consider succession: 'the successive occurrence of plant communities at a given site'. (For similar definitions from other ecological approaches, see Rabotnov 1974 and Miles 1979.) A plant community in turn can be defined (largely after Westhoff & van der Maarel 1978) as a part of a vegetation consisting of interacting populations growing in a uniform environment and showing a floristic composition and structure that is relatively uniform and distinct from the surrounding vegetation. This includes the classical case of a succession from a pioneer community to a climax (or more neutrally: terminal) forest community via a series of developmental communities (be it without Clementsian and neo-Clementsian philosophy as now generally dismissed, e.g. by Miles 1979, 1987).

On the other hand, even if there usually is a clear continuum of transitional stages, this need not be approached with Gleasonal dogmatism, even if modern community ecologists who apparently ignore the modern vegetation-ecological literature would like us to do so (e.g. Crawley 1986). On the contrary, separate communities can be and indeed are recognized, both in modern European and American approaches (cf. Whittaker 1974; Westhoff & van der Maarel 1978).

There is a (confusing) terminology on such series of stages for which Westhoff & van der Maarel (1978) prefer the term climax complex. Although Braun-Blanquet phytosociologists are aware of the dynamical coherence within such a complex (examples in Braun-Blanquet 1964) they did not develop a systematic approach to describing them as units of their own on a higher hierarchical level until Tüxen (1973) introduced the concept of sigma-association and a standard procedure to describe and typify them. Ever since many such sigma-units have been described but usually as ad hoc local community complexes, which are landscape ecological units rather than dynamical entities.

On the basis of this definition pair of community and succession it is relatively easy to distinguish succession s.s. from other forms of vegetation dynamics, i.e. to discern it from dynamics of units below the level of the plant community as well as of community complexes.

This 'community approach' does not preclude the actual study of succession by means of population dynamics. Fekete *et al.* (1988, this volume) and Symonides (1988, this volume) give examples of population characteristics that can be measured and compared: different reproductive allocation and photosynthetic activity of annual versus perennial species in early successional grassland stages, and self-pollination and seed polymorphism of annuals of disturbed environments respectively.

Major forms of vegetation dynamics

There are many ways of classifying dynamics, but few of the classifications presented are comprehensive and an attempt to integrate them all is still lacking. The most useful general classification is probably the one based on the variation in the time-scale, although there is the complication that the relevance of the time-scale is dependent on the average life-span of the prevailing plant species. Major's (1974) scheme of 'kinds of changes in vegetation', in part based on earlier literature, included also ontogenetic and evolutionary changes. Miles (1987) presented a similar scheme, which in a simpler form (e.g. Miles 1979) consists of fluctuation, regeneration and secondary and primary succession. The time-scale classification can easily be integrated with the 'hierarchy of successional causes' of Pickett *et al.* (1987).

Fluctuation. The first and lowest level of vegetation dynamics which we may deduce from these schemes is that of *fluctuation*. Changes in vegetation are considered on the level of the plant individual or shoot and may be either ontogenetical or due to external factors such as predation, competition etc. Pickett *et al.* (1987) give a fairly complete list of processes and include stochastic environmental stress, which equals Miles' (1987) climatic fluctuations as a cause of change. Generally, the changes are rapid in relation to the time interval with which the stress may occur. This is of course more obvious in grasslands and similar communities with species able to rapidly change their biomass production.

According to both American and European ecologists the most important expression of fluctuation is the (quantitative) change in the relative dominance of species. Any quantitative community composition may reappear in the course of time and there is no clear direction of change (evidence for and examples of fluctuation in, e.g., Braun-Blanquet 1964; Rabotnov 1974; Miles 1979). Obviously fluctuation may have a cycle long enough to make it difficult to distinguish it from longer term trends of change.

Moreover, there will always be fluctuations be superimposed on any directional change. Another aspect is that drastic fluctuational stress may easily change the course of longer-term successional changes (e.g. Austin & Williams 1988, this volume).

Multivariate methods have shown to be helpful in discriminating between fluctuation and trend (Austin *et al.* 1981; Cramer & Hytteborn 1987). Wildi (1988, this volume) contributes to this methodology by developing trend analysis further.

Gap dynamics. Ontogenetically determined as well as externally imposed death of plant individuals may imply qualitative changes in vegetation as opposed to the merely quantitative changes with fluctuation. As a consequence species have to regenerate in order to maintain themselves in the community, while at the same time new possibilities for such reestablishment are provided. This aspect of dynamics has become more and more important, especially since Grubb (1977) emphasized the crucial role of the regeneration niche in community dynamics.

Both in forest and grassland dynamics the term gap is used for any opening resulting from the disappearance of plants (or parts of larger plants) and the term *gap dynamics* as used in the forest literature (e.g. Shugart 1984; Pickett & White 1985; Leemans & Prentice 1987) may be chosen to indicate this second level of vegetation dynamics. Gap dynamics in grasslands is thought equally important but more difficult to study. Silvertown & Smith (1988, this volume) show how grassland gaps can be detected more objectively and how one can study their significance for recolonization.

Patch dynamics. A third level of dynamics concerns the disappearance of local populations or patches of one species, or more species together, where the changes continue over a longer period of time. This is roughly equivalent to pattern dynamics as involved in Watt's (1947) classical approach on pattern and process in short vegetation, and to the patch dynamics in forests (e.g., Pickett & White 1985). Such changes may be cyclic in the sense that a given patch structure may re-appear in the course of time. Hence the term replacement cycles used in Major's (1974) scheme. The time-span involved may (but need not) be longer than in the case of gap dynamics, but the main (though gradual) difference is the size of the patch relative to the size of the community. As a lable for this level *patch dynamics* may be suggested. A complicated example of patch dynamics in conjunction with fluctuation is found in the famous Białowieża forest studied by Faliński (1986, 1988, this volume).

Cyclic succession. The fourth level, only gradually differing from the third one concerns the cyclic replacement of the major part of a community. This is particularly the case when the disappearing population is the dominant one. Examples of simply structured forests are found in the early work of Watt (1925) and discussed by Knapp (1974). Miles (1979) presents a scheme of cyclic changes where different dominant species follow each other. Facts about the classical example, the *Calluna* heath are updated in this volume (Gimingham 1988, this volume). Although the various phases in the *Callunetum* cycle have not been formally described as plant community types (part of them are cryptogamic communities for which a formal system has hardly been developed) the structural and floristic differences between the phases are big enough to talk of a succession and hence the well-known term *cyclic succession* (e.g., Braun-Blanquet 1964; Knapp 1974) may be maintained.

Clearly, the transition between patch dynamics and cyclic succession is gradual, not only in the *Callunetum* case. A special complication, of course is the semi-natural character of the heath in most parts of its distribution area. If any of the management practices (sheep grazing, regular mowing, burning, taking sods) would stop, woodland would take over though not beyond the tree limit.

Cyclic succession may be linked to stand dieback, i.e. the massive death of a canopy dominant, either by endogenous processes or by insect attacks and now also air pollution. This phenomenon has recently been claimed to be more universal than hitherto assumed and moreover spreading as a result of direct and indirect effects of air and soil pollution (Mueller-Dombois 1986). Loehle (1988, this volume) extends the theory on dieback by including tree defences against certain attacks. We also have the

Scalesia example mentioned above (Lawesson 1988, this volume). Sherman & Warren (1988, this volume) summarize the factors causing mortality of two tree species in the Yosemite valley forest and show how man, by exluding fire from the area, plays a role in the interplay between *Pinus ponderosa* and *Calocedrus decurrens*, in this case leading to *Calocedrus* taking over dominance from *Pinus*.

Regeneration succession. The next level of dynamics concerns the recovery from a total disturbance of a mature community, by natural agents such as storm, fire and insect attack, or human impact such as burning and clear-cutting. In the various recovery stages communities can easily be recognized and formally we better speak of regeneration succession than just regeneration. The difference with cyclic succession is gradual but clear: in the case of regeneration succession the duration of the mature stage is long on average in relation to the time needed for recovery, whereas in cyclic succession the mature stage is usually relatively short.

Secondary succession. This form of dynamics we will here define as succession towards a mature stage starting from a semi-natural or cultural situation. The starting point is usually a field, but may also be a grassland, whether grazed or mown. The original Clementsian concept takes the in situ developed soil organic matter as a criterion, as available after disturbance of any mature vegetation (see, e.g. Miles 1979; Vitousek & Walker 1987). Hence it includes our regeneration succession. Braun-Blanquet (e.g. 1964) takes previous anthropogenic disturbance as a criterion, which implies a division of regeneration + secondary succession into naturally versus man-induced.

Most of our knowledge of secondary succession stems from 'old-field successions'. And even much of our modern theory does, starting with Egler's (1954) classical paper on initial floristic composition as an important factor in old-field succession. The large number of studies of such secondary successions have learnt us that general conclusions are not easy to draw. The two contributions in this volume are no exception. Bornkamm (1988, this volume) concludes that all three mechanisms included in the Connell & Slatyer approach: facilitation, tolerance and inhibition operate at the same time, on different species.

Most studies deal with fertile soils allowing a rapid return to relatively rich well-structured forest. Clearly, findings from such environments do not necessarily apply to succession on poor soils. Examples are known from the Netherlands where secondary succession on poor sandy soils does not proceed at all through phases with ruderals, tall grasses and shrubs (see e.g., Pickett 1982), but through open short grassland and open *Calluna* heath. Succession on abandoned dense grassland may not even start at all during a number of years, because species of further stages cannot enter the community, or only a slow shrub colonization takes place (e.g., Miles 1979; Schmidt 1988, this volume). Examples of either possibility from one limestone grassland type are found in Rosén (1982), while Rusch (1988) found how the closed canopy of an ungrazed grassland site in this type hinders the germination of most species.

In this connection it does not seem appropriate to compare succession on newly harrowed ground and that on abandoned permanent (be it young) grassland in a search for patterns and strategies during succession in general, as done by Brown & Southwood (1987). A more differentiated approach of secondary (+ regeneration) succession follows from the scheme of van der Maarel (1975) where post-cultural, post-agricultural, post-seminatural and post-natural developments are distinguished.

In the present-day landscape of many areas with an intensive land-use we find post-agricultural and post-seminatural successions in complex patterns, simultaneously with still ongoing anthropogenic degradation of remaining natural woody vegetation. Both Borhidi (1988, this volume) and Poli Marchese *et al.* (1988, this volume) show how on the islands of Cuba and Sicily respectively the natural vegetation has been predominantly forest and diversified in relation to environmental variation, how these forests have disappeared, and in which complex ways certain types can be regained or at least approached through secondary succession.

Primary succession. The next level of dynamics is succession on a virginal substrate, i.e. where no remains of earlier vegetation development occur. Such virginal environments are usually poor in nitrogen, which implies that nitrogen-fixing organisms (often algae prior to phanerogams) are important in early stages, provided phosphorus is available (Vitousek & Walker 1987). Another general characteristic is the absence of diaspores and the relatively large distances to be covered by colonizing species. There are exception to both rules, however. Beaches with driftlines may be locally rich and so are the sand flats in some of the enclosed estuaries in the Southwest Netherlands, well-described by Beeftink and colleagues (van Noordwijk-Puijk *et al.* 1979; Hogeweg *et al.* 1985). Also, diaspores could easily enter a nearby primary succession. Examples are the same sand flats and also new land in land upheaval areas (Cramer & Hytteborn 1987; Vartiainen 1988, this volume). Another special case of primary succession on generally fertile soil on artificially created islands in a Swedish lake is described by Rydin & Borgegård (1988b).

A complicated example of primary/secondary succession has been described for a coastal dune area, the older part of which underwent a rapid secondary succession after being released from overgrazing. However, the multi-pathway succession from open communities via scrub to woodland could not be distinguished from a really primary succession having started simultaneously on dunes which rapidly developed in front of the older dunes (van der Maarel *et al.* 1985; van Dorp *et al.* 1985).

Another phenomenon, regularly discussed in a general way (e.g., Beard 1974; Dansereau 1974) and approached more quantitatively by van der Maarel *et al.* (1985) is that of retrogression or regression, the development during succession of a vegetation structure which is lower and/or less complicated than in some previous stage. Apart from local catastrophies such as fire or mechanical disturbance there are intrinsic mechanisms, mainly the death of a dominating shrub or tree population, e.g. of *Betula pendula*, with a life-span shorter than the time needed to reach the climax stage. This is, in fact, canopy dieback as discussed above, but without any indication of a cyclic succession involved. Piotrowska (1988, this volume) gives an example of regression of dune forest subject to dune abrasion and inblowing sand.

Among the many dichotomies and trichotomies of succession (see e.g. Sjörs 1980) one is important especially for primary succession: the subdivision into autogenic, allogenic and biogenic succession, which goes back to Tansley (see Dansereau 1974). Biogenic succession is nothing else than regeneration (as well as secondary) succession. Autogenic succession is primary succession on a substrate that changes only under the impact of the developing vegetation (soil development). Allogenic succession is development on a substrate that is also changing through input of material from outside. The classical examples are salt marsh succession, where the colonizing first stages of vegetation capture clay and coastal dune succession where the same happens with inblowing sand. In both cases the substrate is changing. Instead of considering this a clear case of facilitation one could also view each subsequent community adapted to the new soil conditions a new terminal stage of a mini-succession.

Secular succession. This is the final level of succession, also called synchronology in phytosociology (Braun-Blanquet 1964). It concerns long-term changes in a landscape as a whole as a result of a long-term changes in the environment, usually climate. The apparent changes are in the distribution area of individual species and through palaeoecology much has become known on postglacial secular changes in forest vegetation, especially in the northern hemisphere (see e.g., Davis 1986 for a survey of papers in the vegetational context).

One fascinating aspect of secular succession is the gradual change in the composition and even the structure of the climax community as a result of changes in soil conditions (Beard 1974). Walker *et al.* (1981) describe a sequence of dune systems in Queensland with ages increasing to probably 400000 yr and considerable leaching of nutrients. The 'climax' community on the older dunes is an open low heathlike community and compared with the taller forest on younger dunes it is a phase of retrogression.

Table 1. Two-dimensional scheme to relate spatial and temporal scales of vegetation dynamics to each other. x: ongoing research; X represented in this volume.

	Individual	Patch	Population	Community	Landscape	Region
Fluctuation	x	XX	XX			
Gap dynamics	x	XX	XX			
Patch dynamics	x	x	x			
Cyclic succession		x	XX	x		
Regeneration succession		XX	XX	XX		
Secondary succession		x	XX	XX	XX	
Primary succession			XX	XX	XX	
Secular succession				x	x	x

Conclusion

The only conclusion to be drawn for this introduction is that there is a threefold variation complicating the discussion on vegetation dynamics: 1) in temporal scale of the dynamical phenomena, 2) in degree of isolation of the piece of vegetation undergoing dynamics, and 3) in level-of-integration on which vegetation is actually studied. Hence it is impossible to create general theories on vegetation dynamics and confusing to apply bits of theory developed in one part of this complex field to other parts. Table 1 lists the various levels of dynamics and of integration and indicates on which cross-roads research on vegetation dynamics takes place.

The present volume covers only a part of these activities, and before starting to learn from its contents one should realize its limitations!

Acknowledgements

I thank David Glenn-Lewin for discussions on vegetation dynamics during a course at Uppsala and John Miles and Wolfgang Schmidt for helpful comments on the manuscript.

References

Austin, M. P. & Williams, O. B. 1988. Influence of climate and community composition on the population demography of pasture species in semi-arid Australia. Vegetatio 77: 43–49.

Austin, M. P., Williams, O. B. & Belbin, L. 1981. Grassland dynamics under sheep grazing in an Australian mediterranean type climate. Vegetatio 47: 201–211.

Beard, J. S. 1974. Vegetational changes on aging landforms in the tropics and subtropics. In: Knapp, R. (ed.), Vegetation dynamics, pp. 219–224. June, The Hague.

Borhidi, A. 1988. Vegetation dynamics of the savannization process on Cuba. Vegetatio 77: 177–183.

Bornkamm, R. 1988. Mechanisms of succession on fallow lands. Vegetatio 77: 95–101.

Bramwell, D. (ed.). 1979. Plants and islands. Academic Press, London.

Braun-Blanquet, J. 1964. Pflanzensoziologie. 3. Aufl. Springer, Wien.

Brown, V. K. & Southwood, T. R. E. 1987. Secondary succession: patterns and strategies. In: Gray, A. J., Crawley, M. J. & Edwards, P. J. (eds), Colonization, succession and stability, pp. 315–337. Blackwell, Oxford.

Burgess, R. L. & Sharpe, D. M. 1981. Forest island dynamics in man-dominated landscapes. Springer, New York.

Colaris, W. J. J. 1983. Application of the theory of island biogeography in the design of a structural concept of nature and landscape conservation. Geol. Mijnbouw 62: 643–652.

Connell, J. H. 1987. Change and persistence in some marine communities. In: Gray, A. J., Crawley, M. J. & Edwards, P. J. (eds), Colonization, succession and stability, pp. 339–352. Blackwell, Oxford.

Connell, J. H., Noble, I. R. & Slatyer, R. O. 1987. On the mechanisms producing successional change. Oikos 50: 136–137.

Connell, J. H. & Slatyer, R. O. 1977. Mechanisms of succession in natural communities and their role in community stability and organisation. Am. Nat. 111: 1119–1144.

Cramer, W. & Hytteborn, H. 1987. The separation of fluctuation and long-term change in vegetation dynamics of a rising seashore. Vegetatio 69: 157–167.

Crawley, M. J. 1986. The structure of plant communities. In: Crawley, M. J. (ed.), Plant ecology, pp. 1–50. Blackwell, Oxford.

Crawley, M. J. 1987. What makes a community invasible? In: Gray, A. J., Crawley, M. J. & Edwards, P. J. (eds), Colonization, succession and stability, pp. 429–453. Blackwell, Oxford.

Dansereau, P. 1974. Classification of succession and of their terminal stages: In: Knapp, R. (ed.), Vegetation dynamics, pp. 123–135. Junk, The Hague.

Davis, M. B. (ed.) 1986. Vegetation - climate equilibrium. Vegetatio 67(2).

Denslow, J. S. 1980. Patterns of plant species diversity during succession under different disturbance regimes. Oecologia (Berl.) 46: 18–21.

Domin, K. 1928. Introductory remarks to the fifth international phytogeographical excursion (I.P.E.) through Czechoslovakia. Acta Bot. Bohem. 6–7:3–76.

Dzwonko, Z. & Loster, S. 1988. Species richness of small woodlands on the western Carpathians foothills. Vegetatio 76: 15–27.

Egler, F. E. 1954. Vegetation science concepts I. Initial floristic composition, a factor in old-field vegetation development. Vegetatio 4: 412–417.

Faliński, J. B. 1986. Vegetation dynamics in temperate lowland primeval forests. Ecological studies in Białowieża forest. Junk, Dordrecht.

Faliński, J. B. 1988. Succession, regeneration and fluctuation in the Białowieża Forest (NE Poland). Vegetatio 77: 115–128.

Fekete, G., Tuba, Z. & Melko, E. 1988. Background processes at the population level during succession in grasslands on sand. Vegetatio 77: 33–41.

Fenner, M. 1987. Seed characteristics in relation to succession. In: Gray, A. J., Crawley, M. J. & Edwards, P. J. (eds), Colonization, succession and stability, pp. 103–114. Blackwell, Oxford.

Finegan, B. 1984. Forest succession. Nature 312: 109–114.

Forman, R. T. T. 1982. Interactions among landscape elements: a core of landscape ecology. In: Tjallingii, S. P. & de Veer, A. A. (eds), Perspectives in landscape ecology, pp. 35–48. Pudoc, Wageningen.

Forman, R. T. T. & Godron, M. 1981. Patches and structural components for a landscape ecology. BioScience 31: 733–740.

Forman, R. T. T. & Godron, M. 1986. Landscape ecology. Wiley, New York.

Gimingham, C. H. 1988. A reappraisal of cyclical processes in *Calluna* heath. Vegetatio 77: 61–64.

Gorman, M. 1979. Island ecology. Chapman and Hall, London.

Gray, A. J., Crawley, M. J. & Edwards, P. J. (eds). 1987. Colonization, succession and stability. Blackwell, Oxford.

Gremmen, N. J. M. 1981. The vegetation of the subantarctic islands Marion and Prince Edward. Junk, The Hague.

Grubb, P. J. 1977. The maintenance of species richness in plant communities: the importance of the regeneration niche. Biol. Rev. 52: 107–145.

Grubb, P. J. 1987. Some generalizing ideas about colonization and succession in green plants and fungi. In: Gray, A. J., Crawley, M. J. & Edwards, P. J. (eds), Colonization, succession and stability, pp. 81–102. Blackwell, Oxford.

Guevara, S. 1986. Plant species availability and regeneration in Mexican tropical rain forest. Thesis Uppsala University.

Guevara, S., Purata, S. E. & van der Maarel, E. 1986. The role of remnant forest trees in tropical secundary succession. Vegetatio 66: 77–84.

Hamann, O. 1981. Plant communities of the Galapagos islands. Dansk Bot. Ark. 34-2, 163 pp.

Hogeweg, P., Hesper, B., van Schaik, C. P. & Beeftink, W. G. 1985. Patterns in vegetation succession, an ecomorphological study. In: White, J. (ed.), The population structure of vegetation, pp. 637–666. Junk, Dordrecht.

Holzner, W., Werger, M. J. A. & Ikusima, I. 1983. Man's impact on vegetation. Junk, The Hague.

Höner, D. & Greuter, W. 1988. Plant population dynamics and species turnover on small islands near Karpathos (South Aegean, Greece). Vegetatio 77: 129–137.

Itow, S. 1988. Species diversity of mainland- and island forests in the Pacific area. Vegetatio 77: 193–200.

Knapp, R. 1974. Cyclic successions and ecosystem approaches in vegetation dynamics. In: Knapp, R. (ed.), Vegetation dynamics, pp. 91–100. Junk, The Hague.

Kolbek, J. 1975. die Festucetalia valesiacae Gesellschaften im Ostteil des Gebirges Českе středohoři (Böhmisches Mittelgebirge). 1. Die Pflanzengesellschaften. Folia Geobot. Phytotax. 10: 1–57.

Kornas, J. 1968. A geographical-historical classification of synanthropic plants. Mater. Zakladu Fitosocjol. Stosow. UW 25: 33–41.

Krahulec, F., Rosén, E. & van der Maarel, E. 1985. Preliminary classification ecology of dry grassland communities on Ölands Stora Alvar. Nord. J. Bot. 6: 797–809.

Lawesson, J. E. 1988. The stand-level dieback and regeneration of forests in the Galapagos islands. Vegetatio 77: 87–93.

Lawton, J. H. 1987. Are there assembly rules for successional communities? In: Gray, A. J., Crawley, M. J. & Edwards, P. J. (eds), Colonization, succession and stability, pp. 225–244. Blackwell, Oxford.

Leemans, R. & Prentice, I. C. 1987. Description and simulation of tree-layer composition and size distribution in a primaeval *Picea-Pinus* forest. Vegetatio 69: 147–156.

Loehle, C. 1988. Forest decline: endogenous dynamics, tree defences, and the elimination of spurious correlation. Vegetatio 77: 65–77.

MacArthur, R. H. & Wilson, E. O. 1967. Island biogeography. Princeton University Press, Princeton.

Major, J. 1974. Kinds and rates of changes in vegetation and chronofunctions. In: Knapp, R. (ed.), Vegetation dynamics, pp. 7–18. Junk, The Hague.

McIntosh, R. P. 1980. The relation between succession and the recovery process in ecosystems. In: Cairns, J. (ed.), The recovery process in damaged ecosystems, pp. 11–62. Ann Arbor Science Publ., Ann Arbor.

Mennema, J. & Weeda, E. J. 1983. Vascular plants. In: Dijkema, K. S. & Wolff, W. J. (eds): Flora and vegetation of the Wadden Sea islands and coastal areas, pp. 38–50. Stichting Veth, Leiden.

Miles, J. 1979. Vegetation dynamics. Chapman and Hall, London.

Miles, J. 1987. Vegetation succession: past and present perceptions. In: Gray, A. J., Crawley, M. J. & Edwards, P. J. (eds), Colonization, succession and stability, pp. 1–29. Blackwell, Oxford.

Mueller-Dombois, D. 1986. Perspectives for an etiology of stand-level dieback. Ann. Rev. Ecol. Syst. 17: 221–243.

Nilsson, S. G. & Nilsson, I. N. 1978. Species richness and dispersal of vascular plants to islands in Lake Möckeln, Southern Sweden. Ecology 59: 473–480.

Nilsson, S. G. & Nilsson, I. N. 1983. Are estimated species turnover rates on islands largely sampling errors? Amer. Nat. 121: 595–597.

Noble, I. R. & Slatyer, R. O. 1980. The use of vital attributes to predict successional changes in plant communities subject to recurrent disturbances. Vegetatio 43: 5–21.

Noy-Meir, I. & van der Maarel, E. 1987. Relations between community theory and community analysis in vegetation science: some historical perspectives. Vegetatio 69: 5–15.

Pickett, S. T. A. 1982. Population patterns through twenty years of oldfield succession. Vegetatio 49: 45–59.

Pickett, S. T. A., Collins, S. L. & Armesto, J. J. 1987. A hierarchical consideration of causes and mechanisms of succession. Vegetatio 69: 109–114.

Pickett, S. T. A. & White, P. S. (eds) 1985. The ecology of natural disturbance and patch dynamics. Academic Press, New York.

Piotrowska, H. 1988. The dynamics of the dune vegetation on the Polish Baltic coast. Vegetatio 77: 169–175.

Podpera, J. 1928. Die Vegetationsverhältnisse im Gebiete der Pollauer Berge Acta Bot. Bohem. 6–7: 77–131.

Poli Marchese, E., di Benedetto, L. & Maugeri, G. 1988. Successional pathways of mediterranean evergreen vegetation on Sicily. Vegetatio 77: 185–191.

Prentice, I. C. & van der Maarel, E. 1987. Theory and models in vegetation science. Vegetatio vol. 69. Junk, Dordrecht.

Rabotnov, T. A. 1974. Differences between fluctuations and successions. In: Knapp, R. (ed.), Vegetation dynamics, pp. 19–24. Junk, The Hague.

Raus, T. 1988. Vascular plant colonization and vegetation development on sea-born volcanic islands in the Aegean (Greece). Vegetatio 77: 139–147.

Risser, P. 1987. Landscape ecology: State of the art. In: Turner, M. G. (ed.), Landscape heterogeneity and disturbance, pp. 3–14. Springer, New York.

Rosén, E. 1982. Vegetation development and sheep grazing in limestone grasslands on South Öland, Sweden. Acta Phytogeogr. Suec. 72: 1–104.

Rosenzweig, M. L. 1987. Community organization from the point of view of habitat selectors. In: Gee, J. H. R. & Giller, P. S. (eds), Organization of communities past and present, pp. 469–490.

Rusch, G. 1988. Reproductive regeneration in grazed and ungrazed limestone grassland communities on Öland. Preliminary results. Acta Phytogeogr. Suec. 76: 113–124.

Rydin, H. & Borgegård, S.-O. 1988a. Plant species richness on islands over a century of primary succession. Ecology (in press).

Rydin, H. & Borgegård, S.-O. 1988b. Primary succession over sixty years on hundred-year old islets in Lake Hjälmaren, Sweden. Vegetatio 77: 159–168.

Schmidt, W. 1988. An experimental study of old-field succession in relation to different environmental factors. Vegetatio 77: 103–114.

Sherman, R. J. & Warren, R. K. 1988. Factors in *Pinus ponderosa* and *Calocedrus decurrens* mortality in Yosemite Valley. Vegetatio 77: 79–85.

Shugart, H. H. 1984. A theory of forest dynamics. Springer, New York.

Silvertown, J. & Smith, B. 1988. Gaps in the canopy: the missing dimension in vegetation dynamics. Vegetatio 77: 57–60.

Sjörs, H. 1980. An arrangement of changes along gradients, with examples from successions in boreal peatland. Vegetatio 43: 1–4.

Slavikova, J. *et al.* 1983. Ecological and vegetational differentiation of a solitary conic hill (Oblik in Česke středohoři Mts.). Vegetace CSSR A13. Academia, Praha.

Soulé, M. E. (ed.). 1987. Viable populations for conservation. Cambridge University Press, Cambridge.

Sukopp, H. 1962. Neophyten in natürlichen Pflanzengesellschaften Mitteleuropas. Ber. Deutsch. Bot. Ges. 75: 193–205.

Sukopp, H. 1969. Der Einfluss des Menschen auf die Vegetation. Vegetatio 17: 360–371.

Symonides, E. 1988. On the ecology and evolution of annual plants in disturbed environments. Vegetatio 77: 21–31.

Tagawa, H., Suzuki, E. & Partomihardjoo, T. 1985. Vegetation and succession on the Krakatau islands, Indonesia. Vegetatio 60: 131–145.

Tüxen, R. 1973. Vorschlag zur Aufnahme von Gesellschaftskomplexen in potentiell natürlichen Vegetationsgebieten. Acta Bot. Sc. Hung. 19: 379–384.

Usher, M. B. 1987. Modelling successional processes in ecosystems. In: Gray, A. J., Crawley, M. J. & Edwards, P. J. (eds), Colonization, succession and stability, pp. 31–55. Blackwell, Oxford.

van der Maarel, E. 1975. Man-made natural ecosystems in environmental management and planning. In: van Dobben, W. H. & Lowe-Mcconnell, R. H. (eds), Unifying concepts in ecology, pp. 263–274. Junk, The Hague & Pudoc, Wageningen.

van der Maarel, E. 1982. Biogeographical and landscape-ecological planning of nature reserves. In: Tjallingii, S. P. & de Veer, A. A. (eds), Perspectives in landscape ecology, pp. 227–235. Pudoc, Wageningen.

van der Maarel, E. 1988. Floristic diversity and guild structure in the grasslands of Öland's Stora Alvar. Acta Phytogeogr. Suec. 76: 53–65.

van der Maarel, E., Boot, R., van Dorp, D. & Rijntjes, J. 1984. Vegetation succession on the dunes near Oostvoorne, the Netherlands; a comparison of the vegetation in 1959 and 1980. Vegetatio 58: 137–187.

van der Maarel, E. & Werger, M. J. A. 1978. On the treatment of succession data. Phytocoenosis 7: 257–277.

van der Werff, H. 1983. Species numbers, area and habitat diversity in the Galapagos islands. Vegetatio 54: 167–175.

van Dorp, D., Boot, R. & van der Maarel, E. 1984. Vegetation succession on the dunes near Oostvoorne, the Netherlands, since 1934, interpreted from air photographs and vegetation maps. Vegetatio 58: 123–136.

van Noordwijk-Puijk, D., Beeftink, W. G. & Hogeweg, P. 1979. Vegetation development of salt-marsh flats after disappearance

of the tidal factor. Vegetatio 39: 1–13.
Vartiainen, T. 1988. Vegetation development on the outer islands of the Bothnian Bay. Vegetatio 77: 149–158.
Vitousek, P. M. & Walker, L. R. 1987. Colonization, succession and resource availability: ecosystem-level interactions. In: Gray, A. J., Crawley, M. J. & Edwards, P. J. (eds), Colonization, succession and stability, pp. 207–223. Blackwell, Oxford.
Walker, J., Thompson, C. H., Fergus, I. F. & Tunstall, B. R. 1981. Plant succession and soil development in coastal sand dunes of subtropical eastern Australia. In: West, D. C., Shugart, H. H. & T. Botkin, D. B. (eds), Forest succession: concepts and applications, pp. 107–131. Springer, Berlin.
Walker, L. R. & Chapin III, F. S. 1987. Interactions among processes controlling successional change. Oikos 50: 131–135.
Westhoff, V. & van der Maarel, E. 1978. The Braun-Blanquet approach. In: Whittaker, R. H. (ed.), Classification of plant communities, pp. 289–399. Junk, The Hague.
Watt, A. S. 1925. The development and structure of beech forests in the Sussex Downs. J. Ecol. 13: 27–73.
Watt, A. S. 1947. Pattern and process in the plant community. J. Ecol. 35: 1–22.
White, J. 1985. The census of plants in vegetation. In: White, J. (ed.), The population structure of vegetation, pp. 33–88.
Whittaker, R. H. 1974. Climax concepts and recognition. In: Knapp, R. (ed.), Vegetation dynamics, pp. 137–154. Junk, The Hague.
Whittaker, R. H. & Levin, S. A. 1977. The role of mosaic phenomena in natural communities. Theor. Pop. Biol. 12: 117–139.
Wildi, O. 1988. Linear trend in multi-species time series. Vegetatio 77: 51–56.
Williamson, M. H. 1987. Are communities ever stable? In: Gray, A. J., Crawley, M. J. & Edwards, P. J. (eds), Colonization, succession and stability, pp. 353–371. Blackwell, Oxford.

Vegetatio 77: 21–31, 1988
© Kluwer Academic Publishers, Dordrecht – Printed in the Netherlands

On the ecology and evolution of annual plants in disturbed environments

Ewa Symonides
Department of Phytosociology and Plant Ecology, Institute of Botany, Warsaw University, Al. Ujazdowskie 4, 00-478 Warsaw, Poland

Accepted 28.6.1988

Keywords: Amphicarpy, Life history, Seed polymorphism, Self-pollination

Abstract

This paper is concerned with the effect of disturbance on some crucial characteristics of annual plants. The theoretically optimal life-history traits that maximize individual fitness in disturbed environments are described and critically evaluated. It seems that none of them holds for all annual species.

Self-pollination and especially seed polymorphism are considered important adaptations to life in unpredictable environments. The thesis is put forward that amphicarpic annuals, which exhibit both self-pollination and extreme seed polymorphism, are best adapted to life in hazardous habitats. The hypothetical course of the evolution of amphicarpy is demonstrated on the grounds of the comparison of contemporary annual species producing chasmogamous and cleistogamous flowers on a single individual.

Nomenclature: follows Flora Europaea (Tutin *et al.* 1964–1980), except for non-European species where the nomenclature used in the papers cited has been followed.

Introduction

Disturbance has been defined as mechanisms which limit the plant biomass by causing its partial or total destruction (Grime 1979), as relatively discrete events in time that disrupts ecosystem, community or population structure and changes resources, substrate availability or physical environment (Pickett & White 1985), or even wider as unusual events relative to a normal situation (Begon *et al.* 1986). Disturbance can differ in type, frequency, extent and intensity; nevertheless, it is a crucial aspect of ecological reality for most organisms regardless of their life-history patterns (see e.g. Bazzaz 1983; Grubb 1985).

From recent papers it may be inferred that the search for correlations between life-history traits and features of disturbed environments is especially interesting in the case of annual plants. On the one hand, due to their life cycle restricted to only a few months or weeks, even short disturbance in their environment may considerably affect a plant's chance of survival and reproduction. On the other hand, the short life cycle of many contemporary annual species has been probably formed in response to disturbance, and favoured by natural selection. It is well proved by the fact that even amongst semelparous (monocarpic) species with undetermined life length the fraction of individuals terminating their life cycle in one year is clearly higher in disturbed than in undisturbed environments, as in *Poa annua* (Fig. 1). Moreover, in populations of typical winter annuals like *Erophila verna* and *Galium aparine* new ecotypes originating due to disturbance have been observed to terminate their cycle within spring and summer (Symonides 1983a, b; Groll & Mahn 1986). It is also worth noting that periodically and continu-

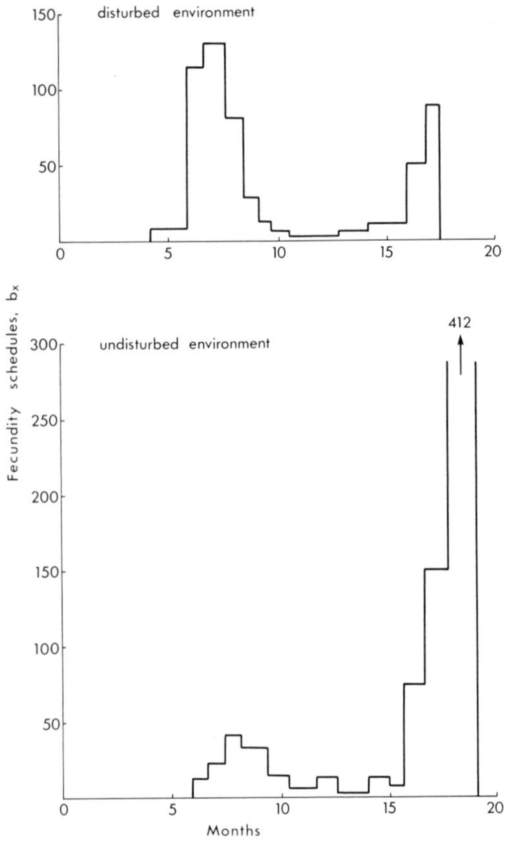

Fig. 1. Fecundity schedules for two populations of *Poa annua* derived from contrasting environments (From Law *et al.* 1977).

ally disturbed environments are usually colonized, and often dominated in late successional stages, by annual plants.

This paper will only survey the crucial traits of annual plants in environments that are disturbed in time and space, mainly physiological and behavioural properties that enable them to minimize loss of energy for reproduction and (or) minimize offspring mortality.

Optimal life history: models and reality

Theoretically, high fecundity will be typical of annual plants. Optimal life-history models make some important predictions about the life-history traits that might be expected to maximize individual fitness in short-lived organisms. These include: (a) a rapid switch from vegetative to reproductive growth late in the life cycle (Vincent & Pulliam 1980; Kermit 1983); (b) translocation of nutrients from vegetative to reproductive organs (Chapin 1980); (c) a high reproductive effort (King & Roughgarden 1982); (d) maximization of seed number within available resources, resulting in low individual seed weight (Baker 1974); (e) a long-term seed dormancy and the presence of a seed bank as bet-hedging tactics in an unpredictable environment (Cohen 1966; Leon 1985), or alternatively a strong tendency to disperse in space (Venable & Lawlor 1980). Finally, semelparity should also be associated with high survival of juvenile individuals (Silvertown 1987).

A question arises whether the actual strategy of annual plants, especially those from disturbed environments, corresponds to that which is theoretically optimal. The recent papers show that this is only partly true, and only for some species.

Against theoretical predictions a rapid switch from vegetative to reproductive growth in reality is rather rare. Maybe it is true only for species with genetically determined timing of flowering, e.g. for *Plantago erecta* (King & Roughgarden 1983), *Echinocystis lobata* (Silvertown 1985), and *Aira caryophyllea* (Rozijn & van der Werf 1986). Some annuals may shorten their life cycle and fairly rapidly start flowering in response to various perturbance agents like excessive increase in the CO_2 concentration, as in the case of *Layia platyglossa* and *Clarkia rubicunda* (Omer & Horvath 1983), essential increase in water temperature, as in *Ludwigia leptocarpa* growing in a reactor sewage (Christy & Sharitz 1980), or strong oscillations in water level, as in the case of *Limnanthes floccosa* (Ritland & Jain 1984). However, species that continue vegetative growth during the reproductive phase, and hence early allocate a part of their resources for the production of generative organs, are much more numerous. Doubtless, such resource partitioning increases the fitness of an organism in periodically disturbed environments. It is typical of, e.g. many *Veronica* species (Harper 1977), *Polygonum minimum* (Jackson & Bliss 1984), *Soliva pterosperma* (Maxwell *et al.* 1986), *Aira praecox* (Rozijn & van der Werf 1986), *Urtica urens* (Boot *et al.* 1986), *Lesquerella gordonii* (Delph 1986), and other species both from natural and man-made ecosystems. Most annuals of this group produce seeds gradually over an extended period or in a number of discrete flushes. In a sense they are uniseasonally iteroparous (polycarpic) species (Kirkendall & Stenseth 1985). Amphicarpic species i.e. plants producing two kinds of fruits on a single individual, are classical iteroparous annual.

It also seems that translocation of nutrients from vegetative to reproductive organs is not a rule amongst annuals, although it occurs in some species from disturbed, unstable habitats e.g. in *Bromus mollis* (Woodmansee & Duncan 1980), *Aristida oligantha* (Adams & Wallace 1985), as well as *Phleum arenarium* and *Erodium glutinosum* (Ernst 1983). However it has not been observed in *Cassia fasciculata* (Kelly 1986), and may also be true for other annuals of disturbed environments if their generative organs are capable of photosynthesis (as in *Cassia*), or if generative and vegetative phases overlap.

Reproductive effort is also extremely variable amongst annuals (Table 1). It may vary significantly among local populations of a single species and in successive generations (e.g. Symonides 1974; Kawano 1975). Morphs diversified, e.g. in respect of resources provided for reproduction may occur within one generation (van Andel *et al.* 1986). Sometimes even within one cohort, that represents one form, individuals exhibit a highly different reproductive effort; in populations of *Erophila verna, Veronica triphyllos* and *Arabidopsis thaliana* from arable fields the reproductive effort between neighbours may differ by more than 30% (Fig. 2). In some species, e.g. *Polygonum cascadense* and *Trifolium hirsutum,* an increase in the reproductive effort occurs with disturbance (Hickman 1975; Jain 1983). In others the opposite phenomenon takes place which seems to be correlated with an increased level of ploidy, like in some closely related *Setaria* species (Kawano & Miyake 1983). It should be noted, however, that a reproductive effort as discussed above has been evaluated on the grounds of biomass or energy

Table 1. The range of reproductive effort (RE) in annual species.

Species	RE (%)	References
Chenopodium album	≥ 60	Maillette 1985
Spergula arvensis		Maillette 1985
Polygonum cascadense	50 – 60	Hickman 1975
Plantago insularis		Bell *et al.* 1979
Spergula vernalis		Symonides 1974
Chaenactis carphoclinia	41 – 50	Bell *et al.* 1979
Plantago indica		Symonides 1979
Phleum arenarium		Ernst 1981
Astragalus sabulanum	31 – 40	Bell *et al.* 1979
Schismus arabicus		Bell *et al.* 1979
Leavenworthia stylosa		Solbrig & Rollins 1977
Oenothera deltoides	21 – 30	Bell *et al.* 1979
Amaranthus powelli		Weaver *et al.* 1982
Leavenworthia crassa		Solbrig & Rollins 1977
Androsace septentrionalis	11 – 20	Symonides 1979
Polygonum minimum		Hickman 1977
Cerastium semidecandrum		Symonides unpubl.
Medicago laciniata	≤ 10	Friedman & Elberse 1976
Abutilon theophrasti		Abrahamson 1979
Erodium cicutarium		Jaksic & Montenegro 1979

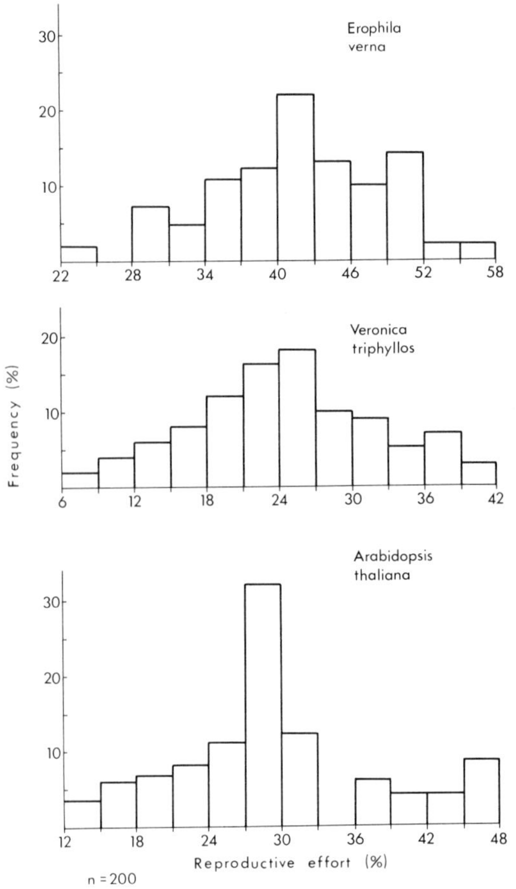

Fig. 2. Variation in reproductive effort within a single cohort in some annuals from arable fields (orig.).

allocation. It might be that nutrient allocation is a more appropriate measure and is differently allocated.

The total seed number per individual plant may also vary, ranging from 1–3 diaspores in *Vulpia fasciculata* (Watkinson & Harper 1978) to hundreds of thousands of seeds in *Chenopodium album* (Maillette 1985). In addition, many species both in stable and unpredictable environments have evolved towards an increase in the weight of seeds, not in their number. The cases of *Capsella bursa-pastoris, Arabidopsis thaliana* and *Sesbania vesicaria* indicate that annual plants may respond to current disturbance (e.g. herbivore pressure) by an increase in the seed weight and a decrease in the seed number (Hurka & Benneweg 1979; Marshall *et al.* 1985). Also plasticity in partitioning of resources between seeds and fruits is a common phenomenon in species that experience the stress of biotic and abiotic perturbance. From a theoretical viewpoint it is surprising that annuals from unstable habitats often maintain high numbers of fruits, while the seed number per fruit decreases (see e.g. Delph 1986). This may lessen the resource pool potentially allocated to increase seed production. However, such plasticity raises the fitness of an organism in an unpredictable environment. In summary, it seems that the total seed number, the weight of a single seed, as well as the partitioning of resources between fruits and seeds vary with variable environmental conditions. One may presume that the theoretically assumed tendency to produce the highest seed number within the available resource pool does not occur in nature. Such a strategy would inevitably lead to a reduction of seed weight, which amongst many annuals is already fairly low. However it may characterize some species under highly unfavourable conditions when a decrease in seed weight associated with an increase in seed number, in spite of all, raises the chance of reproductive success (cf. Watkinson 1981; Kelly 1984).

Again, following the cited authors, we may conclude that annuals are highly diversified in respect of both longevity and dispersability of seeds regardless of the environment in which they grow and disturbance types which have been selectional agents in the course of their evolution. This problem has been discussed by Symonides (1988). It is also worth noting that the thesis of Venable & Lawlor on negative correlation between dispersal in space and that in time is highly disputable. The correlation appears to hold for species producing two dispersal morphs on a single plant, but as might be expected, does not always hold when comparisons are made between species. For example, the dispersing seeds of *Spergula vernalis* form a persistent bank, whilst non-dispersing seeds of its neighbour *Erophila verna* are short-lived (Symonides 1974, 1984).

Finally, from the data published to date it may be concluded that both winter and summer annuals show great variety of survivorship patterns between germination and seed set ranging from Deevey type I to extreme Deevey type III. It also appears that there is a lack of correlation between the number of seeds produced per plant and the shape of survivorship curve, as well as between survivorship pattern and the environment of individual species (cf. Symonides 1988, and references therein). For example, in a single community of psammophilous annuals in the Toruń Basin an extreme variety of survivorship curves of plant populations have been found although they experience similar environmental conditions (Fig. 3). Again the survivorship curve cannot be considered as characteristic of a given species. It may vary markedly depending upon the habitat conditions and population density (Symonides 1974, 1983; Klemow & Raynal 1983; Beeftink 1985).

Self-pollination and seed polymorphism as important adaptations

Self-pollination

Stebbins (1970) suggested that the transition from

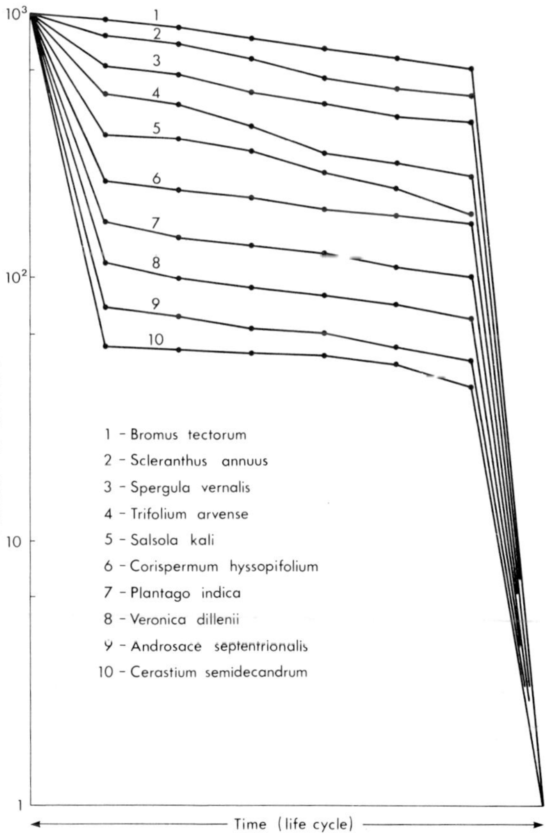

Fig. 3. Survivorship curves in ten annual species in a disturbed grassland, Toruń Basin (orig.).

predominant outcrossing to predominant selfing is more common in the angiosperms than any other evolutionary change. The change in the pollination mode and ability to produce offspring 'single-handed' is crucial for short-lived semelparous organisms, especially those living in hazardous environments. No wonder this is a widespread phenomenon among both wind- and insect-pollinated annual plants, although only some species are as well self-fertile. Self-pollination has been observed among many annual species of *Arenaria, Gilia, Setaria, Erophila, Veronica, Bromus, Ipomoea, Vicia, Limnanthes* etc.

So far the results of the studies have not answered satisfactorily the question when, how and why self-pollination evolved. But the relation between disturbance in the environment and transition from outcrossing to selfing seems undoubted. According to Cruden (1977) the evolutionary shift from xenogamy to obligate autogamy and cleistogamy is associated with a significant decrease in the mean pollen-ovule ratio, which is low in disturbed ecosystems and high at late successional stages. Most authors suggest that the evolution of self-pollination in different species has resulted from strong selection to ensure seed production under the conditions that make outcrossing difficult or impossible due to the effect of unfavourable, abiotic and biotic agents (e.g. Solbrig 1972; Arroyo 1975; Solbrig & Rollins 1977; Lloyd 1979; Schoen 1982a, 1982b; Gouyon *et al.* 1983). For example, competition for pollinators may lead to the evolution of self-pollination, as in *Ipomoea hederacea* (Stucky 1984) and *Arenaria uniflora* (Wyatt 1986).

Lack of data makes it impossible to evaluate the ecological consequences of the change in pollination mode in annuals. Its effect is only known to be either negative or positive. For example, in *Leavenworthia crassa* and *L. alabamica* self-fertilization is associated with lower seed production and lower seed quality (Lloyd 1965). This, however, is not common amongst annuals (cf. Solbrig 1980, and references therein). The individuals of self-pollinated populations of *Arenaria uniflora,* although they are smaller and produce smaller numbers of flowers which are also devoid of nectar, have a higher reproductive effort than those in cross-pollinated populations (Wyatt 1984). In *Gilia achilleifolia* selfed progeny from a normally cross-pollinated population shows unexpectedly high fitnesses relative to outcrossed progeny, suggesting that if such a mutant should arise it would increase in frequency (Schoen 1983). It has also been shown that self-pollinated populations are highly polymorphic and usually consist of a large number of specialized genotypes (Jain 1983; Zangerl & Bazzaz 1984a, 1984b). Hybridization is also a common phenomenon amongst annuals of disturbed environments. Due to self-pollination these new lines may be fixed in just a few generations (Zohary & Plitmann 1979).

Seed polymorphism

Seed polymorphism is a crucial characteristic amongst annuals that have evolved in variable, disturbed environments, although to a different extent in various species. In the simplest case, morphologically identical seeds have diversified dormancy, as in *Lactuca serriola* (Marks & Prince 1981), *Erucastrum gallicum* (Klemow & Raynal 1983) and *Papaver dubium* (Arthur *et al.* 1973). Thus, from one seed pool two or more cohorts appear having different size, mortality and seed production, as a result of the different growth conditions experienced by the individuals in different cohorts.

More often somatic polymorphism of seeds, originating in a single fruit, is associated with different seed weight, like in *Ludwigia leptocarpa* (Dolan & Sharitz 1984) or *Raphanus raphanistrum* (Stanton 1984), and in some species with different seed structure, as in *Hypochoeris glabra, Heterotheca latifolia* and some other composites (Baker & O'Dowd 1982; Venable & Levin 1985a, b, c). Sometimes somatic polymorphism of seeds is associated with their physiological polymorphism. The phenomenon has been analysed in a few annual halophytes: *Salicornia europaea, S. patula, Atriplex triangularis* (Ungar 1987). For example, in *Salicornia patula* large seeds may germinate at high soil salinity and over a wide range of temperature and light conditions, whilst small seeds are less tolerant to unfavourable environmental conditions (Berger 1985).

In all annuals with polymorphic seeds only small and light seeds are capable of germinating for a number of years. It has already been stated that long-term seed dormancy can be seen in terms of bet-hedging of an individual plant in disturbed environment with a high probability of seedling failure in any given year (Cohen 1966). However the presence of a large reservoir of seeds in the soil also has the consequence of stabilizing population fluctuations from year to year and preventing rapid directional selection (Epling *et al.* 1960).

From the above examples it may be concluded that many annual species adopt a mixed strategy by equiping some seeds for long-distance dispersal and others for staying near the mother plant, as well as for synchronous and long-term germination. The question of how many large and how many small seeds will be produced is a problem of Evolutionary Stable Strategy, since the production of large numbers of non-dispersing seeds would lead to sib competition.

Hypothetical evolution of amphicarpy and its ecological consequences

Amphicarpic annual plants seem to be best adapted to life in disturbed unpredictable environments. They also exhibit both self-pollination and extreme seed polymorphism. They have evolved not only towards dual resource allocation but also towards a dual reproductive strategy. The latter consists in the production of potentially cross-pollinated chasmogamous, aerial flowers and self-pollinated cleistogamous, subterranean flowers. The phenomenon is fairly rare, although it has evolved independently in seven different families (Cheplick 1987).

From the comparison of contemporary plants with both cleistogamous and chasmogamous flowers on a single individual one may guess about a possible evolution of amphicarpy (Fig. 4). Cleistogamous flowers have probably derived from chasmogamous ones, which is well illustrated by *Collomia grandiflora:* within a single inflorescence flowers exhibit a gradient from large, open, potentially cross-pollinated to small, closed, self-pollinated. The second evolutionary stage is characterized by the lack of transient flowers. A single plant produces only chasmogamous-, only

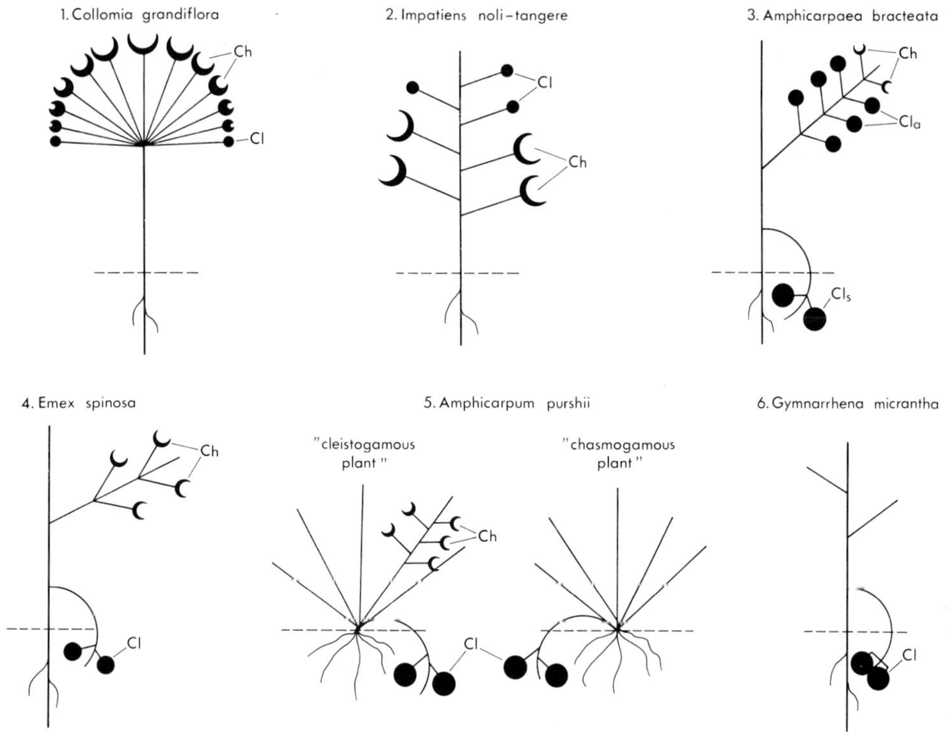

Fig. 4. The hypothetical stages in the evolution of amphicarpy. Ch - chasmogamous flowers, Cl_a - cleistogamous, aerial flowers, Cl_s - cleistogamous, subterranean flowers.

cleistogamous-, or both types of flowers. The process depends upon environmental conditions: the more unfavourable they are the higher is the fraction of cleistogamous flowers (Schemske 1978; Falencka 1983). It should be noted that, like in *Collomia,* both types of flowers develop here only on aerial shoots. The next step is the production of three types of flowers: chasmogamous, aerial cleistogamous and subterranean cleistogamous. Subterranean flowers have required remodelling of the system of reproductive organs, especially the change in the geotropism of a pedicel from negative to positive. *Amphicarpaea bracteata* is a classical example (Schnee & Waller 1986). At the further stage plants produce cleistogamous flowers and fruits only below the ground, like in *Emex spinosa* (Weiss 1980) and *Cardamine chenopodifolia* (Cheplick 1983). At last, an intriguing form of amphicarpy has evolved, called pessimistic by Zeide (1978). Its classical form occurs in *Amphicarpum purshii,* a grass from disturbed habitats. Its populations are usually dominated by the individuals of cleistogamous origin which allocate about 40% of their biomass to reproduction: 37% to early, heavy, subterranean diaspores and only 3% to a few light aerial seeds which develop one month later. Whilst the reproductive effort of the plants of chasmogamous origin is only 22%, they produce solely subterranean seeds and fruits, which arise two weeks later than in their cleistogamous neighbours (Cheplick & Quinn 1982).

An extreme form of pessimistic reproductive strategy that consists in almost total extinction of dual resource allocation is the last stage in the hypothetical scheme of evolution of amphicarpy. It has been found in the small desert plant *Gymnarrhena micrantha.* From the beginning of their growth each plant allocates most resources to the production of 1-3 very heavy, subterranean diaspores, and only under exceptionally favourable conditions it takes the risk of chasmogamous reproduction (Koller & Roth 1964).

Cheplick (1987) discusses some disturbance

agents which probably have been responsible for the evolution of amphicarpy (drought, fire, herbivore pressure), and some selective advantages of this strategy. In fact, there are some arguments that the life history of amphicarpic annuals is optimal in disturbed environments: (a) minimum energy loss resource allocation that enables plants to carry over the biomass stored in subterranean seeds; (b) the chance of undisturbed reproduction below the ground in spite of possible disturbance above, followed by the ultimate reproductive success; (c) germination of subterranean 'stay-at-home' seeds and growth of seedlings in a safe site already experienced by a parent plant; (d) large size of seedlings growing from heavy, subterranean seeds, and in consequence having high tolerance to unfavourable abiotic conditions and to the competition stress; (e) the chance to increase both offspring number and its genetic variability under favourable conditions due to chasmogamous reproduction.

Conclusions

From these data it may be concluded that there are many patterns of life history of annual species that enable them to exist in disturbed environments, although bet-hedging, risk-avoiding tactics are common in their evolution. The tolerance to unfavourable habitat conditions on the one hand, and low competitive ability of short-living plants on the other, cause that most annuals colonize hazardous and pioneer environments, which are permanently or periodically disturbed both by natural and anthropogenic factors. Some authors emphasise that the species which are capable of living without mycorrhiza are especially well adapted to colonize disturbed environments, in spite of their weaker growth and lower competitive ability in comparison with mycorrhizal species (Allard 1965; Moorman & Reeves 1979; but see also Medre 1984).

The life-history traits of annuals, described in this paper, explain often their dominance in communities subjected to natural and anthropogenic disturbance (Joenje 1985; Lee & Ignaciuk 1985; Nakagoshi 1985; Watkinson & Davy 1985; Hatton & West 1987). Such plant communities are characterized by rapid, often irregular fluctuations both in species composition and dominance structure (e.g. Symonides 1979; Pineda *et al.* 1987). Partly they result from a specific response of single species to various disturbance agents (see Mahn 1984; Davy & Smith 1985). In this case the presence of some species in a community, as well as the sizes of their populations may play the role of indicators of a form and intensity of disturbance (Hobbs & Hobbs 1987). However, some studies show that fluctuations in species composition and quantitative relations between populations may result from non-interactive structure of an annual community, which is variable in time and depends on the intensity of disturbance (Hobbs & Hobbs 1987; Pineda *et al.* 1987).

One may currently observe rapid changes in the life-history traits of plants due to man-made disturbance. New biotypes originate which are resistant to herbicides, air pollutants or heavy metals. Short life-cycle and self-pollination may cause new lines to be fixed over a few generations (see e.g. Zohary & Plitmann 1979). Moreover, new forms may differ in traits that on the surface are unrelated to their tolerance to disturbance like a lower reproductive effort, delayed flowering or lower competitive ability, observed for biotypes of *Amaranthus retroflexus* resistant to atrozine (Conard & Radosevick 1979; Weaver *et al.* 1982). Probably, further studies will show whether such response occurs in other species and due to other disturbance agents.

Acknowledgements

I am very grateful to Professor Eddy van der Maarel for valuable remarks concerning the text and Mrs Anna Mazur for the English translation. I also recognize the help of three reviewers.

References

Abrahamson, W. G. 1979. Pattern of resource allocation in wildflower populations of fields and woods. Am. J. Bot. 66: 71–79.

Abrahamson, W. G. & Hershey, B. J. 1977. Resource allocation and growth in *Impatiens capensis* (Balsaminaceae) in two habitats. Bull. Torrey Bot. Club 104: 160–164.

Adams, D. E. & Wallace, L. L. 1985. Nutrient and biomass allocation in five grass species in a Oklahoma tallgrass prairie. Am. Mid. Nat. 113: 170–181.

Allard, R. W. 1965. Genetic systems associated with colonizing ability in predominately self-pollinated species. In: Baker, H. G. & Stebbins, G. L. (eds), The genetics of colonizing species, pp. 49–75. Academic Press, New York.

Arroyo, M. T. K. de. 1975. Electrophoretic studies of genetic variation in natural populations of allogamous *Limnanthes alba* and autogamous *Limnanthes floccosa* (Limnanthaceae). Heredity 35: 153–164.

Arthur, A. E., Gale, J. S. & Lawrence, K. J. 1973. Variation in wild populations of *Papaver dubium*. VII. Germination time. Heredity 30: 189–197.

Baker, H. G. 1974. The evolution of weeds. Ann. Rev. Ecol. Syst. 5: 1–24.

Baker, G. A. & O'Dowd, D. J. 1982. Effects of parent plant density on the production of achene types in the annual *Hypochoeris glabra*. J. Ecol. 70: 201–215.

Bazzaz, F. A. 1983. Characteristics of populations in relation to disturbance in natural and man-modified ecosystems. In: Mooney, H. A. & Gordon, M. (eds), Disturbance and ecosystems. Components of response, pp. 259–275. Springer, Berlin.

Beeftink, W. G. 1985. Population dynamics of annual *Salicornia* species in the tidal salt marshes of the Oosterschelde, The Netherlands. Vegetatio 61: 127–137.

Bell, K. L., Hiatt, H. D. & Niles, W. E. 1979. Seasonal changes in biomass allocation in eight winter annuals of the Mojave desert. J. Ecol. 67: 781–787.

Begon, M., Harper, J. L. & Townsend, C. R. 1986. Ecology. Individuals, populations and communities. Blackwell Scientific Publications, Oxford.

Berger, A. 1985. Seed dimorphism and germination behaviour in *Salicornia patula*. Vegetatio 61: 137–143.

Boot, R., Raynal, D. J. & Grime, J. P. 1986. A comparative study of the influence of drought stress on flowering in *Urtica dioica* and *U. urens*. J. Ecol. 74: 485–495.

Chapin, F. S. 1980. The mineral nutrition of wild plants. ARES 11: 233-260.

Cheplick, G. P. 1983. Differences between plants arising from aerial and subterranean seeds in the amphicarpic annual *Cardamine chenopodifolia* (Cruciferae). Bull. Torrey Bot. Club 110: 442–448.

Cheplick, G. P. 1987. The ecology of amphicarpic plants. Trends in Ecol. and Evol. 2: 97–101.

Cheplick, G. P. & Quinn, J. A. 1982. *Amphicarpum purshii* and the pessimistic strategy in amphicarpic annuals with subterranean fruits. Oecologia (Berl.) 52: 327–332.

Christy, E. J. & Sharitz, R. R. 1980. Characteristics of three populations of a swamp annual under different temperature regimes. Ecology 61: 454–460.

Cohen, D. 1966. Optimizing reproduction in a randomly varying environment. J. Theor. Biol. 12: 119–129.

Conard, G. & Radosevick, S. R. 1979. Ecological fitness of *Senecio vulgaris* and *Amaranthus retroflexus* biotypes susceptible or resistant to atrozine. J. Appl. Ecol. 16: 171–177.

Cruden, R. W. 1977. Pollen-ovulae ratios: a conservative index of breeding systems in flowering plants. Evolution 31: 22–46.

Davy, A. J. & Smith, H. 1985. Population differentiation in the life-history characteristics of salt-marsh annuals. Vegetatio 61: 117–127.

Delph, L. F. 1986. Factors regulating fruit and seed production in the desert annual *Lesquerella gordonii*. Oecologia (Berl.) 69: 471–476.

Dolan, R. W. & Sharitz, R. R. 1984. Population dynamics of *Ludwigia leptocarpa* (Onagraceae) and some factors affecting size hierarchies in a natural population. J. Ecol. 72: 1031–1041.

Epling, C., Lewis, H. & Bell, F. M. 1960. The breeding group and seed storage: a study in population dynamics. Evolution 14: 238–255.

Ernst, W. H. O. 1981. Ecological implication of fruit variability in *Phleum arenarium* L., an annual dune grass. Flora 171: 387–398.

Ernst, W. H . O. 1983. Mineral nutrition of two contrasting dune annuals, *Phleum arenarium* and *Erodium glutinosum*. J. Ecol. 71: 197–209.

Falencka, M. 1983. Struktura i dynamika populacji *Impatiens noli-tangere* L. w róznych warunkach ekologicznych. Ph.D. thesis, Białowieza (mscr.).

Friedman, J. & Elberse, W. T. 1976. Competition between two desert varieties of *Medicago laciniata* (L.) Mill. under controlled conditions. Oecologia (Berl.) 22: 321–339.

Gouyon, P. H., Lumaret, R., Valdeyron, G. & Vernet, P. 1983. Reproductive strategies and disturbance by man. In: Mooney, H. A. & Godron, M. (eds), Disturbance and ecosystems. Components of response, pp. 214–225. Springer, Berlin.

Grime, J. P. 1979. Plant strategy and vegetation processes. Wiley, New York.

Groll, U. & Mahn, E.-G. 1986. Zur Entwicklung ausgewählter Populationen des Kletten-Labkrautes (*Galium aparine* L.). Flora 176: 93 – 110.

Grubb, P. J. 1985. Plant population and vegetation in relation to habitat, disturbance and competition: problem of generalization. In: White, J. (ed.), Handbook of vegetation science. 3. The population structure of vegetation, pp. 595–622. Junk, Hague.

Harper, J. L. 1977. Population biology of plants. Academic Press, London.

Hatton, T. J. & West, N. E. 1987. Early seral trends in plant community diversity on a recontoured surface mine. Vegetatio 73: 21–29.

Hickman, J. C. 1975. Environmental unpredictability and plastic energy allocation strategies in the annual *Polygonum cascadense* (Polygonaceae). J. Ecol. 63: 689–701.

Hobbs, R. J. & Hobbs, V. J. 1987. Gophers and grassland: a model of vegetation response to patchy soil disturbance. Vegetatio 69: 141–146.

Hurka, H. & Benneweg, M. 1979. Patterns of seed size variation in populations of the common weed *Capsella bursa-pastoris* (Brassicaceae). Biol. Zbl. 98: 699–709.

Jackson, L. E. & Bliss, L. C. 1984. Phenology and water relations of three life forms in a dry tree-line meadow. Ecology 65: 1302–1314.

Jain, S. 1983. Genetic characteristics of populations. In: Mooney, H. A. & Godron, M. (eds), Disturbance and ecosystems. Components of response, pp. 240–258. Springer, Berlin.

Jaksic, F. M. & Montenegro, G. 1979. Resource allocation of Chilean herbs in response to climatic and microclimatic factors. Oecologia (Berl.) 40: 81–89.

Joenje, W. 1985. The significance of waterfowl grazing in the primary vegetation succession on embanked sandflats. Vegetatio 62: 399–407.

Kawano, S. 1975. The productive and reproductive biology of flowering plants. II. The concept of life history strategy in plants. J. Coll. Liberal Arts, Toyama Univ. 8: 51–86.

Kawano, S. & Miyake, S. 1983. The productive and reproductive biology of flowering plants. X. Reproductive energy allocation and propagule output of five congeners of the genus *Setaria* (Gramineae). Oecologia (Berl.) 57: 6–13.

Kelly, D. 1984. Seeds per fruit as a function of fruits per plant in depauperate annuals and biennials. New Phytol. 96: 103–114.

Kelly, C. A. 1986. Extrafloral nectaries, ants, herbivores and fecundity in *Cassia fasciculata*. Oecologia (Berl.) 69: 600–605.

Kermit, R. 1983. The joint evolution of seed dormancy and flowering time in annual plants living in variable environments. Theor. Popul. Biol. 24: 213–343.

King, D. & Roughgarden, J. 1982. Multiple switches between vegetative and reproductive growth in annual plants. Theor. Popul. Biol. 21: 194–204.

King, D. & Roughgarden, J. 1983. Energy allocation patterns of the California grassland annuals *Plantago erecta* and *Clarkia rubicunda*. Ecology 64: 14–24.

Kirkendall, L. R. & Stenseth, N. C. 1985. On defining 'breeding once'. Am. Nat. 125: 189–204.

Klemow, K. M. & Raynal, D. J. 1983. Population biology of an annual plant in a temporally variable habitat. J. Ecol. 71: 691–703.

Koller, D. & Roth, N. 1964. Studies on the ecological and physiological significance of amphicarpy in *Gymnarrhena micrantha* (Compositae). Am. J. Bot. 51: 26–35.

Law, R., Bradshaw, A. D. & Putwain, P. D. 1977. Life history variation in *Poa annua*. Evolution 31: 233–246.

Lee, J. A. & Ignaciuk, R. 1985. The physiological ecology of strandline plants. Vegetatio 62: 319–327.

Leon, J. A. 1985. Germination strategies. In: Greenwood, P. J., Harvey, P. H. & Slatkin, M. (eds), Evolution: essays in honour of John Maynard Smith, pp. 129–142. Cambridge Univ. Press, Cambridge.

Lloyd, D. G. 1965. Evolution of self compatibility and racial differentiation in *Leavenworthia* (Cruciferae). Contrib. Gray Herb. Harv. Univ. 195: 3–134.

Lloyd, D. G. 1979. Some reproductive factors affecting the selection of self-fertilization in plants. Am. Nat. 113: 67–79.

Mahn, E. G. 1984. Structural changes of weed communities and populations. Vegetatio 58: 79–85.

Maillette, L. 1985. Modular demography and growth patterns of two annual weeds (*Chenopodium album* L. and *Spergula arvensis* L.) in relation to flowering. In: White, J. (ed.), Studies on plant demography, pp. 239–255. Academic Press, London.

Marks, M. & Prince, S. 1981. Influence of germination date on survival and fecundity in wild lettuce *Lactuca serriola*. Oikos 36: 326–330.

Marshall, D. L., Levin, D. A. & Fowler, N. L. 1985. Plasticity in yield components in response to fruit predation and date of fruit initiation in three species of *Sesbania* (Leguminosae). J. Ecol. 73: 71–81.

Maxwell, C. D., Jacob, N., Bollard, S. & Lovell, P. 1986. Factors affecting establishment and survival of *Soliva* (Onchung weed) at Auckland, New Zeland. New Zeal. J. Bot. 24: 79–87.

Medre, R. J. 1984. The mycorrhizae of pioneer species in disturbed ecosystems in western Pennsylvania. Am. J. Bot. 71: 787–714.

Moorman, T. & Reeves, F. B. 1979. The role of endomycorrhizae in revegetation practices in the semi-arid West. II. A bioassay to determine the effect of land disturbance on endomycorrhizal populations. Am. J. Bot. 66: 14–18.

Nakagoshi, N. 1985. Buried viable seeds in temperate forests. In: White, J. (ed.), The population structure of vegetation, pp. 551–570. Junk, Dordrecht.

Omer, L. S. & Horvath, S. M. 1983. Elevated carbon dioxide concentrations and whole plant senescence. Ecology 64: 1311–1314.

Pickett, S. T. A. & White, P. S. 1985. Natural disturbance and patch dynamics: an introduction. In: Pickett, S. T. A. & White, P. S. (eds), The ecology of natural disturbance and patch dynamics, pp. 3–13. Academic Press, New York.

Pineda, F. D., Casado, M. A., Peco, B., Olmeda, C. & Levassor, C. 1987. Temporal changes in therophytic communities across the boundary of disturbed-intact ecosystems. Vegetatio 71: 33–41.

Ritland, K. & Jain, S. 1984. The comparative life histories of two annual *Limnanthes* species in a temporally variable environment. Am. Nat. 124: 656–679.

Rozijn, N. A. M. G. & van der Werf, D. C. 1986. Effect of drought during different stages in the life-cycle on the growth and biomass allocation of two *Aira* species. J. Ecol. 74: 507–523.

Schemske, D. W. 1978. Evolution of reproductive characteristics in *Impatiens* (Balsaminaceae). The significance of cleistogamy and chasmogamy. Ecology 59: 596–613.

Schnee, B. K. & Waller, D. M. 1986. Reproductive behavior of *Amphicarpaea bracteata* (Leguminosae), an amphicarpic annual. Am. J. Bot. 73: 376–386.

Schoen, D. J. 1982a. The breeding system of *Gilia achilleifolia*: variation in floral characteristics and outcrossing rate. Evolution 36: 352–360.

Schoen, D. J. 1982b. Genetic variation and the breeding system of *Gilia achilleifolia*. Evolution 36: 361–370.

Schoen, D. J. 1983. Relative fitness of selfed and outcrossed progeny in *Gilia achilleifolia* (Polemoniaceae). Evolution 37: 292–301.

Silvertown, J. 1985. Survival, fecundity and growth of wild cucumber, *Echinocystis lobata*. J. Ecol. 73: 841–849.

Silvertown, J. W. 1987. Introduction to plant population ecology,

2nd ed. Longman, Harlow.

Solbrig, O. T. 1972. Breeding system and genetic variation in *Leavenworthia.* Evolution 26: 155–160.

Solbrig, O. T. 1980. Demography and evolution in plant populations. Blackwell, Oxford.

Solbrig, O. T. & Rollins, R. C. 1977. The evolution of autogamy in species of the mustard genus *Leavenworthia.* Evolution 31: 265–281.

Stanton, M. L. 1984. Developmental and genetic sources of seed weight variation in *Raphanus raphanistrum* L. (Brassicaceae). Am. J. Bot. 71: 1090–1098.

Stebbins, G. L. 1970. Adaptative radiation in angiosperms. I. Pollination mechanisms. An. Rev. Ecol. Syst. 1: 307–326.

Stucky, J. M. 1984. Forager attraction by sympatric *Ipomoea hederacea* and *I. purpurea* (Convolvulaceae) and corresponding forager behaviour and energetics. Am. J. Bot. 71: 1273–1244.

Symonides, E. 1974. Populations of *Spergula vernalis* Willd. on dunes in the Toruń Basin. Ekol. pol. 22: 379–416.

Symonides, E. 1979. The structure and population dynamics of psammophytes on inland dunes. III. Populations of compact psammophyte communities. Ekol. pol. 27: 235–257.

Symonides, E. 1983a. Population size regulation as a result of intra-population interactions. I. Effect of density on the survival and development of individuals of *Erophila verna* (L.) C. A. M. Ekol. pol. 31: 839–882.

Symonides, E. 1983b. Population size regulation as a result of intra-population interactions. II. Effect of density on the growth rate, morphological diversity and fecundity of *Erophila verna* (L.) C. A. M. individuals. Ekol. pol. 31: 883–912.

Symonides, E. 1984. Population size regulation as a result of intra-population interactions. III. Effect of *Erophila verna* (L.) C. A. M. population density on the abundance of the new generation of seedlings. Summing-up conclusions. Ekol. pol. 32: 557–580.

Symonides, E. 1988. Population dynamics of annual plants. In: Davy, A. J., Hutchings, M. J. & Watkinson, A. R. (eds), Plant population biology, Blackwell, London (in press).

Ungar, I. A. 1987. Population ecology of halophyte seeds. Bot. Rev. 53: 301–334.

van Andel, J., Rozijn, N. A. M. G., Ernst, W. H. O. & Wellissen, H. J. M. 1986. Variability in growth and reproduction in F_1 families of an *Erophila verna* population. Oecologia (Berl.) 69: 79–85.

Venable, D. L. & Lawlor, L. 1980. Delayed germination and dispersal in desert annuals: escape in space and time. Oecologia (Berl.) 46: 272–282.

Venable, D. L. & Levin, D. A. 1985a. Ecology of achene dimorphism in *Heterotheca latifolia.* I. Achene structure, germination and dispersal. J. Ecol. 73: 133–145.

Venable, D. L. & Levin, D. A. 1985b. Ecology of achene dimorphism in *Heterotheca latifolia.* II. Demographic variations within populations. J. Ecol. 73: 743–755.

Venable, D. L. & Levin, D. A. 1985c. Ecology of achene dimorphism in *Heterotheca latifolia.* III. Consequences of water availability. J. Ecol. 73: 757–763.

Vincent, T. L. & Pulliam, H. R. 1980. Evolution of life history strategies for an asexual plant model. Theor. Pop. Biol. 17: 215–231.

Watkinson, A. R. 1981. Interference in pure and mixed populations of *Agrostemma githago.* J. Appl. Ecol. 18: 967–976.

Watkinson, A. R. & Davy, A. J. 1985. Population biology of salt marsh and sand dune annuals. Vegetatio 62: 487–499.

Watkinson, A. R. & Harper, J. L. 1978. The demography of a sand dune annual: *Vulpia fasciculata.* I. The natural regulation of populations. J. Ecol. 66: 15–33.

Weaver, S. E., Warwick, S. I. & Thompson, B. K. 1982. Comparative growth and atrazine response of resistant and susceptible populations of *Amaranthus* from southern Ontario. J. Appl. Ecol. 19: 611–620.

Weiss, P. W. 1980. Germination, reproduction and interference in the amphicarpic annual *Emex spinosa* (L.) Campd. Oecologia (Berl.) 45: 244–251.

Woodmansee, R. G. & Duncan, D. A. 1980. Nitrogen and phosphorus dynamics and budgets in annual grassland. Ecology 61: 893–904.

Wyatt, R. 1984. The evolution of self-pollination in granite outcrop species of *Arenaria* (Caryophyllaceae). III. Reproductive effort and pollen-ovule ratios. Syst. Bot. 9: 432–440.

Wyatt, R. 1986. Ecology and evolution of self-pollination in *Arenaria uniflora* (Caryophyllaceae). J. Ecol. 74: 403–418.

Zangerl, A. R. & Bazzaz, F. A. 1984. Effects of short term selection among environmental gradients on variation in populations of *Amaranthus retroflexus* and *Abutilon theophrasti.* Ecology 65: 207–217.

Zangerl, A. R. & Bazzaz, F. A. 1984b. Niche partitioning between two phosphoglucoisomerase genotypes in *Amaranthus retroflexus.* Ecology 65: 218–222.

Zeide, B. 1978. Reproductive behaviour of plants in time. Am. Nat. 112: 636–639.

Zohary, D. & Plitmann, H. 1979. Chromosome polymorphism, hybridization and colonization in the *Vicia sativa* group (Fabaceae). Pl. Syst. Evol. 131: 143–156.

Vegetatio 77: 33–41, 1988
© Kluwer Academic Publishers, Dordrecht – Printed in the Netherlands

Background processes at the population level during succession in grasslands on sand

G. Fekete, Z. Tuba[1] & E. Melkó[2]
Institute for Ecology and Botany of the Hungarian Academy of Sciences, Vácrátót, Hungary, H-2163
[1]Present address: Department of Botany, Agricultural University, Gödöllö, Hungary, H-2100; [2]Present address: 'Rozmaring' Agricultural Co-operative Farm, Solymár, Hungary, H-2083

Accepted 24.5.1988

Keywords: *Festucetum vaginatae,* Path analysis, Photosynthetic intensity, Reproductive allocation, Successional stage

Abstract

Two population characteristics, reproductive allocation (RA) and the intensity of photosynthesis, were investigated in the successional stages of the sandy grassland communities of the Great Hungarian Plain. Most of the species studied changed their allocational response as succession advanced. Compound path schemes, as casual models, were constructed and analysed for describing intra-individual effects (influence of assimilating organs and root mass of the plant sampled) as well as the phytosociological effects on RA. In pioneer stages, the high values of path coefficients show that the effect of assimilating organs is responsible for the variability of RA. It is also shown that annuals and perennials exhibit different behaviour as to the regulation of RA. It seems that – in terms of path analysis – the reproduction of annuals is influenced by internal factors only. In case of perennials, an external (namely phytosociological) regulation was also observed. It is suggested that successional stages play a role in regulating the photosynthetic intensity of both the dominant species and the whole plant assemblage.

Abbreviations: RA = Reproductive Allocation

Nomenclature: Soó, R. 1980. Taxonomic and plant geographical manual of Hungarian flora and vegetation (Hung.) Budapest.

Introduction

An increasingly promising approach to a deeper understanding of succession utilizes population characteristics. This 'reductionist' alternative to the Clementsian 'organismic' concept was revisited in Drury & Nisbet's paper (1973) which evoked intense reaction among students of vegetation. The recently popular species replacement concept is largely based on events at the population level. The models suggested by Connell & Slatyer (1977; Noble & Slatyer 1980) are also associated with population strategies. There is a vast literature attempting to interpret succession in terms of population characteristics (Fekete 1985). Certain overall trends in adaptive strategies performed by both plants and animals along a successional gradient, are summarized by Brown & Southwood (1987).

In many population-based succession models it is assumed, explicitly or implicitly, that characteristics of the same population are invariant during succession. However, the possibility that at least some population characteristics are subject to change in the consecutive stages of succession cannot be excluded completely. This accords with the view that the regulation of populations is subject to small changes during succession. These changes themselves may become the driving force of succession

This paper is concerned with two population attributes of very different nature. It is examined - in the same species - whether they remain unchanged or suffer some modifications during succession.

Reproduction and succession

It is well-known that the resources available to an organism are partitioned among its activities necessary for survival and successful reproduction. The basic types of matter allocation are strategy-like adaptations of plants (Harper & Ogden 1970). The knowledge on how plants allocate their resources between reproductive and vegetative activities (reproductive allocation, RA, Cody 1966) is important for a general life history theory (Reekie & Bazzaz 1987a, b). In environments where juvenile survival is high, an increase in the proportion of resources allocated to propagules is to be expected. On the other hand a decrease is expected, where juvenile survival is low. This is in agreement with observations suggesting that matter and energy allocation to reproductive organs decreases along with succession of vegetation (Abrahamson & Gadgil 1973; Gaines *et al.* 1974; Newell & Tramer 1978; Abrahamson 1979; Grace & Wetzel 1981). Some counterarguments were presented by Luftensteiner (1980) and Stewart & Thompson (1982). Differences of RA in *Andropogon scoparius* populations from old fields of different ages were found by Ross & Quinn (1977). Similar findings were reported by Scheiner & Goodnight (1984) concerning *Danthonia spicata* populations from stands of different successional ages. In both cases changes occurred due to phenotypic plasticity. Gray (1987) discusses this question in connection with genetic changes during succession. Brown & Southwood (1987) found a strong decrease of RA in *Holcus lanatus* populations growing in three sites of different ages.

In the studies cited above, RA is expressed in terms of biomass allocation. The same method is used in the present study, as well. Reekie & Bazzaz (1987a) proved carbon determination to be a more appropriate method, that can be used as a common currency to assess allocation pattern.

Photosynthesis and succession

There are very few papers available examining the ecophysiology of plants in the course of succession. Photosynthesis is of particular importance among the ecophysiological processes (Bazzaz 1979). The differences found in various parameters (light saturation curves of photosynthesis, the light compensation point, etc.) are due to the sun-adapted and shade-adapted character of species confined to the early and late successional stages, respectively. Similarly, Bazzaz & Carlson (1982) pointed to another relevant characteristic, the photosynthetic flexibility. There are only a few data concerning the photosynthetic behaviour of a species in the consecutive stages of succession. Nearly nothing is known about the photosynthetic productivity of the stands representing these stages.

Material

The sampling site is on sandy tableland situated 25 km west of the town of Kecskemét in the Hungarian Great Plain (Kiskunság National Park, near Fülöpháza). Most sand hills are covered by grasslands, but there are some sand dunes still moving. Here the succession did not reach the forest stage for reasons of vegetation history and ecology. Small clumps of *Populus alba* and its forerunner, *Salicetum rosmarinifoliae* intermingled with *Festuca vaginata* grassland, can only be found in depressions among the dunes. *Juniperus communis* shrubs have colonized occasionally.

A successional scheme concerning the whole xeroseries on the neighbouring sand areas is suggested by Hargitai (1940). Additional important information on the local vegetation is presented by Szodfridt (1969).

Some large moving dunes and colonized sand hills

gave us a good opportunity for succession studies. We were able to distinguish some distinct stages of vegetation development. Generally, the existence of stages is connected to the moving of the sand and the period after colonization. Note that the stage studied represent only the first part of the xeroseries; later stages were not studied.

The dominant grassland in the area is the *Festucetum vaginatae danubiale.* This is the typical plant community on the arid calcareous sand of the Hungarian Great Plain, rich in continental, Pontic, and Pannonian species.

The first stage on moving sand is either a loose assemblage of some annuals (*Secale silvestre, Bromus squarrosus, Silene conica, Corispermum* spp., *Cenchrus tribuloides,* and the perennial *Centaurea arenaria*) or large tufts of *Festuca vaginata* with *Centaurea arenaria* in between. An open perennial grassland forms the second stage. It is relatively rich in species, for example, *Festuca vaginata, Stipa sabulosa, Alyssum montanum* ssp. gmelinii, Koeleria glauca, *Euphorbia seguieriana, Dianthus serotinus, Scabiosa ochroleuca,* etc. Then, in the third stage, the cover of *Festuca* increases causing numerous species to disappear and, on some places, species belonging to the next, shrubby stage occur.

Sampling and methods

For studying RA some 'target' species were chosen. Perennials: the hemicryptophyte *Festuca vaginata, Alyssum gmelinii, Euphorbia seguieriana;* annuals: *Polygonum arenarium, Silene conica, Arenaria serpyllifolia, Erigeron canadensis.* Plants with nearly ripe fruits were collected only.

In the case of *Festuca vaginata* whole tufts (with fertile shoots) were taken together with the upper 30 cm of the root system. In the case of the other two perennials, entire individuals were collected with fruits as well. The *Euphorbia* individuals sampled, usually consisted of both fertile and sterile shoots. In order to obtain the rhizomes of this species, the upper 50 cm of underground parts were removed. For sampling the root system of *Alyssum* a depth of 30 cm was sufficient.

Thirty to fifty individuals of each target species were taken in each successional stage (Fekete & Melkó 1981; Melkó 1984).

In the case of every target individual data were also collected concerning surrounding individuals belonging to either the same or other species. Circular sampling plots with the target individual in the centre were used. Cover was estimated in a species-dependent way, or biomass was measured by removing the individuals together with their root system down to 30–50 cm in depth depending on the species. In the case of annuals individuals were counted. Each species was treated in its specific way in every successional stage. The diameter of the circular sampling plot was dependent on the target species (*Festuca:* 200 cm; *Alyssum:* 50 cm; annuals 10 or 20 cm). The size of the sample area was determined after estimating the influence area of the sampled (target) plant and its competitors (considering rhizosphere volume, mutual shading).

After clearing and removal of the foreign materials the samples were fractionized and the fertile part, assimilation organs and root mass were separated in each species.

In *Festuca* the fertile part means the spica together with the floral axis, the assimilation part includes all parts excluding the fertile part and the roots. The fertile part of *Alyssum* is the raceme, its assimilation part is the ligneous stem together with the leaves. In *Euphorbia* the fertile part consists of the whole umbel with the involuclar braclets and the ripe fruits. The remaining leafy stem of the fertile shoot and the green parts of the sterile shoots belong to the assimilation part, whereas the root system and the rhizome belong to the underground part. The fertile part of *Arenaria* is the capsule with seeds in it, its assimilating part is the remaining aboveground part. The fractions of *Silene* are formed in the same way. The three parts of *Erigeron* are distinguished: (1) the whole raceme (peduncles, capitulum with the ripe fruits; (2) leafy stem and basal leaf; (3) root system. In *Polygonum:* crop; stem with leaves; roots.

The fractions obtained were dried and weighed. The data set for every individual of the target species includes weights of the three main plant parts (as 'intra-individual characteristics') and the scores of sociological components. We note here that for species of minor importance, the total number of individuals or the total cover was recorded within the sample plot, without identification of taxa.

For evaluating the data concerning RA we used path analysis, a multivariate statistical method based on multiple regression. Here, the main task is the determination of path coefficients as standardized partial regression coefficients. The path coefficients reflect the standard deviation of dependent variables explained by one of the independent variables while the others kept constant. The mutual relations and those of cause and effect between variables are shown in a path diagram. The method is applicable to estimate relationships between variables located far apart in the system (path chains). Path analysis was introduced by Wright (1921, 1934) to analyse genetical relationships. Li (1955) and Le Roy (1960) may be consulted for more details of the method. See Hermy (1987) for an example in vegetation ecology.

Photosynthetic activity was measured by a portable equipment on the site. The C_{14} labelled CO_2 content of detached leaves was measured. A detailed description of equipments and measurements is given in Fekete & Tuba (1982).

Results

Changes in relative RA with succession on sand

It seems to be adequate to express the allocation in relative terms (dry weight of reproductive part/dry weight of other parts). Concerning absolute values we characterize only the behaviour of dominant species. There are one-two magnitude differences between the stages. The response of total biomass of *Festuca* is more pronounced between stages 1 and 2, whereas the reproduction is changed considerably, i.e., RA strongly decreases, in stage 3 (Fig. 1).

The relative RA value of the perennials, as seen in Table 1, decreases during succession (in *Festuca,* in the second stage a transitional increase was found).

The response of annuals is less equivocal, but the effect of stages is significant in most cases (Table 1).

For further details see Fekete & Melkó (1981) and Melkó (1984).

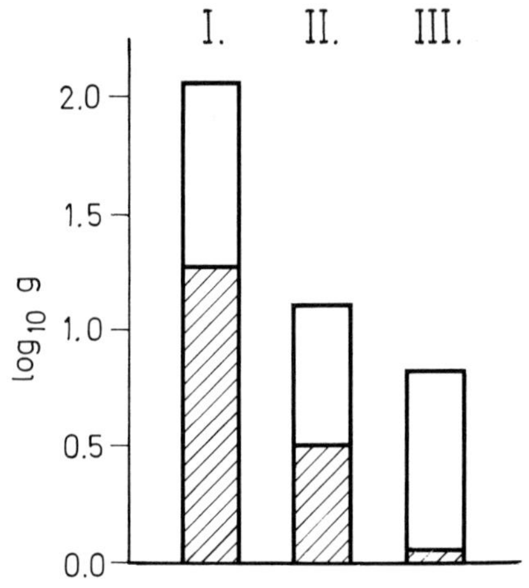

Fig. 1. Dry weight of an average individual of *Festuca vaginata* in successional stages. Columns are proportional to the weight of whole plants, shaded sections indicate the proportion of reproductive organs. I: open annual-, II: semi-closed perennial-, and III: closed perennial stages.

*Intra-individual regulation of reproduction (*Euphorbia seguieriana*)*

The main considerations in constructing our descriptive models (path models) are as follows.

Reproduction is internally influenced and it is to be expressed quantitatively as well (allocational property). So, the internal regions of regulation are the individual (intra-individual) variables – in our case – the mass (weight) of assimilating organs and roots. In a given environment there is generally a definite correlation between reproductive and other main organs of plants. The internal regulation is not independent of external factors.

The simple response model for the rhizomatous *Euphorbia seguieriana* is restricted to internal effects, and the sociological variables are excluded. Observations were made in three stages and in a stand of *Festucetum vaginatae salicetosum rosmarinifoliae* developed in depressions among dunes (Table 1). The latter community is greatly influenced by soil water and its closure is the highest.

The path coefficients (Fig. 2) indicate pronounced trends. At the beginning of colonization in the pioneer open grassland - according to the values of path coefficients - the assimilating organs alone

Table 1. Averages of the relative RA data according to the successional stages and levels of significance of their differences of the species examined (Results of ANOVA) $^*P \leq 5\%$, $^{**}P \leq 1\%$, $^{***}P \leq 0.1\%$.

Festuca vaginata

Successional stage	I	II	III
I	0.183		
II	**	0.261	
III	**	***	0.063

Euphorbia seguieriana (see also Fig. 2.)

Successional stage	I	II	III	IV
I	0.111			
II	NS	0.103		
III	*	*	0.070	
IV	***	***	**	0.049

Silene conica

Successional stage	I	II
I	0.293	
II	***	0.383

Alyssum gmelinii

Successional stage	I	II
I	0.232	
II	***	0.171

Arenaria serpyllifolia

Successional stage	I	II	III
I	0.288		
II	*	0.346	
III	NS	***	0.254

Erigeron canadensis

Successional stage	I	II
I	0.640	
II	**	0.440

Polygonum arenarium

Successional stage	I	II
I	0.334	
II	NS	0.330

are responsible for the variability of reproduction. In the half-closed grassland the influence of rhizome mass is still low, but there is an abrupt change in the closed community. Competition is obviously strong and *Festuca vaginata* supersedes many other species. Reproduction becomes influenced by rhizomes. The *salicetosum rosmarinifoliae* subassociation fits into this sequence; in this community the assimilating organs negatively influence the reproductive allocation.

*Intra-individual and sociological regulation of reproduction (*Festuca vaginata, Erigeron canadensis*)*

The path schemes to be discussed below will express in some sense the relationship between the individu-

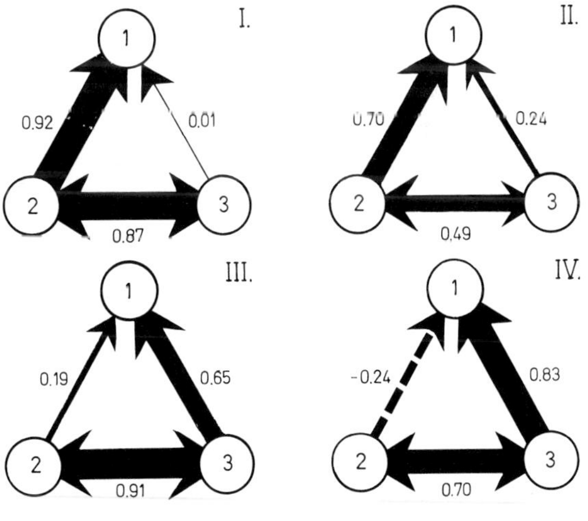

Fig. 2. Intra-individual responses of *Euphorbia seguieriana* to successional stages. One-headed arrows represent path coefficients, double-headed arrows show correlation coefficients. The thickness of lines is proportional to the corresponding value. Solid lines: positive values, dotted lines: negative values. In circles: 1. weight of reproductive parts, 2. weight of assimilating parts, 3. weight of roots. States as in Fig. 1 plus IV: *Festucetum vaginatae salicetosum rosmarinifoliae.*

al and supra-individual levels of organisation. The sociological variables reflect the 'external region', thus expressing the 'density dependence'. (In the models applied the abiotic external world remains unspecified, although abiotic variables, such as soil and other parameters, could have been included.)

Of the perennials examined, the dominant species of the community, *Festuca vaginata,* is discussed here (Fig. 3). On the top of moving dunes (Stage 1) the direct effect of green assimilating parts is decisive on the variability of RA (high value of path coefficient). In the half-closed community (Stage 2), the direction and degree of the two internal components are interchanged. In the closed stage 3 the internal effects are negligible and the coenological (mainly intraspecific) effects become controlling factors. Especially informative is the comparison of stages 2 and 3. Although all compartments are similar (even compartment 7, the 'other species', includes the same taxa), the difference is obvious and important. In a static sense, considering only species association based on presence/absence data, high similarity is expected, but there are significant changes in the abundance and population dynamics of species.

Let us consider now the behaviour of an annual plant, *Erigeron canadensis.* This species is an offensive weed in Hungary. It is the pioneer colonizing species of abandoned vineyards, orchards and ploughed land. In the pioneer assemblages of moving sand *Erigeron canadensis* has also established. In perennial sand steppe communities, it is very scattered and occurs only on bare ground among the *Festuca* tufts. Using the two schemes discussed here we attempt to examine whether the quantity of reproduction of this species is influenced by a sociological variable in stage 2.

The information summarized in Fig. 4 is sufficient to show that in an annual grassland the direct effects, which influence reproduction, are compensated by negative common effects. The direct effect of assimilating organs is very strong and, like for other annuals studied, positive. Sociological impact comes primarily from the 'other species', and reproduction is affected indirectly through influences on roots and green parts. The reproduction of *Erigeron* penetrating into the perennial grassland is most influenced through the assimilating organs. A difference is that the effect of roots is no longer

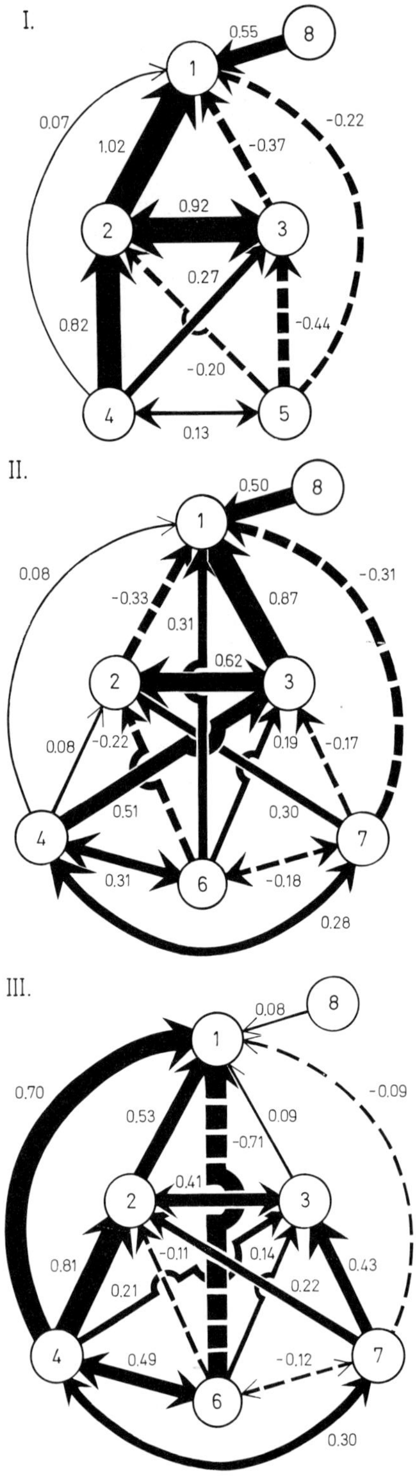

Fig. 3. Path schemes reflecting causes that affect the RA of *Festuca vaginata* in pioneer- (I), semiclosed- (II) and closed (III) stage. Circles 1–3 as in Fig. 2, 4: density of *Festuca vaginata,* 5: density of *Centaurea arenaria,* 6: density of *Euphorbia seguieriana,* 7: density of other species, 8: error.

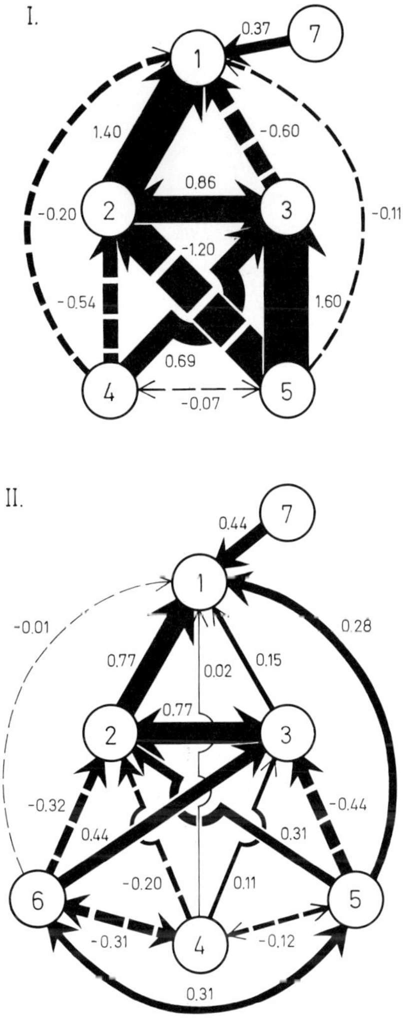

Fig. 4. Path schemes reflecting causes that effect the RA of *Erigeron canadensis* in pioneer annual- (I) and semiclosed perennial (II) stage. Circles 1–3 as in Fig. 2, 4: density of *Erigeron,* 5: density of other species, 6: density of *Festuca vaginata,* 7: error.

negative. The direct sociological effects are negligible. Common effects are of minor importance.

The sociological influences on reproduction provide no explanation for the observation that *Erigeron* usually does not establish in perennial sand grasslands. It is likely that this species is generally weakened if compared to its status in the annual grassland. The mean weight of an individual is six times lower, whereas that of reproductive organs is nine times lower than the respective figures recorded in the annual grassland.

The regulation of photosynthesis (The response of dominant species. Collective responses)

In the analysis of photosynthesis *Festuca vaginata* receives priority over all the other species. Since the absolute magnitude of measurements differed with the data of field work, whereas those obtained at the same time were comparable, the stages will be characterized by relative scores: 100, 52, and 40 (as averages of relative values of measurements taken on three different dates). For details see Fekete & Tuba (1982)). As seen, most significant is the sudden decrease of photosynthetic activity between stages 1 and 2.

It is essential to examine the intensity of photosynthesis at a level of organisation higher than the population level. To do this, five stands of 1 m^2 size were selected in each successional stage and the photosynthetic intensity of each species found in the stand was measured. In every stand, the biomass of green parts was measured for each species. The value of photosynthetic intensity of each species was weighed by biomass, thus yielding the productivity of the stands (Fig. 5). The photosynthetic production calculated for 1 g biomass of the stand was also calculated.

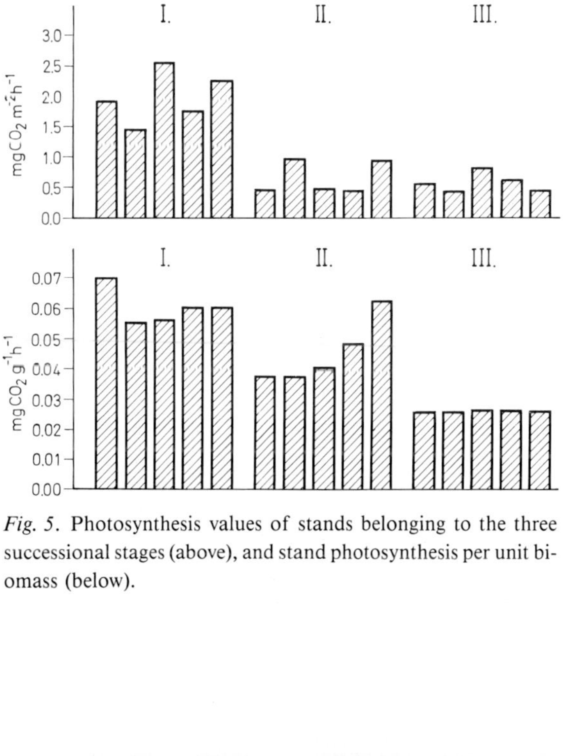

Fig. 5. Photosynthesis values of stands belonging to the three successional stages (above), and stand photosynthesis per unit biomass (below).

Conclusions

The fact that in some species the relative values of RA differ significantly in different stages of succession on sand, implies an external control. Therefore, our aim was to develop a causal scheme for modelling this regulation. During succession both the sociological and abiotic environment may change. In the present work only sociological data were estimated and incorporated into the path-models used. In the pioneer stages, both the annuals and the perennials require large amounts of assimilates for colonisation. (All species examined have small seeds with limited nutrient storage.) Competition is not strong yet and the root system is of subordinate importance. This is in accordance with the fact that in the first stage the assimilating parts are important (in terms of path analysis), more important than the roots in determining RA of annuals and perennials, as well. It may be assumed that the behaviour of perennials and annuals is similar during colonization. (The K-strategist characteristics are not yet manifest.) In annuals, the root will not influence RA even in the later stages. This is true also in the case of other annual species (Fekete & Melkó 1981; Melkó 1984). On the contrary, there is heavy competition among perennials for the nutrient resources in the soil so root mass becomes decisive. The most important finding is that sociological regulation of RA was indicated for perennials only and not for annuals.

An appropriate extension of the path model may be used to evaluate the effect of density on survivorship as well on natality. Such a scheme, including both RA and demographic events, is shown in a simple form in Fig. 6. Of course, the sampling conditions adapted to this suggested model differ from those applied above. Instead of using one individual per sampling unit, the reproductive mass of all the individuals of the target plant within the sampled area should be included in category D_T. Thus, in a conspecific sense, the individual and sociological levels become simultaneously expressed.

Another – population centred – approach to characterizing successional stages is the measurement of photosynthetic productivity. The only species present in every stage is *Festuca vaginata*. The others were different. This is why the data of each population were weighed by biomass, which made the comparison of stands possible. Figure 5 clearly shows that as succession advances, photosynthesis is reduced and in stage 3 exactly the same value is reached consistently. As far as the interpretation is concerned, two alternative explanations appear straightforward, one emphasizing physiology and the other stressing the sociological aspect. The first is the aging of individuals, especially those of perennial species. No doubt that in the pioneer stage several species have dark green, vigorous and large individuals. As maturity advances, the individuals become smaller in size and altered in colour. Leaf marcescence is accelerated, leading to a reduced photosynthetic production in intact leaves. As pointed out earlier, RA decreases with maturity so that fewer individuals are produced, contributing much to a more rapid aging of the stand. Succession induces a shift of age distribution and a uniformisation of ages. Table 2 refers indirectly to this fact; in lieu of an appropriate method of measuring the age of individuals, the size of individuals is used assuming a correlation between age and size.

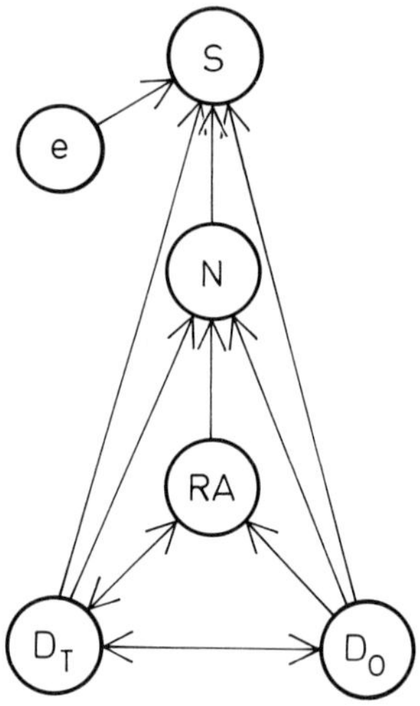

Fig. 6. Proposed elementary demographic model. The path scheme takes into consideration the RA, the natality (N) producing a new cohort, survivorship (S) of a given target plant and the dependence of these parameters on density (D_T, density of target plant; D_O: density of other species), e = error.

The second explanation is that reduced production is a consequence of increased closure in the community. If photosynthesis is measured in stands, there is a striking disproportionateness: even though the green biomass is higher, the photosynthetic

Table 2. Percentage distribution of diameter values of *Festuca vaginata* tufts in stages 1 – 3, on the basis of 200 measurements in each stage.

diameter, cm	1	2	3
0 – 2	9	2	2
2 – 4	4	2	0
4 – 6	7	0	1
6 – 8	11	0	0
8 – 10	12	35	23
10 – 12	27	50	70
12 – 14	27	11	4
14 –	3	0	0
H_e	1.841	1.113	0.840

production per unit weight rapidly decreases. In addition, there is an indication that root mass greatly increases as succession goes on. Its significance is obvious in the competition for nutrients. The extra energy utilized for root growth and respiration, as an allocational feature, probably influences - negatively - the photosynthetic production.

References

Abrahamson, W. G. 1979. Patterns of resource allocation in wildflower populations of fields and woods. Amer. J. Bot. 66: 71–79.

Abrahamson, W. G. & Gadgil, M. D. 1973. Growth form and reproductive effort in goldenrods (*Solidago*, Compositae). Am. Nat. 107: 651–661.

Bazzaz, F. A. 1979. The physiological ecology of plant succession. Ann. Rev. Ecol. Syst. 10: 351–371.

Bazzaz, F. A. & Carlson, R. W. 1982. Photosynthetic acclimatisation to variability in the light environment of early and late successional plants. Oecologia (Berlin) 54: 313–316.

Brown, V. K. & Southwood, T. R. E. 1987. Secondary succession: patterns and strategies. In: Gray, A. J., Crawley, M. J. & Edwards, P. J. (eds), Colonization, succession and stability, pp. 315–337. Blackwell, Oxford.

Cody, M. L. 1966. A general theory of clutch size. Evolution 20: 174–184.

Connell, J. H. & Slatyer, R. O. 1977. Mechanisms of succession in natural communities and their role in community stability and organization. Am. Nat. 111: 1119 – 1144.

Drury, W. H. & Nisbet, I. C. T. 1973. Succession. J. Arn. Arb. 54: 331–368.

Fekete, G. 1985. A terresztris vegetáció szukcessziója: elméletek, modellek, valóság (Succession of terrestrial vegetation: theories, models, reality). In: Fekete, G. (ed.), A cönológiai szukcesszió kérdései (Problems of coenological succession), pp. 31–63. Akadémiai Kiadó, Budapest.

Fekete, G. & Melkó, E. 1981. Reproductive allocation in the stages of sandy succession. Acta Bot. Hung. 27: 351–364.

Fekete, G. & Tuba, Z., 1982. Photosynthetic activity in the stages of sandy succession. Acta Bot. Hung. 28: 291–296.

Gaines, M. S., Vogt, K. J., Hamrick, J. L. & Caldwell, J. 1974. Reproductive strategies and growth patterns in sunflower (*Helianthus*). Am. Nat. 108: 889–894.

Grace, J. B. & Wetzel, R. G. 1981. Phenotypic and genotypic components of growth and reproduction in *Typha latifolia:* experimental studies in marshes of differing successional maturity. Ecology 62: 789–801.

Gray, A. J. 1987. Genetic change during succession in plants. In: Gray, A. J., Crawley, M. J. & Edwards, P. J. (eds), Colonization, succession and stability, pp. 273 – 293. Blackwell, Oxford.

Hargitai, L. 1940. Nagykörös növényvilága II. A homoki növényszövetkezetek (Plant life in Nagykörös II. The plant communities on sand). Bot. Közl. 37: 205–240.

Harper, J. L. & Ogden, J. 1970. The reproductive strategy of higher plants. I. The concept of strategy with special reference to *Senecio vulgaris.* J. Ecol. 58: 681–698.

Hermy, M. 1987. Path analysis of standing crop and environmental variables in the field layer of two Belgian riverine forests. Vegetatio 70: 127–134.

Le Roy, H. L. 1960. Statistische Methoden der Populationsgenetik. Birkhauser, Basel.

Li, C. C. 1955. Population genetics. Univ. of Chicago Press, Chicago.

Luftensteiner, H. W. 1980. Der Reproduktionsaufwand in vier Mitteleuropäischen Pflanzengemeinschaften. Pl. Syst. Evol. 135: 235–251.

Melkó, E. 1984. Reproductive allocation in the stages of sandy succession II. Acta Bot. Hung. 30: 129–137.

Newell, S. J. & Tramer, E. J. 1978. Reproductive strategies in herbaceous plant communities during succession. Ecology 59: 228–234.

Noble, I. R. & Slatyer, R. O. 1980. The use of vital attributes to predict successional changes in plant communities subject to recurrent disturbances. Vegetatio 43: 5–21.

Reekie, E. G. & Bazzaz, F. A. 1987a. Reproductive effort in plants. 1. Carbon allocation to reproduction. Am. Nat. 129: 876–896.

Reekie, E. G. & Bazzaz, F. A. 1987b. Reproductive effort in plants. 3. Effect of reproduction on vegetative activity. Am. Nat. 129: 907–919.

Ross, F. H. & Quinn, J. A. 1977. Phenology and reproductive allocation in *Andropogon scoparius* (Gramineae) populations in communities of different successional stages. Am. J. Bot. 64: 535–540.

Scheiner, S. M. & Goodnight, C. J. 1984. A comparison of phenotypic plasticity and genetic variation in populations of the grass *Danthonia spicata.* Evolution 38: 845–855.

Stewart, A. J. A. & Thompson, K. 1982. Reproductive strategies of six herbaceous perennial species in relation to a successional sequence. Oecologia Berl. 52: 269–272.

Szodfridt, I. 1969. Borókás-nyárasok Bugac környékén (Juniper-poplar stands in the environs of Bugac). Bot. Közl. 56: 159–165.

Wright, S. 1921. Correlation and causation. J. Agr. Res. 20: 557–585.

Wright, S. 1934. The method of path coefficients. Ann. Math. Stat. 5: 161–215.

Vegetatio 77: 43–49, 1988
© Kluwer Academic Publishers, Dordrecht – Printed in the Netherlands

Influence of climate and community composition on the population demography of pasture species in semi-arid Australia

M. P. Austin & O. B. Williams
CSIRO Division of Wildlife and Ecology, P.O. Box 84, Lyneham, A.C.T. 2602, Australia

Accepted 2.9.1988

Keywords: *Astrebla, Callitris,* Demography, El Niño/Southern Oscillation, *Sclerolaena*

Abstract

Substantial recruitment of *Callitris glaucophylla* in woodland, *Sclerolaena birchii* in cleared woodland, and *Astrebla lappacea* in grassland is related to catastrophic events of the past century in the form of interactions between climate, the impact of European land use (sheep, cattle, rabbits) and the rabbit myxoma epizootic. The direct effect of rainfall on the demography of these species and its indirect effect through competition via suites of accompanying plant species are examined. Major long-term changes in plant populations are generated by extreme sequential events rather than by random isolated events. One of the most potent climatic agents for change in eastern Australia is the El Niño/Southern Oscillation phenomenon.

Introduction

Vegetation dynamics are influenced by factors that operate either simultaneously or sequentially (Austin 1981). These factors include both the intrinsic properties of the component species populations as expressed in primary succession and regeneration cycles, and extrinsic forcing factors such as climate. In semi-arid environments the occurrence of extreme rainfall events in successive years is a potent extrinsic factor and the long-term dynamics of plant communities initiated by sequential extreme events may bear little resemblance to predicted communities offered by models based on random climatic events.

In this paper we select three Australian examples which illustrate the consequences of sequential climatic extremes. The first example describes the direct impact of climate on the dynamics of *Callitris glaucophylla* Thompson and Johnson, a conifer tree in grazed, semi-arid woodlands in Australia south of the Tropic of Capricorn (Thompson & Johnson 1986), and the two other examples describe the indirect effect of climate on the dynamics of *Sclerolaena birchii* (F. Muell.) Domin in former *Eucalyptus populnea* F. Muell. woodland and *Astrebla lappacea* (Lindl.) Domin in *Astrebla* tussock grassland through its impact on the establishment and survival of associated species.

Callitris glaucophylla

An area of 400000 ha known as the Pilliga scrub and now mostly a State Forest was open grazing country 120 years ago (Rolls 1981; Chiswell 1982). When the first Europeans moved into the area in the 1830s, they found an open grassy 'forest' with large trees of *Eucalyptus crebra* F. Muell. and *Callitris glaucophylla* at a density of about 8 trees/ha. This woodland had developed under an Aboriginal fire regime and grazing by indigenous marsupials. Figure 1 summarizes the events which then took place. Less

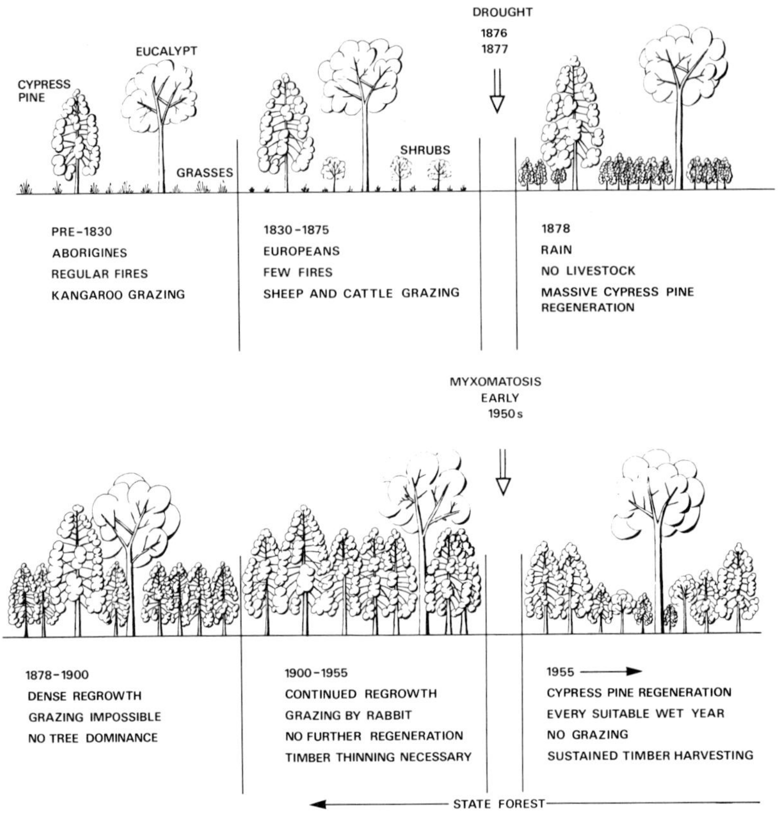

Fig. 1. Events leading to the transformation of forest under Aboriginal hunting and gathering followed by European pastoralism into the Pilliga Scrub and sustainable timber harvesting.

frequent fires and increased grazing by cattle and sheep quickly resulted in an increase in shrubs (the 'woody weed' problem which now plagues many of Australia's grazing lands). These shrubs and other associated species apparently had been, and still are, characteristic of the ridges but over a century ago they began to invade the broad valleys. The resumption of burning by the settlers to control these invaders was unsuccessful. Droughts in 1876 and 1877 forced destocking of the pastures, and when the drought broke in 1878 abundant regeneration of *Callitris, Eucalyptus* and other woody species occurred. *Callitris* does not self-thin very rapidly, and the dense thickets of slow-growing saplings that developed were so firmly entrenched by 1900 that grazing was nearly impossible (Chiswell 1982).

A second catastrophic event in the late 1880s occurred when the European rabbit (*Oryctolagus cuniculus*) invaded the area and stopped any further regeneration of the conifer. As a consequence, by 1950 ecologists had come to regard *C. glaucophylla* as a relict species under the current climate (Lacey 1972). The third catastrophic event was the myxoma epizootic which reached the area in the early 1950s and almost eliminated rabbits. This event, which was aided by abundant rainfall, high rivers and high insect vector populations allowed prolific regeneration of *C. glaucophylla*, completely altering its earlier ecological status under Aboriginal and European cultures. Over 2.5 million stems per ha were recorded locally. Recruitment is now a common phenomenon.

Different types of grazing by native marsupials, domestic livestock and rabbits have interacted with different fire regimes, disease and extreme climatic events to totally change the appearance, ecology and economics of a large area (Rolls 1981; Chiswell 1982). To quote Adamson & Fox (1982), 'The European invasion was a watershed beyond which Australian ecosystems are permanently changed. A revolution has occurred and no steady state is in sight'. It is this turbulent environment with which Australian ecologists have to contend.

Detailed demographic studies of such major events are rare; evidence is anecdotal and interpretations are often limited by confounding factors. Conversely long-term experiments on grazing effects may be confounded by climatic effects; e.g. Austin *et al.* (1981) found that successional trend, seasonal fluctuations in winter rainfall and soil type differences totally obscured any effect of the designed experimental treatment of grazing intensity in a 20 year grazing trial.

Scleroleana birchii in southern Queensland

In undisturbed *Eucalyptus populnea* woodland this species is a sparse stunted (to 10 cm) spiny shrub. Following tree clearing and sheep grazing *S. birchii* exhibits periodic abrupt increases in abundance (Fig. 2) and it has been proclaimed a noxious weed under state legislation (Menz & Auld 1977). It can develop into wide-spread and dense monospecific communities of plague proportions, with individual hemispherical plants to 1 m diameter. By three months of age *S. birchii* plants produce viable seed within spined woody monocarpic fruits even on heavily grazed plants; ultimately the plants decay and the fruits break-down in the soil over a period of years to produce a fluctuating pool of germinable seed (Auld 1981). R. Roe established an experiment near St George, Queensland (28°10′ 148°54′) using permanent quadrats under various grazing treatments over the period 1937–47 to examine the population dynamics of the species. The results reported here are quoted from a collaborative but unpublished re-analysis of the original data by Austin, Roe, Williams and Werner. The full results will be published shortly.

The detailed records enabled cohort life tables to be prepared. These showed germination occurring in all seasons, but with major recruitment usually in the cool season. Figure 3 shows the survival of the cohorts from the years 1936 to 1947 in a permanent quadrat within the light grazing treatment. The cohorts from August–December 1936, March 1937 and August 1942 were longer lived than all other cohorts, due to precipitation which briefly arrested mortality. From early 1940, severe drought was punctuated by a few small rainfall events which were adequate to initiate cohorts, but insufficient to prolong their existence. From 1937 to 1945 the density of all species on the permanent quadrat was

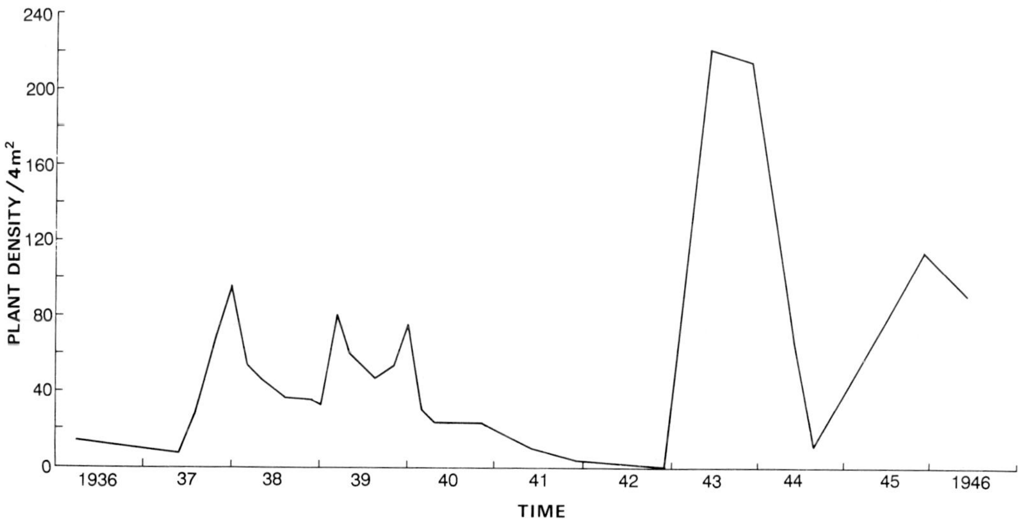

Fig. 2. Population dynamics of mature plants (>6 months age per 4 m^2) of *Sclerolaena birchii* 1936–1946 under light grazing by Merino sheep.

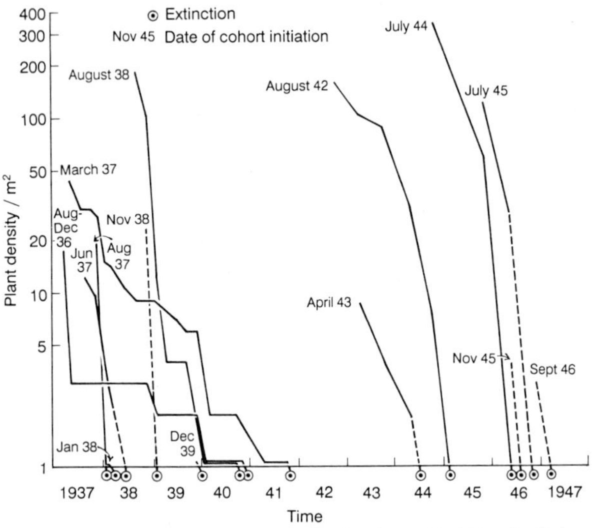

Fig. 3. Survivorship of cohorts of *Sclerolaena birchii* under light grazing by Merino sheep 1936–1947 (per 1 m^2). Extinction o; Date of cohort initiation.

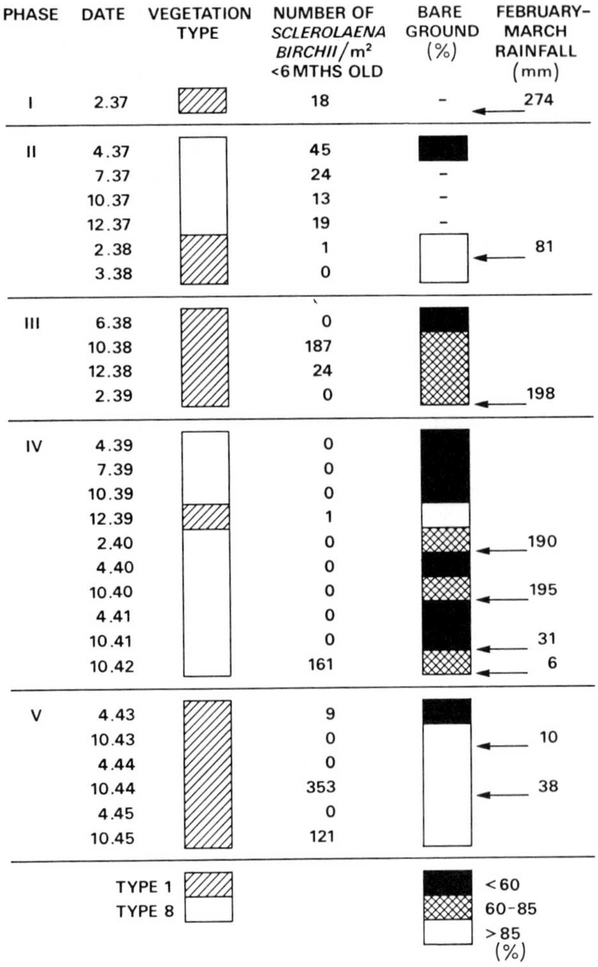

Fig. 4. Changes in vegetation type and numbers of *Sclerolaena birchii* juveniles (>6 months age) under light grazing by Merino sheep in relation to bare ground over the period February–March 1937–1945 (per 1 m^2). Type 1 is erect, perennial, small shrub. Type 8 is prostrate, perennial and herbaceous.

recorded and this enabled us to complete a floristic analysis of changes in the total vegetation of the quadrats (cf. Austin *et al.* 1981); no pattern or sequence of relevant changes was detected. Grazed semi-arid pastures in Australia are a mixture of native grasses, herbs and small shrubs with many introduced species of annual weeds and grasses with similar habitat requirements. Because many of the species have similar phenological life histories and morphological growth patterns it was hypothesized that they might well substitute for each other in the open grazed community. The species were then classified into a number of groups with similar morphologies. The observations for each quadrat at each recording were then classified on the basis of the presence of these species groups.

Figure 4 shows the behaviour of the vegetation through time for the quadrat in the light grazed treatment. Abundant February–March rainfall triggers a change in vegetation type and a decrease in bare ground. There is relatively little change in *S. birchii* numbers. With insubstantial late summer rainfall the vegetation type switches – autumn germinating species are replaced by winter and spring germinators, bare ground increases, and there is a spring flush of *S. birchii* juveniles. Abundant late summer rain occurred in 1939 and 1941. Without a preceding spring flush these late summer rainfall events produce a massive growth of autumn annuals. Dense stands of small plants, e.g. *Tripogon loliiformis* (F. Muell.) C. E. Hubbard with mats of warm season species, e.g. *Portulaca oleracea*, L. suppress the *S. birchii* plants that germinate in the following cool season. Note that even though winter rains were generally satisfactory for recruitment in this period and a substantial population of adult plants of *S. birchii* persisted, no plague outbreak of *S. birchii* occurred. After 1941 a general drought occurred, characterized by a lack of late summer rains which increased the amount of bare ground. Sporadic rainfalls initiated 2 cohorts in 1943 (Fig. 3). The drought killed the perennial grass species of genera such as *Stipa* and *Aristida* that partially control *S. birchii* when there is abundant summer rainfall; although *S. birchii* recruitment occurred in the absence of this competition on ungrazed plots, the recruits were both fewer and smaller than with grazing.

Our age-specific analysis of the reproductive status of *S. birchii* during 1937–48 on the southwestern Queensland site suggest that depletion of the labile seed pool in the soil is unlikely (cf. Auld 1981 for a more southerly site). There is a 'window' for *S. birchii* outbreaks through the agency of suitable winter and spring rains, but *only* if there are no prior heavy rains in late summer. Indeed, the average rainfall in February–March for four consecutive years in this early experimental period when establishment did not occur was at least three times the long-term average. Drought opens this 'window' for *S. birchii* on all types of grazing management in this type of degraded plant community. Although rabbits do not have a formative role as with *Callitris glaucophylla* in the Pilliga example, their activities do maintain heavy fruiting populations of *S. birchii* around warrens in ungrazed treatments at the time of maximum *S. birchii* control through the summer rainfall – perennial grass phase described earlier.

In this degraded plant community with its short-lived herbaceous vegetation, the changes are transient but have important impacts and implications for pasture management and research. Climate overwhelms the effects of management practices imposed on a suite of plant species none of which are capable of exerting a long-term dominant role either singly or in combination. *S. birchii*, like *Callitris glaucophylla*, is playing a role that it did not have under Aboriginal management. Recruitment in both species depends on occasional rainfall events.

Astrebla grasslands

Grasslands dominated by *Astrebla lappacea, A. pectinata* (Lindl.) F. Muell. ex Benth, and *A. elymoides* F. Muell. ex Benth. occur on heavy clay soils in Queensland and the Northern Territory. Debate over the decline of these grazing lands and concern at

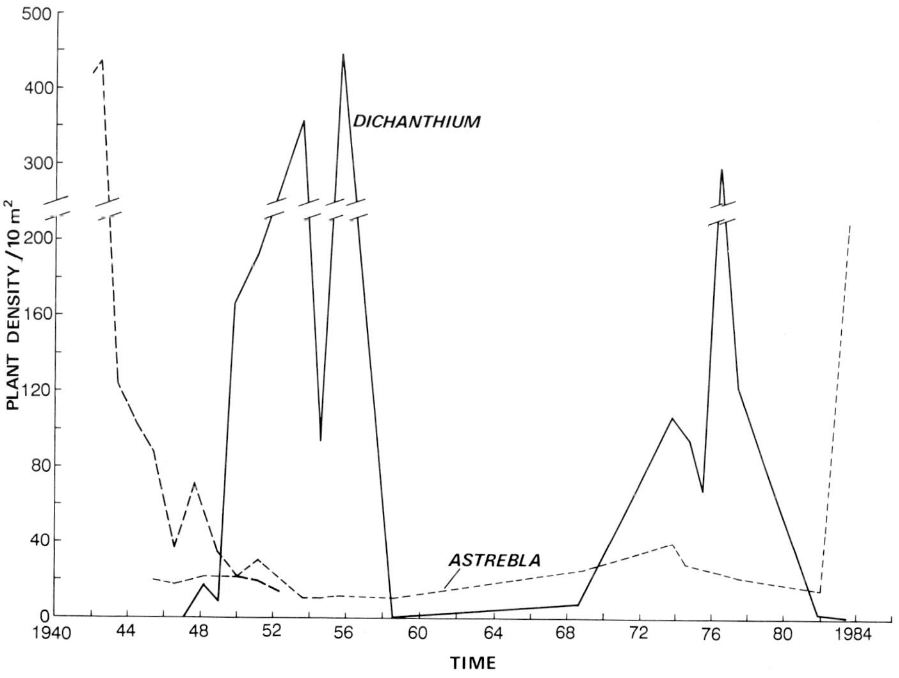

Fig. 5. Population dynamics of *Astrebla lappacea* and *Dichanthium sericeum* 1940–1984 (per 10 m²).

overstocking is a periodic phenomenon. In addition, early rather limited ecological studies (Blake 1939) had suggested that major community changes could take place on these heavy clay soils from an *Astrebla* dominated grassland to one dominated by *Dichantheum sericeum* (R. Br.) A. Camus (the so-called 'shifting climax'). However, observations by R. Roe and O. B. Williams on the long-term dynamics and demography of *Astrebla lappacea*, using permanent plots and careful recording within a simple experimental design provide clear evidence of what has occurred (Fig. 5). Grazing has been shown to have little impact on recruitment and survival of *A. lappacea* plants. Although seedlings establish in small numbers every few years when it rains, few plants persist. Records from 1941–1983 (Fig. 5) show only two major seedling establishment events. For *A. lappacea* to establish in any numbers requires either (a) at least 100 mm of rain on one occasion in spring followed by consistent summer rains or (b) a similar autumn rainfall with suitable rain in the following winter.

When populations of *Astrebla* spp. fall to a hundred plants or less per ha, exceptional rainfall sequences are required in order to produce a seed-bearing cohort which in turn can generate a substantial cohort in the following year (Roe 1941). The community in south-western Queensland appears to get this rainfall sequence once in every fifteen to twenty years, with anecdotal evidence suggesting previous events around 1916 and 1934. The long gap from 1941 to 1984 may be attributed to competition from *S. sericeum* cohorts in the wet 1950s, drought in the middle 1960s and infrequent cool-season rainfall in the 1970s (see also Roe & Davies 1985). Both *D. sericeum* and, to a lesser extent, the annual grass *Isielema membranacea*, will germinate and establish in dense populations under warm-season rainfall conditions to out-compete *A. lappacea* spp. seedlings. Further, the large suite of cool-season species that grow when the occasional winter rainfalls occur can also outcompete *A. lappacea*, hence the significance of the late spring and autumn rains in providing the limited 'window' for *Astrebla* recruitment.

Records of other associated species suggest (Fig. 5) that the observations in the literature about *D. sericeum* dominance were probably made during visits in years when there had been a pulse of establishment and growth by this species which behaves as a short-lived perennial in semi-arid Queensland.

Conclusion

Knowledge of Australian environment and the changes which have taken place since the invasion of Europeans and their animals and plants is still poorly documented and poorly understood (Williams 1985). Mechanisms determining community structure and their relative importance are much discussed and speculated upon at present. Without observations and understanding of what actually happens in a series of communities from different parts of the Australian environment such theoretical debate is unfruitful. Conclusions based on single locations and for short time periods would be risky in the light of the examples discussed here and the current knowledge of north-south demographic clines in eastern Australia for species such as *Enteropogon* (Williams 1970; Michalk & Herbert 1978) and *Astrebla* spp. (D. M. Orr pers. comm.).

Knowledge of climate, global weather systems and local weather can aid demographic analysis. Central to this knowledge is the operation of the El Niño/Southern Oscillation (ENSO) phenomenon (Nichols 1987), formerly studied as two separate entities, but now recognised as linked parts of the same atmosphere-ocean climate system. A strong occurrence of ENSO affects weather on the west coast of South and North America, Australia, New Zealand and Indonesia (World Climate Data Programme 1987). Each of the examples given can be placed in the context of ENSO climatic events.

We can now appreciate that the Pilliga scrub developed in the rainy aftermath of the severe ENSO event of 1876–78, and the myxoma virus epizootic recruitment of *Callitris glaucophylla* and *Eucalyptus crebra* occurred in the rainy aftermath of the 1951 ENSO. The converse appears to operate with *Sclerolaena birchii* with the warm-season rainy phase of ENSO discriminating against *S. birchii* recruitment in the following cool season. Significant *Astrebla lappacea* recruitment can be shown to be the result of either the rainy phase of an ENSO event, or substantial warm-season precipitation in an off-ENSO year. Further establishment events occurred when a 'window' was opened briefly following the ENSO events of 1957–58, 1965–66, 1972–73 and 1982–83.

We suggest that vegetation responses at both coarse and fine scale, whether under the full power of a strong ENSO event, a mild event, or a substantial rainfall event in a non-ENSO year, are capable of prediction to an extent not appreciated hitherto. Such prediction could have a substantial impact on how we model plant demography in semi-arid regions and conduct future pasture research and management.

Acknowledgements

We gratefully acknowledge the contribution to this subject made by R. Roe, P. A. Werner, D. M. Orr and B. A. Auld. The editorial group of the CSIRO Division of Water Resources, Canberra, prepared the illustrations and manuscript for publication.

References

Adamson, D. A. & Fox, M. D. 1982. Change in Australasian vegetation since European settlement. In: Smith, J. M. B. (ed.), A history of Australasian vegetation. McGraw-Hill Book Co., Sydney.

Auld, B. A. 1981. Aspects of the population ecology of Galvanised Burr (*Sclerolaena birchii*). Aust. Rangel, J. 3: 142–148.

Auld, B. A. 1983. Historical aspects of the Galvanised Burr problem in New South Wales. J. Aust. Inst. Agric. Sci. 49: 17–19.

Austin, M. P. 1981. Permanent quadrats: An interface for theory and practice. Vegetatio 46: 1–10.

Austin, M. P., Williams, O. B. & Belbin, L. 1981. Grassland dynamics under sheep grazing in an Australian Mediterranean type climate. Vegetatio 47: 201–211.

Blake, S. T. 1938. The plant communities of western Queensland and their relationships, with special reference to the grazing industry. Proc. Roy. Soc. Queensland 49: 156–204.

Chiswell, E. L. 1982. The Pilliga Story. Forest and Timber 18: 23–24.

Lacey, C. J. 1972. Factors influencing occurrence of Cypress Pine regeneration in New South Wales. Forestry Comm. of New South Wales Techn. Paper 21.

Menz, K. M. & Auld, B. A. 1977. Galvanised Burr, control and public policy towards weeds. Search 8: 281–287.

Michalk, D. L. & Herbert, P. K. 1978. The effects of grazing and season on the stability of *Chloris* spp. (windmill grasses) in natural pasture at Trangie, New South Wales. Aust. Rangeland J. 1: 106–111.

Nicholls, N. 1987. The El Niño/Southern Oscillation phenomenon. In: Glaztz, M., Katz, R. & Krenz, M. (eds), The societal impacts associated with the 1982–83 worldwide climate anomalies, pp. 2–10. National Centre for Atmospheric Research, Boulder, Colorado.

Roe, R. 1941. Studies of the Mitchell Grass Association in southwestern Queensland. 1. Some observations on the response of Mitchell grass pastures to good summer rains following the 1940 drought. J. Coun. Sci. Industr. Res. 14: 253–259.

Roe, R. & Davies, H. I. 1985. Long-term variation in the density of *Astrebla* spp. in the Warrego region of Queensland. Tropical Grasslands 19: 87–95.

Rolls, E. C. 1981. A million wild acres. Nelson, Melbourne.

Thompson, J. & Johnson, L. A. S. 1986. *Callitris glaucophylla*, Australia's 'White Cypress Pine' – a new name for an old species. Telopea 2: 731–736.

Williams, O. B. 1970. Population dynamics of two perennial grasses in Australian semi-arid grassland. J. Ecol. 58: 869–875.

Williams, O. B. 1985. Population dynamics of Australian plant communities, with special reference to the invasion of neophytes. In: White, J. (ed.), The population structure of vegetation, pp. 623–625. Junk, Dordrecht.

World Climate Data Programme 1987. The global climate system, autumn 1984–April 1986. CSM R 84/86. W.M.O., Geneva.

Vegetatio 77: 51–56, 1988
© Kluwer Academic Publishers, Dordrecht – Printed in the Netherlands

Linear trend in multi-species time series

Otto Wildi
Swiss Federal Institute of Forestry Research, Landscape Division, CH-8903 Birmensdorf, Switzerland

Keywords: Classification, Monitoring, Noise, Ordination, Permanent plot, Regression, Smoothing

Abstract

Investigation of permanent plots is the traditional approach to detect changes in species performance and floristic composition. When the time reserved for investigations is limited and statistically independent replicate samples for normal time series analysis do not exist, ordination of multi-species series is often applied. The approach is further developed here with time series data from wetland communities over six consecutive years. Random fluctuation and linear trend are the two mechanisms which can explain the observed changes. Trend analysis of species scores allows to smooth the data and hence the resulting ordination pattern. The expected scores are a conservative measure for trend, taking into account all the recorded time states of the system.

Nomenclature: Hess, H. E., Landolt, E. & Hirzel, R. 1976–1980. Flora der Schweiz und angrenzender Gebiete. (2nd ed.). Birkhäuser, Basel. Vol. I–III.

Introduction

Investigations of natural systems in community ecology yield data sets in which uninterpretable noise often represents the main source of variation. Most theories on vegetation on the other hand, deal with functional relations or distinct structures. Examples are Clement's (1916) organismal view of plant communities, Gleason's (1926, 1939) concept of gradients in space and time, Watt's (1947) theory on pattern and process and other recent approaches reviewed by Crawley (1986) and Noy-Meir & van der Maarel (1987). Even though random variation is accepted as a natural phenomenon and discussed in the context of uniformity (Dahl 1960) or homogeneity (Tüxen 1977), it is generally considered to limit the interpretability of results.

Multivariate analysis is an efficient tool to summarize noisy and complex data (Orlóci 1978), within which ordinations were soon considered an effective means to unfold trends (van der Maarel 1969; Austin 1977). Although some of the information is thus disregarded, random noise tends to get partly suppressed in favour of systematic trends. However, Swaine & Greig-Smith (1980) remark: 'The success of this approach depends on the assumption that the change in vegetation will lead towards other species-compositions already represented in the ordinations' (p. 33). Distortion may occur when the main configuration cannot be represented in two or three dimensions. 'Background noise' is a second limiting factor for the interpretability of the results, and here Swaine & Greig-Smith (1980) suggest constraint oridination, in which between-stand variation is first removed and only temporal changes are considered. This, however, only works when treatments based on true replicate samples are compared. As our case shows, not even the trace of an impact is known and each sample plot represents an individual stand developing in an unpredictable direction.

That persisting changes are hidden by uninterpretable variation occurs frequently in plant ecology (Green 1979). As long as sufficient data are available for the analysis, canonical analysis of changing species frequencies (Orlóci 1981) often separates monotonous trends from uninterpretable noise (Feoli & Orlóci 1985). In single plots, curve fitting procedures may be applied to species time-response curves (Austin *et al.* 1985). The more complex the functions, the more data points in time are needed to achieve significant results. Thus, if only very few time steps are known, the only remaining option is linear regression.

Data and method

The data discussed here are from 25 permanent plots on a former lake bottom of the Aargovian Reuss valley, south of Bremgarten, Switzerland. The area, about 6400 ha, is now mainly agricultural with scattered nature reserves representing remnants of a time when the former lake became silted up and early settlers extensively used the periodically flooded plain. The vegetation types range from dry *Bromus erectus* stands and *Molinia* dominated meadows through small sedge meadows, *Carex elata* communities and *Phragmites* swamps to open backwater areas (Klötzli 1969). The plots have been marked with iron stakes, which are easily located with a precision theodolite. Relevés have been made each year from 1980–1986. The abundance was estimated using the Braun-Blanquet scale, and the occurrence of all species within five subplots was recorded. The latter measure is an unbiased ordinal frequency measure with a range from 0–5 and has been used in the present analysis.

As the selection of appropriate ordination methods is crucial (Kenkel & Orlóci 1986; Minchin 1987a), alternatives have been considered. Firstly, a variant of principal component analysis was used. The frequency counts are square-root transformed, the relevé vectors normalized as suggested in the papers mentioned above. Secondly, the often differing untrended version of correspondence analysis was used. A third attempt with nonmetric multidimensional scaling was given up after the algorithms failed to converge sufficiently when applied to the complex field data.

Many different methods exist to detect a trend. In order to operate at a level close to the raw data, the increase or decrease of species is determined individually. We compute a regression between time step x and species performance y:

$$y = ax + b + e \qquad (1)$$

where a and b are regression coefficients and e is a randomly distributed variable containing the residual. The correlation coefficient r is a measure of goodness of fit. The expectations for the ith species score $\hat{y}_i$ at any given time step x are trended values without noise:

$$\hat{y}_i = a_i x + b_i \qquad (2)$$

For practical reasons, the scores have to be limited to the maximum range of the observed values, i.e. 0–5. From this procedure simulated relevés for each time step result. They represent input for further ordinations and other analyses. In order to measure the rate of change, the similarity ratio of relevés is computed between one or several time steps. Taking the entire period of time as a reference (Fig. 1), the similarity ratio between the first and the last simulat-

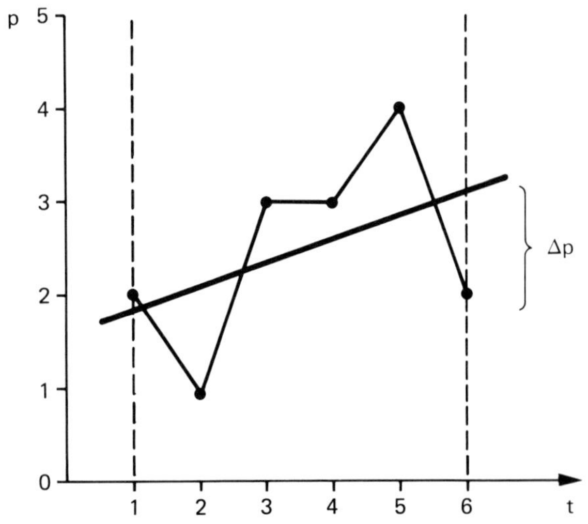

Fig. 1. Linear trend (heavy line) of one species score p with range 0–5 over 6 time steps. Δp is considered the optimal estimate for the persisting change in performance.

ed relevé is easily interpreted in terms of 'typical' plant communities (Gauch 1982, p. 102): If any resemblance drops below about 0.5, it can be assumed that the type of vegetation has changed fundamentally.

For all analyses, a revised version of the program package by Wildi & Orlóci (1983) has been used.

Results

In a first step, ordination is performed with all relevés including sites and time steps. The results are shown in Fig. 2. The arrows indicate the change of plots in time. It is evident that in some cases there is much variation or even a shift occurring earlier or later during the time of investigation. Examples are the plots 5, 6, 7, 10 and 21. Others exhibit minor fluctuations, such as 2, 14 and 23. But in most cases it is difficult to decide whether a true shift occurred or not. The performance of the two ordination methods applied, differs greatly. Correspondence analysis resolves the extreme types (fertilized) best, yet squeezes all the others. Principal component analysis offers a more even representation, but leaves the extremes difficult to identify. Considering more axes (not shown here) does not help. Consequently, it appears that both ordinations are needed for the interpretation.

After computing the regression the resulting data (expectations) which are free of noise are subjected to the same type of analysis. The results are shown in Fig. 3, where the similarity between the first and the last relevé in each vector is added in front of the plot label. Obviously, the trend is much easier to interpret than in the original data. In both ordinations, plots 5, 6 and 7 as well as 2, 15 and 24 converge. The correspondence analysis produces more coherent results than achieved with the original, more varied data (Fig. 2). On the other hand some rather short arrows represent plots with low overall similarity, indicating a high degree of change (e.g. plots 9 and 10). Even though the direction of the arrows appears to be meaningful, this kind of distortion means that the amount of change cannot be deduced from this ordination. In the principal component analysis ordination, the length of the arrows is a more significant, yet still not perfect indication. As a disadvantage, they are curved along the edges of the graph, an effect well known from the analysis of artificial data (c.f. Kenkel & Orlóci 1986). Some types are superimposed, such as plots 10 and 11 with 5, 6 and 7. Compared with correspondence analysis, more axes would have to be considered to resolve the main trends.

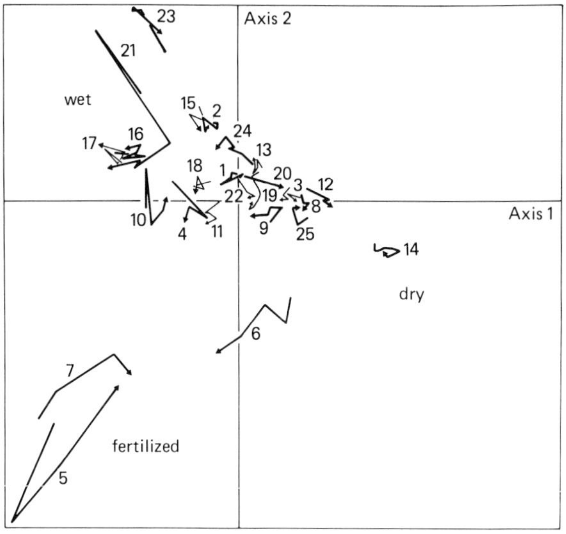

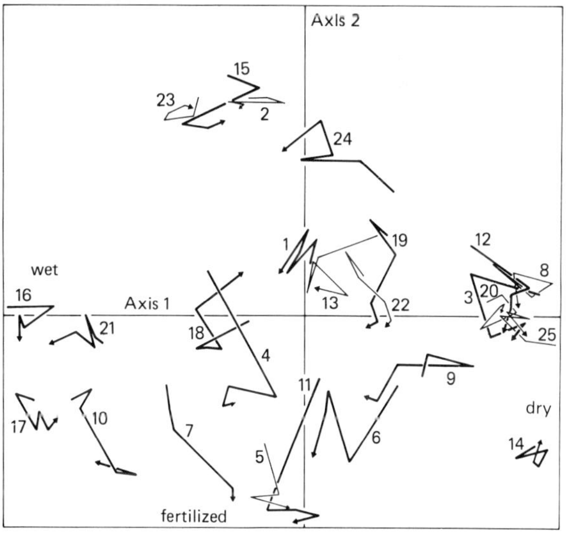

Fig. 2. Ordination of the time series data. Upper graph: Correspondence analysis. Lower graph: A variant of principal component analysis.

Although the ordinations of smoothed data showed the main trends, an other representation, not

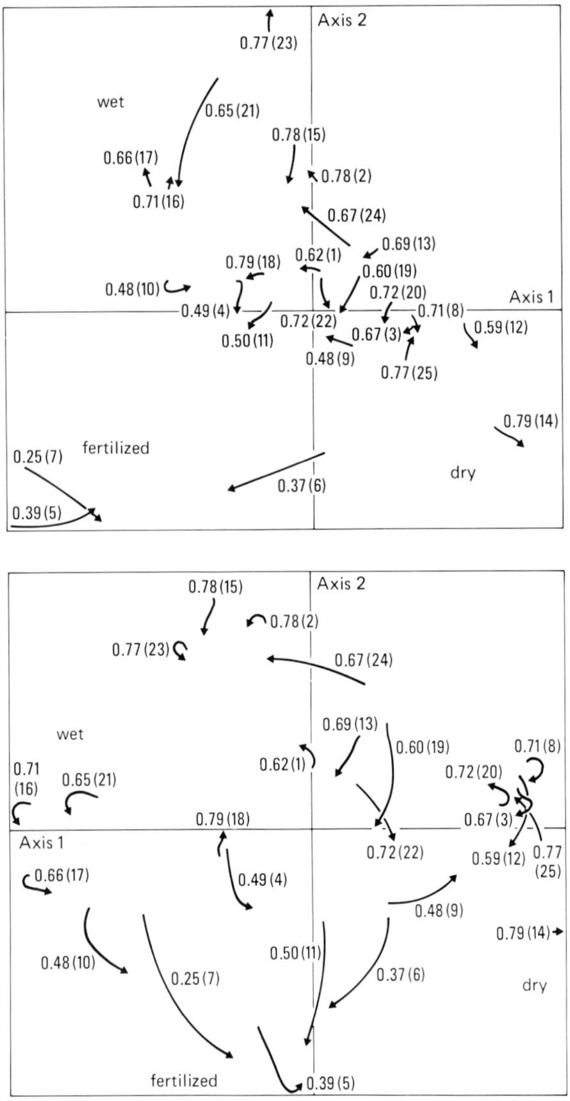

Fig. 3. Ordination of the time series data, smoothed by linear regression. Upper graph: Correspondence analysis. Lower graph: A variant of principal component analysis. Plot numbers (in parentheses) correspond with those in Fig. 2. The amount of change can be deduced from the similarity between the first and last simulated relevé.

suffering from any distortion would be desirable. The similarity of the first and last relevé in each plot is a measure which can easily be interpreted in phytosociological terms (Fig. 4). As a result of minimum variance cluster analysis (Pielou 1984), the plots are ordered in groups. This helps to interpret the observed changes. The rather stable group A

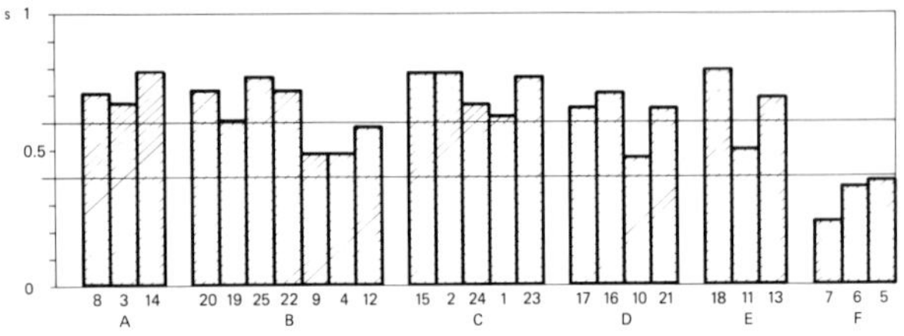

Fig. 4. Vegetation change in 25 plots, measured by the similarity ratio S between the first and the last simulated relevé. Plots with $S > 0.6$ are considered 'stable', between 0.4 and 0.6 'doubtful', below 0.4 they are 'unstable'. The plots are ordered according to the result of a cluster analysis.

summarizes dry sites with *Briza media, Carex flacca, Linum catharticum* and *Colchicum autumnale* as differentiating species. Group B is of the *Molinia coerulea* type, with *Festuca rubra, Betonica officinalis* and *Holcus lanatus* occurring in all stands. While most are fairly stable, plots 4 and 9 tend towards persisting changes. Group C, characterized by *Juncus alpinus, Mentha aquatica* and *Ranunculus flammula,* again, is less variable. Group D (*Carex elata* and *Galium palustre* as typical species) as well as Group E (*Molinia coerulea, Lysimachia vulgaris, Vicia cracca, Galium album* and *Primula elatior*) consist of mostly stable stands, with only one of each being questionable. The area of group F with the typical species *Poa annua* and *Plantago lanceolata*, merely cut since 1980, had originally been fertilized. This impact shows as a clear trend with similarities of the extremes below 0.4.

In order to help to interpret the results, the scale in Fig. 4 is divided into three arbitrary classes. Plots with similarity S above 0.6 usually belong to the same type of vegetation, and changes within this range are assumed to represent 'normal' variation. With S 0.4–0.6, the situation is doubtful. Only a prolonged period of investigation could show if the observed trend persists. With $S < 0.4$ the trend is strong and considered likely to continue. In our example, this has been observed only where the human impact is known to have changed. In all other plots, the yearly variation is still dominating.

Discussion and conclusion

Trend and noise are properties of central interest to plant ecology. While the former is widely discussed in the context of gradient analysis, there is a 'low level of interest attached to noise' (Gauch 1982). An exception is gradient simulation where species response curves are constructed in one or more dimensions (Lagonegro 1984; 1986; Minchin 1987b, and others). Since the data represent functional relationships of given site models noise is generated to achieve more realistic conditions for data analysis. Different multivariate analysis methods are then tested in their ability to unfold trends under complex and specifically disturbed conditions. Lagonegro (1984) refers to this as 'calibration of methods'. Most of the methods used in community ecology have been tested using artificial data (cf. Minchin 1987a) or well known field data. The ordinations in Fig. 3 represent an intermediate attempt. The data used are neither real and therefore noisy, nor are they entirely artificial with a priori defined distribution. Figure 3 shows how correspondence analysis performs much better under conditions, where the variation has been reduced, than it does with unembellished field data. In principal component analysis, the configurations resulting from noisy and smoothed data are more alike. Consequently, this method is considered more robust with regard to minor changes in the data, even though the ordinations suffer from the well known distortions (Kenkel & Orlóci 1986).

A critical point is the response of the species to changes in time. Linearity, as in this example, is only justified for small trends. Otherwise a Gaussian or similarly curved function would be more appropriate (Austin *et al.* 1984; ter Braak & Looman 1986). With ten or more time steps, the method would become a Gaussian regression. The approach described in this paper is only justified where the number of time steps lies between four and about ten, and where mainly continuous and no drastic changes occur in species compositions.

Of course simulation of data has much wider applications than shown in this particular example. Investigation of homotoneity (Tüxen 1977) is a case in point. Whenever a vegetation table is classified, group centroids can be determined. A new set of relevés, generated by introducing random noise instead of computing trend, may be compared with the field data. The interpretation may help to answer one of the most difficult questions in plant ecology: is an unexpected combination of species caused by a trended environmental effect or just triggered by random events?

Acknowledgements

The author thanks the Aargovian Government and Dr E. Leupi, ANL, for financial support and collaboration.

References

Austin, M. P. 1977. Use of ordination and other multivariate methods to study succession. Vegetatio 35: 165–175.

Austin, M. P., Cunningham, R. B. & Fleming, P. M. 1984. New approaches to direct gradient analysis using environmental scales and statistical curve fitting procedures. Vegetatio 55: 11–27.

Clements, F. E. 1916. Plant succession and analysis of the development of vegetation. Carnegie Inst. Wash. Publ. 242: 1–512.

Crawley, M. J. (ed.) 1986. Plant ecology. Blackwell Scientific Publications, Oxford.

Dahl, E. 1960. Some measures of uniformity in vegetation analysis. Ecology 41: 805–808.

Feoli, E. & Orlóci, L. 1985. Species dispersion profiles of anthropogenic grasslands in the Italian eastern Pre-Alps. Vegetatio 60: 113–118.

Gauch Jr, H. G. 1982. Multivariate analysis in community ecology. University Press, Cambridge.

Gleason, H. A. 1926. The individualistic concept of the plant association. Bull. Torrey Bot. Club 53: 7–26.

Gleason, H. A. 1939. The individualistic concept of the plant association. Amer. Midl. Nat. 21: 91-100.

Green, R. H. 1979. Sampling design and statistical methods for environmental biologists. Wiley, New York.

Kenkel, N. & Orlóci, L. 1986. Applying metric and nonmetric multidimensional scaling to ecological studies: some new results. Ecology 67: 919-928.

Klötzli, F. 1969. Die Grundwasserbeziehungen der Streu- und Moorwiesen im nördlichen Schweizer Mittelland. Beitr. Geobot. Landesaufn. 52.

Lagonegro, M. 1984. SPAGHET: A coenocline simulator useful to calibrate software detectors. Stud. Geobot. 4: 63–99.

Lagonegro, M. 1986. Performances of a proximity index defined on a dendrogram table or a minimum spanning tree graph. Universita degli studi di Trieste, 63 pp.

Minchin, P. R. 1987a. An evaluation of the relative robustness of techniques for ecological ordination. Vegetatio 69: 89-107.

Minchin, P. R. 1987b. Simulation of multidimensional community patterns: towards a comprehensive model. Vegetatio 71: 145-156.

Noy-Meir, I. & van der Maarel, E. 1987. Relations between community theory and community analysis in vegetation science: Some historical perspectives. Vegetatio 69: 5–15.

Orlóci, L. 1978. Multivariate analysis in vegetation research. 2nd ed., Junk, The Hague.

Orlóci, L. 1981. Probing time series vegetation data for evidence of succession. Vegetatio 46: 31–35.

Pielou, E. C. 1984. The interpretation of ecological data. A primer on classification and ordination. Wiley, New York.

Swaine, M. D. & Greig-Smith, P. 1980. An application of principal components analysis to vegetation change in permanent plots. J. Ecol. 68: 33–41.

ter Braak, C. J. F. & Looman, C. N. W. 1986. Weighted averaging, logistic regression and the Gaussian response model. Vegetatio 65: 3–11.

Tüxen, R. 1977. Zum Problem der Homogenität von Assoziations-Tabellen. Doc. Phytosoc. 1: 305–320.

van der Maarel, E. 1969. On the use of ordination models in phytosociology. Vegetatio 39: 97–114.

Watt, A. S. 1947. Pattern and process in the plant community. J. Ecol. 35: 1–22.

Wildi, O. & Orlóci, L. 1983. Management and multivariate analysis of vegetation data. (2nd revised ed.) Eidg. Anst. Forstl. Versuchswes. Ber. 215. 139 pp.

Vegetatio 77: 57–60, 1988
© Kluwer Academic Publishers, Dordrecht – Printed in the Netherlands

Gaps in the canopy: the missing dimension in vegetation dynamics

Jonathan Silvertown & Bridget Smith
Department of Biology, Open University, Walton Hall, Milton Keynes, MK7 6AA, Great Britain

Accepted 24.5.1988

Keywords: Canopy gap, Grassland, Plant demography

Abstract

Canopy gaps are important as entry points for new genotypes and new species into many types of vegetation, yet little is known about them in any type of vegetation but forests. Forest gaps are too large for manipulative experiments to be readily undertaken, and hitherto grassland gaps have been too small to be easily mapped. Preliminary results from mapping small (> 1 cm) grassland gaps with a new fibre-optic device suggest that experiments need to be performed at a smaller physical scale than has hitherto been achieved.

Introduction

In most types of vegetation outside the arid regions of the world, gaps in the canopy appear to be essential for new genotypes and new species to enter plant communities (e.g. Brokaw 1985; Goldberg & Werner 1983; Collins & Pickett 1987; Martinez-Ramos & Alvarez-Buylla 1986; Miles 1974; Wells & Haggar 1984). Furthermore, it has been suggested that different species have different requirements for the kind of gap they are able to colonize and that this provides one or more dimensions in which niche separation may take place (Denslow 1980; Grubb 1977; Silvertown & Wilkin 1983). Irrespective of whether equilibrium or non-equilibrium conditions apply, gaps are a fundamental feature of the working mechansm (*sensu* Watt 1947) of plant communities. A considerable amount is now known about the creation, size frequency distribution, timing of appearance and disapearance of gaps – or what might be called 'gap demography' – in forests (e.g. Faille *et al.* 1984a & b; Foster & Reiners 1986; Lang & Knight 1983) but there is almost no experimental test of the role played by gaps in determining the species composition or community structure of forests. The reason for this lack is obvious: the physical scale of forest vegetation makes experimental treatments difficult to replicate sufficiently and the long lifespan of trees requires extended periods of study. Grasslands offer an alternative system in which to study the role of gaps in plant communities, but here we have the converse problem: there are many experimental studies of colonization (e.g. Fenner 1978; Gross 1980; Hillier 1986; Rusch 1988; Silvertown & Wilkin 1983) and almost no information on the demography of gaps. The reason for this is that the small physical scale of grassland plants demands a scale of measurement beneath the vegetation canopy which has not hitherto been achieved in the field. In this paper we describe preliminary results obtained from the use of a new instrument designed to map gaps in grassland.

Methods

Maps of grassland gaps

Leaves selectively absorb red light and transmit near infra-red. Measurements of the ratio of light intensity in non-overlapping wavebands in the red and near infra-red (R/IR) made beneath the grassland canopy are directly correlated with leaf area index (Jordan 1969; Frankland & Poo 1980). A field-portable, microcomputer-controlled instrument (Silvertown *et al.* 1988) with a fibre-optic probe was used to record the R/IR ratio at 1 cm intervals in a grid pattern in grassland quadrats at the campus of the Open University in Milton Keynes, Buckinghamshire. Twelve quadrats 25 cm × 25 cm each in a ca 6 cm tall grassland of *Lolium perenne* were mapped in this fashion on 18 August 1987. Contour maps of R/IR ratio were produced on a VAX computer system using the UNIRAS graphics package. These provide a quantitative picture of the distribution of gaps in each quadrat.

Because R/IR data provide a quantitative picture of the grassland canopy, exactly what is meant by a 'gap' can be defined quantitatively too. Gaps were identified and counted by a computer algorithm which operated on the matrix of raw R/IR ratios as follows. Each point in the matrix was checked in turn. Points with a R/IR ratio above a predetermined threshold value for a gap were identified and all adjacent points checked. The coordinates of adjacent points which qualified were entered in a list of points belonging to the same gap. This procedure was repeated on all coordinates in the list until the boundaries of the gap were encountered.

Germination in grassland gaps

Seeds of *Geranium dissectum* L. were sown individually with forceps at 2 cm intervals into 16 25 cm × 25 cm quadrats, 121 seeds per quadrat. Before sowing, each quadrat was mapped as described above. Four of the quadrats were at Milton Keynes (MK) and the remaining 12 in a 20-year-old grassland ley on a nature reserve at Little Wittenham, Oxfordshire (LW). The quadrats at Little Wittenham were situated in nine 0.25 ha paddocks grazed by sheep. Sheep grazing in each paddock is controlled to produce differences in sward height between paddocks. Sowing locations in each quadrat were reinspected and germination was scored one month after sowing.

Correlations between germination and R/IR ratio measured at individual sowing positions were determined using the GLIM statistical package (Payne 1986) with models appropriate for binomial data. The following model was run for each of the 16 quadrats:

$$\text{GERMINATION} = \text{QUADRAT}/(\text{ROW} + \text{COLUMN}) + \text{RATIO}$$

in which the term QUADRAT/(ROW + COLUMN) determines the effect of spatial correlation (if any) on the germination of seeds sown in the same row or column of the planting grid within a quadrat.

Results

R/IR maps and gaps

An example of a contour map of R/IR is shown in Fig. 1. Few entirely bare areas (R/IR > 1) were present in the twelve quadrats: only 13 gaps in the

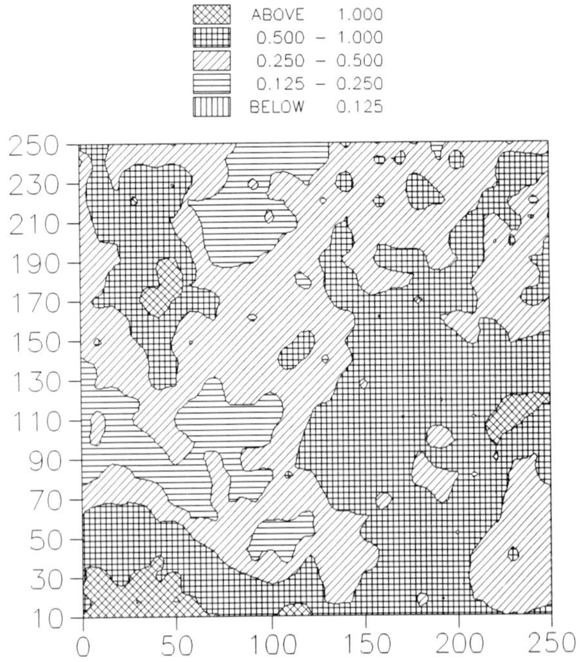

Fig. 1. Contour map of R/IR ratio in a representative quadrat mapped in Milton Keynes.

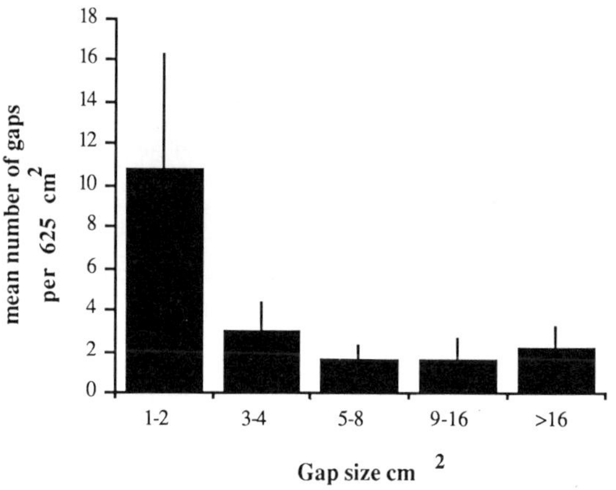

Fig. 2. Mean number of gaps per 625 cm^2 in five size classes at Milton Keynes. Bars show standard deviations, $n=12$.

1-2 cm^2 size-class and one gap in the size-class 3–4 cm^2 were recorded in the entire sampling area of 0.75 m^2. Defining gaps with a threshold R/IR ratio > 0.5 yielded 130 gaps and 36 gaps in these two size-classes respectively. The full size-class distribution of gaps (R/IR > 0.5) in twelve quadrats is shown in Fig. 2.

Germination in grassland gaps

Laboratory germination of *Geranium dissectum*

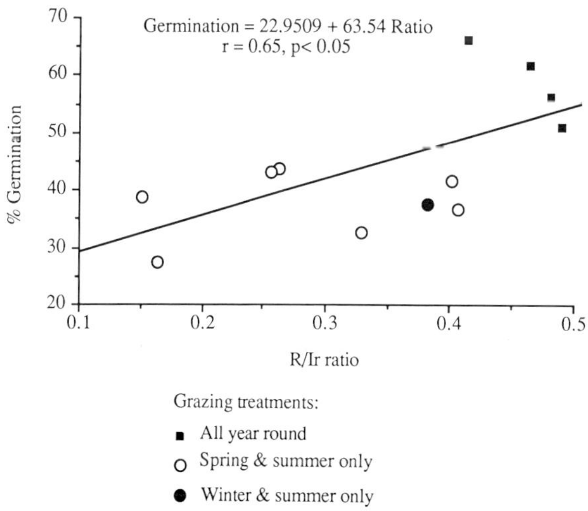

Fig. 3. Relationship between percent germination of *Geranium dissectum* and mean R/IR ratio in twelve grazed quadrats at Little Wittenham.

was 100%, but in the field germination ranged between 25 and 70% per quadrat. There was significant spatial correlation in one of four quadrats at MK and in 5 of 12 quadrats at LW. There was no significant relationship between R/IR ratio and germination in any quadrat at MK and a signficant positive relationship (χ^2 = 9.9, d.f. = 1, $p<0.005$) in only one of the twelve quadrats at LW.

Percent germination in quadrats at LW was significantly positively correlated with mean R/IR ratio per quadrat and was highest in paddocks with the heaviest grazing (Fig. 3).

Discussion

The size distribution of gaps clearly depends upon how a 'gap' is defined and upon the scale at which measurements are made. Size distributions of gaps in forests often show a skew towards smaller gap-sizes (e.g. Brokaw 1982; Faille *et al.* 1984a; Foster & Reiners 1986), similar to the pattern observed in our grassland quadrats at Milton Keynes (Fig. 2). The modal size-class of gaps at this site was 1–2 cm^2 which is significantly smaller than the size of gap generally created in experimental studies in grasslands. If the results of this preliminary study prove to be representative of other grasslands, then the gap dynamics of grasslands need to be studied at a much smaller physical scale than has hitherto been attempted. Gaps may be the missing dimension that is needed to understand the mechanism of vegetation dynamics.

The question of what the appropriate physical scale is on which to study grassland gap dynamics is crucial if we are to uncover the mechanism(s) of grassland vegetation dynamics. In order to answer this question, we first need to know how plants themselves respond to gaps of different size. Our results indicate that, in the case of *Geranium dissectum,* germination is not sensitive to plant cover at the scale of 1 cm^2 *within* quadrats, though *between* quadrats there is a correlation between percent germination and mean R/IR ratio (Fig. 3). It appears that this species, which has relatively large seeds weighing 2.75 mg, is responding to a coarser grain of environmental variation than the 1 cm^2 resolution of our R/IR measurements. *Geranium dissectum* is a ruderal species which, although found in grasslands in our study areas, is perhaps not typical of grassland dicots in general. Further studies of germination and survival in relation to R/IR ratio should reveal precisely where the threshold lies between invadable and non-invadable microsites for this species. Our ultimate objective is to identify this threshold for a group of grassland species and to be able to predict the potential for invasion into pastures from a knowledge of the gap dynamics of the grassland and the behaviour of individual species.

Acknowledgements

We thank Jo Treweek and Trudy Watt for their assistance at Little Wittenham Nature Reserve and the Northmoor Trust for permission to work there. We thank Kevin McConway for statistical advice and we acknowledge financial support from the Natural Environment Research Council and the Open University Research Committee.

References

Brokaw, N. V. L. 1982. The definition of treefall gap and its effect on measures of forest dynamics. Biotropica 14: 158–160.

Brokaw, N. V. L. 1985. Gap-phase regeneration in a tropical forest. Ecology 66: 682–687.

Collins, B. S. & Pickett, S. T. A. 1987. Influence of canopy openings on the environment and herb layer in a northern hardwood forest. Vegetatio 70: 3–10.

Denslow, J. S. 1980. Gap partitioning among tropical rainforest trees. Biotropica (Suppl.) 12: 47–55.

Faille, A., Lemée, G. & Pontailler, J. Y. 1984a. Dynamique des clairières d'une forêt inexploitée (reserves biologiques de la forêt de Fontainebleau) 1. Origine et état actuel des ouvertures. Acta. Oecol. Oecol. Gen. 5: 35–51.

Faille, A., Lemée, G. & Pontailler, J. Y. 1984b. Dynamique des clairières d'une forêt inexploitée (reserves biologiques de la forêt de Fontainebleau) 2. Fermeture des clairières actuelles. Acta. Oecol. Oecol. Gen. 5: 181–199.

Fenner, M. 1978. A comparison of the abilities of colonizers and closed turf species to establish from seed in artificial swards. J. Ecol. 66: 953–963.

Foster, J. R. & Reiners, W. A. 1986. Size distribution and expansion of canopy gaps in a northern Appalachian spruce-fir forest. Vegetatio 68: 109–114.

Frankland, B. & Poo, W. K. 1980. Phytochrome control of seed germination in relation to natural shading. pp. 357–366. In: J. De Greef. (ed.), Photoreceptors and plant development. Antwerpen University Press, Antwerpen, Belgium.

Goldberg, D. & Werner, P. A. 1983. The effects of size of opening in vegetation and litter cover on seedling establishment of goldenrods (*Solidago* spp.). Oecologia 60: 149–55.

Gross, K. L. 1980. Colonization by *Verbascum thapsus* (Mullein) of an old field in Michigan: experiments on the effects of vegetation. J. Ecol. 68: 919–928.

Grubb, P. J. 1977. The maintenance of species richness in plant communities. The importance of the regeneration niche. Biol. Rev. 52: 107–145.

Hillier, S. H. 1986. A quantitative study of gap recolonization in two contrasted limestone grasslands. Ph. D. Thesis Univ. Sheffield.

Jordan, C. F. 1969. Derivation of leaf-area index from quality of light on the forest floor. Ecology 50: 663–666.

Lang, G. E. & Knight, D. H. 1983. Tree growth, mortality, recruitment, and canopy gap formation during a 10-year period in a tropical moist forest. Ecology 64: 1075–1080.

Martinez-Ramos, M. & Avarez-Buylla, E. 1986. Seed dispersal, gap dynamics and tree recruitment: the case of *Cecropia obtusifolia* at Los Tuxtlas, Mexico. In: Estrada, A. & Fleming, T. H. (eds), Frugivores and seed dispersal.

Miles, J. 1974. Effects of experimental interference with stand structure on establishment of seedlings in Callunetum. J. Ecol. 62: 675–687.

Payne, C. D. (ed.) 1986.The Generalized Linear Interactive Modelling system. Release 3.77. Numerical Algorithms Group, Oxford.

Rusch, G. 1988. Reproductive regeneration in grazed and ungrazed limestone grassland communities on Öland. Preliminary results. Acta Phytogeogr. Suec. 76: 113–124.

Silvertown, J. & Wilkin, F. R. 1983. An experimental test of the role of micro-spatial heterogeneity in the co-existence of congeneric plants. Biol. J. Linn. 19: 1–8.

Silvertown, J., Prince, S. D. & Smith, B. A. 1988. A field-portable instrument for mapping the micro-environment within grass canopies. Functional Ecology 2: (in press).

Watt, A. S. 1947. Pattern and process in the plant community. J. Ecol. 35: 1–22.

Wells, G. J. & Haggar, R. J. 1984. The ingress of *Poa annua* into perennial ryegrass swards. Grass Forage Sci. 39: 297–303.

Vegetatio 77: 61–64, 1988
© Kluwer Academic Publishers, Dordrecht – Printed in the Netherlands

A reappraisal of cyclical processes in *Calluna* heath

C. H. Gimingham
Department of Plant Science, University of Aberdeen, St. Machar Drive, Aberdeen, AB9 2UD, Scotland, Great Britain

Accepted 24.5.1988

Keywords: *Calluna vulgaris*, Canopy gap, Cyclical change, Succession

Abstract

Small-scale, short-term cycles constitute an important aspect of vegetation dynamics. It has been widely accepted that such cycles are characteristic of unmanaged heath communities where the dominant, *Calluna vulgaris*, undergoes an age-related series of changes in its growth and morphology. This is thought to give rise to a repetitive sequence of changes in the occupancy of gaps which form in the *Calluna* canopy and are first colonized by other species, with *Calluna* re-establishing later. Recently, doubts have been cast as to whether the changes are in fact often repetitive. A new study has shown much greater variability than was previously suspected in the transitions occurring in gaps. There is evidence that, in some instances though by no means all, *Calluna* can re-establish and thereby initiate a genuine cycle. It seems, however, that such re-entry is seldom by means of seedling establishment, but more frequently by development of adventitious shoots and roots on stems which have come to lie across the gap and have been covered by moss, moist litter or humus.

Nomenclature: Clapham, A. R., Tutin, T. G. & Warburg, E. F. (1981). Excursion Flora of the British Isles, 3rd ed. Cambridge University Press.

Introduction

Most of the papers in this series are concerned with plant succession, treated as a non-repetitive series of changes in vegetation. However, during the past 40 years it has become increasingly apparent that small-scale, short-term cycles of change constitute an important aspect of vegetation dynamics. It is evident that in many plant communities, although the overall floristic composition may be relatively constant over considerable periods of time (of the order of tens or even hundreds of years), the occupancy of any particular spot may pass through a series of phases characterised by different species. This has an important bearing on the species richness of the community as a whole and on its horizontal heterogeneity or 'patchiness'.

The first thorough treatment of this concept of cyclical processes in plant communities was that by Watt (1947). He described a number of examples of vegetational cycles in communities which were assumed to be 'relatively stable over a reasonable period of time'. A feature common to all these examples was the development of gaps in the cover of a dominant species, which could be formed either by the death of individuals (e.g. trees in a forest) or through some intrinsic feature of the plants' growth and morphology (e.g. radial spread of rhizomes accompanied by degeneration at the clone centre). Observations, and recordings of permanent quadrats, indicated that the dominant seldom recolonized gaps immediately. Instead, a sequence of other species usually occupied the patch for a time before re-establishment of the dominant took place. Where the population of the dominant was uneven-aged, gaps appeared irregularly in the canopy and the community structure took the form of a mosaic of patches of limited area, each at a different stage in the cycle. Hence, at any one time the composition of the patches represents a series of phases which are 'dynamically related to each other'.

Heathland communities dominated by *Calluna vulgaris* figured among the examples Watt described in 1947, and he developed the idea of cyclical processes in heath vegetation in a subsequent paper (1955). Here the terms he had used previously to distinguish four phases in the morphological life-history of bracken (*Pteridium aquilinum*) were applied to *Calluna*: pioneer, building, mature, and degenerate. Although these phases are not sharply differentiated one from another, subsequent workers have had little difficulty in separating them in *Calluna* (except in certain rather extreme habitats), as follows:

Pioneer. Early stages of establishment and growth. Regular branching from the axis of a single leading shoot, at first. Height up to ca 6 cm, shape pyramidal, cover incomplete. Up to 6 years of age.

Building. Becoming bushy, branches radiating from the centre, productivity of peripheral shoots high and flowering vigorous. Height up to ca 0.5 m (or more), shape hemispherical, canopy dense, cover approaching 100%. Up to about 15 years of age (depending on habitat).

Mature. Extension growth declining, though green shoot production and flowering still high. Shape still hemispherical, but central branches inclined to spread sideways, initiating gap formation; cover slightly reduced. Up to 20 years of age (or more).

Degenerate. Central branches dying, creating gap. Some other branches which have become partially buried in litter etc. may remain alive (because of adventitious rooting), sustaining a ring of foliage-bearing twigs. Cover much reduced. Eventually whole plant may die. Death at 30–40 years of age.

The behaviour of a *Calluna* individual is such that during the mature and degenerate phases a gap forms in the canopy. Watt's observations in a sandy heath in the Breckland region of S.E. England indicated that the accumulated litter and mor humus beneath the gap was normally colonized by lichens and mosses, but that eventually the mor would decay or become dispersed, leaving a humus-stained mineral soil surface. In addition to cryptogams, a limited number of vascular plant species were noted as capable of occupying the site, and Watt added 'Seedling *Calluna* may become established on the mor but its chances of survival are greater on the mineral soil. Even then a number of years may elapse before an effective colonization initiates a new cycle. A new cycle may also be initiated by vegetative lateral spread from neighbouring bushes'.

The *Calluna* stands examined by Watt had not been managed by burning for at least 25 years, probably more, and individuals at all phases of their development were present side by side, together with patches occupied by other plants. Accordingly, he interpreted the patchiness in terms of the life history of the *Calluna* plant and clearly took the view that repetitive cyclical processes were operating. Other heathland ecologists in Britain, working in different localities, also found it possible to recognise the growth phases of *Calluna* and, in examples of undisturbed heath, to find patches occupied by lichens, mosses and vascular plants such as *Deschampsia flexuosa, Vaccinium myrtillus, Empetrum nigrum* or *Arctostaphylos uva-ursi*. The idea that cyclical change is a fundamental and probably universal feature of heathland communities became widely accepted, and remained almost unquestioned until recently.

Questioning the theory

Among the first to express scepticism about the universality of these ideas were Dutch ecologists who indicated that in the heaths of the Netherlands they could not find evidence of the complete sequence of growth-phases in *Calluna*, as described by Watt (1947, 1955) and Barclay-Estrup & Gimingham (1969). Furthermore, they could find no evidence of the return of *Calluna* in gaps in the stands. Their records of permanent quadrats over extensive time-periods suggested that mosaic patterns were caused not by cyclical processes, but by periodic catastrophic events such as severe drought or attacks of heather beetle, *Lochmaea suturalis* Thomson, leading to the death of individual *Calluna* bushes (Prentice *et al.* 1987). Under these circumstances *Calluna* may be replaced by species such as *Deschampsia flexuosa, Molinia caerulea* or *Empetrum nigrum*, and may fail to re-establish on the sites it formerly occupied. In Britain, Marrs (1986) has re-examined some of Watt's study areas, and found that *Calluna*, instead of maintaining repeated cycles, has been replaced either by trees or by bracken. Indications such as these, together with his own observations, led Miles (1981) to suggest that the pattern of intrinsic cycles put forward by Watt for the East Anglian breckland might not be the norm, and that extrapolation to other heaths might be invalid. He also pointed out that the litter and humus which accumulate under heather, or the cover of lichens and mosses such as that typical of most gaps, do not form favourable regeneration niches for *Calluna*. He states that despite widespread search he had not at the time seen *Calluna* seedlings in the gaps formed at the centre of degenerate bushes. (He noted, however, the possibility of re-establishment of *Calluna* by vegetative spread from neighbouring bushes, also referred to by Watt).

A new search for evidence

In view of this challenge to the accepted theory, a reappraisal and a renewed search for evidence has been undertaken, with the help of research students and other associated workers at the University of Aberdeen. In the first place, ample evidence has been obtained in undisturbed locations in Scottish heaths of the passage of *Calluna* through the four growth phases as described by Watt. In a detailed study by Barclay-Estrup & Gimingham (1969) and Barclay-Estrup (1970, 1971), a number of permanent quadrats were set up to contain *Calluna* bushes in each of the four phases, and were mapped at yearly intervals over a period of years. The passage from one phase to the next was clearly documented in all cases. It is suggested that in those instances where this does not obtain, as for example in the Netherlands and parts of southern England, the *Calluna* may have been subject to stresses leading habitually to the death of the older individuals before they reach the mature or degenerate phases.

The observation that *Calluna* individuals pass through the sequence of phases as originally described does not of itself constitute evidence for a cyclical process. Only if *Calluna* eventually re-establishes in the gap and the sequence is repeated would the theory hold. Surveys have revealed numerous instances where this is not the case. In addition to Marrs' data (mentioned above), studies on Dinnet Moor near Aberdeen have shown that *Betula* spp. can establish in gaps in the centre of mature or degenerate *Calluna* (Gong & Gimingham 1984). As the trees mature they eventually eliminate *Calluna* beneath their crowns and lead to permanent changes in the ground flora. In cases like this, Watt's assumption of relative stability over a reasonable period of time does not apply, and the heath vegetation is evidently seral. Hence, the theory of cyclical process may be invoked only when there is no potential for invasion by a more permanent and long-lived species such as a tall shrub or tree.

Repetitive cycles remain a possibility where tree seed-parents or aggressive grass species are absent. Direct evidence may be sought in undisturbed stands by quantitative surveys of gaps to establish the frequency of young *Calluna*, whether derived from seedling establishment or vegetative layering. In one such study, although recently-germinated seedlings were fairly abundant, survival was poor and few instances were found of gap colonization by seedlings (de Hullu & Gimingham 1984). On the other hand, in a transect of 25 contiguous 50 cm × 50 cm quadrats in an undisturbed, uneven-aged *Calluna* stand, Scandrett (pers. comm.) found clusters of young shoots derived from layered stems in 16 of them. It seems that *Calluna* may reappear in gaps as a result of trailing branches from nearby bushes falling across the gap, becoming partially buried in moss or litter and rooting adventitiously.

The problem has also been approached experimentally by creating artificial 'gaps' in continuous *Calluna* stands. In some seasons, *Calluna* seedlings appear in numbers where bare ground free of deep litter is available. However, very few survive to a second or third year, and since most gaps are litter-covered or occupied by other vegetation, establishment of *Calluna* from seed must be a rare event. On the other hand, even quite old *Calluna* stems can be induced to root adventitiously by packing them round with moist humus or peat. Lateral vegetative spread appears, therefore, to provide a mechanism by which *Calluna* may re-establish in gaps.

Watt's work suggested that, in any particular locality, the changes proceeded in a regular, orderly sequence, which was repetitive, and that this accounted for the persistence of pattern in the plant community. Our quadrat records, however, do not support this notion of an orderly sequence. Although they demonstrate that cycles may occur, there is certainly much more variation than was previously recognised. The species taking part in the sequences vary from patch to patch, and while *Calluna* may on some occasions reappear in a gap, on others it may be replaced for long periods of time by another species. Some gaps persist apparently indefinitely in the lichen-moss condition. Instead of describing the community in terms of a mosaic representing the stages of an orderly sequence, it becomes necessary to describe it in terms of a matrix of probabilities of change from one vegetation state to another.

Conclusions

This reappraisal of Watt's concept of cyclical processes in heathland vegetation in no way detracts from his achievement in formulating it. The idea has been of the utmost value in many ways. It provided a strong stimulus to the investigation of pattern in plant communities, by offering an explanation to be considered where habitat heterogeneity was insufficient to account for the mosaic. It provided a framework on which to base detailed research on the dynamics of even-aged stands of *Calluna* created by burning. In so far as it concerns the morphological life-history of *Calluna* it has been enormously valuable in interpreting the varying responses of *Calluna* to environmental influences ranging from climatic and edaphic factors to pests and pollutants.

Nevertheless, like many biological theories as at first propounded, it has proved too simple to account for all subsequent observations. In the first place, in lowland heaths *Calluna* is not a climax dominant and most *Calluna*-dominated communities are inherently unstable. Except where severe environments exclude tall shrubs and trees, there is always the likelihood of invasion by species typical of later successional stages. By their nature, repetitive processes can operate only where there is no opportunity for more permanent replacement. Hence, cyclical change in heathland can be expected only if further successional change fails to occur.

Secondly, recent investigations (supported by simulation studies, e.g. van Tongeren & Prentice 1986) have proved that the details of the small-scale dynamic processes in heathland communities are much more variable than was once thought. In this respect, our conclusions parallel other recent reappraisals of theory in the field of vegetation dynamics. However, the doubts cast on the validity of the concept of cyclical change in heathland have been valuable in stimulating re-investigation, as a result of which firm evidence has been obtained for the existence of cycles in some situations.

References

Barclay-Estrup, P. 1970. The description and interpretation of cyclical processes in a heath community. II. Changes in biomass and shoot production during the *Calluna* cycle. J. Ecol. 58: 243–249.

Barclay-Estrup, P. 1971. The description and interpretation of cyclical processes in a heath community. III. Microclimate in relation to the *Calluna* cycle. J. Ecol. 59: 143–166.

Barclay-Estrup, P. & Gimingham, C. H. 1969. The description and interpretation of cyclical processes in a heath community. I. Vegetational change in relation to the *Calluna* cycle. J. Ecol. 57: 737–758.

Gong, Wooi Khoon & Gimingham, C. H. 1984. Birch regeneration in heath vegetation. Proc. Roy. Soc. Edinb. 85B: 73–81.

de Hullu, E. & Gimingham, C. H. 1984. Germination and establishment of seedlings in different phases of the *Calluna* life cycle in a Scottish heathland. Vegetatio 58: 115–121.

Marrs, R. H. 1986. The role of catastrophic death of *Calluna* in heathland communities. Vegetatio 66: 109–115.

Miles, J. 1981. Problems in heathland and grassland dynamics. Vegetatio 46: 61–74.

Prentice, I. C., van Tongeren, O. & de Smidt, J. T. 1987. Simulation of heathland vegetation dynamics. J. Ecol. 75: 203–219.

van Tongeren, O. & Prentice, I. C. 1986. A spatial simulation model for vegetation dynamics. Vegetatio 65: 163–173.

Watt, A. S. 1947. Pattern and process in the plant community. J. Ecol. 35: 1–22.

Watt, A. S. 1955. Bracken versus heather, a study in plant sociology. J. Ecol. 43: 490–506.

Vegetatio 77: 65–78, 1988
© Kluwer Academic Publishers, Dordrecht – Printed in the Netherlands

Forest decline: endogenous dynamics, tree defenses, and the elimination of spurious correlation

Craig Loehle
Environmental Sciences Division, Savannah River Laboratory, Savannah River Plant, Aiken, SC 29808-0001, USA

Accepted 28.4.1988

Keywords: Acid rain, Dieback, Pollution, Senescence, Tree defense, Tree ring width

Abstract

Two important unresolved problems in pollution impact research are early detection of changes in tree and forest growth and determination of the causes of observed growth declines. Consideration of tree defensive systems can shed light on dieback phenomena. Trees possess both chemical and structural defenses, with both types having both active and passive dimensions. When stress or pollution causes a reduction in vigor, defenses may become ineffective, resulting in insect pest and pathogen attack. This leads to a positive feedback cycle of declining vigor. Level of defenses differs by species, site, and life-stage. Consideration of these factors leads to better statistical methods for correlating pollution dose with tree growth response. It is shown that several recent studies are potentially flawed by inadequate consideration of species type, tree age, or stand density. It is shown that forest growth models can be used to correct for endogenous changes in tree growth rates as long as the trees used to derive parameters for the model were not themselves affected by pollution.

Introduction

Tree stands exhibiting synchronous dieback have been observed for many decades. They have become particularly prominent in industrial nations where it is speculated that mortality is due to air pollution. Natural stand dieback can result from a combination of tree aging and stresses. It is important to recognize these two causes and their interactions because a great deal of effort is going into research on mechanisms of pollutant effects. Failure to consider the effects of life-stage and differential susceptibility confound attempts to correlate decline with doses of pollutants.

Mueller-Dombois (1986) has put forth an integrated theory to explain natural stand-level dieback. In this paper I extend this theory to include the role of tree defenses. This new model is used to predict the types of forests in which dieback is likely to occur and the characteristics of species likely to be involved. The existing literature is examined and shown to be flawed by inadequate consideration of tree defenses, stand age and density, and site factors. Failure to consider these factors leads to both spurious correlation and inconclusive studies.

Decline etiology

Early studies of forest dieback focused on the proximate causes of mortality, such as specific pathogens (Mueller-Dombois 1986). Investigation usually revealed that 1) single pathogens were not an adequate explanation, 2) boundaries of dieback stands were often sharp and bordered stands with no dieback, and 3) understory regeneration of the same species was vigorous under dying adults. These factors led Mueller-Dombois to propose the following 'cohort senescence theory' for stand-level dieback (items 1–6 below taken from Mueller-Dombois (1986); bracketed statements added for clarity):

1. A perturbation sets the stage for cohort invasion of a potentially dominant canopy species. The perturbation may either be a catastrophic stand-leveling disturbance, such as a lava flow or hurricane, or it may be a canopy dieback with trees remaining upright. The subsequent invasion will probably not be uniform over very large areas but may result in a mosaic of cohorts developing at somewhat unequal rates on the same and different habitats.
2. Stand development will proceed under the influence of recurring physiological upsets due to extreme [climate] events. These may force cohort stands on harsh sites into premature senescence while those on more favorable sites may grow with less impeded vigor into old age. Moreover, stands on harsh sites may exhibit [more intense growth declines] and thus become more synchronized in their terminal life stage, while those on more favorable sites may go through [less intense growth declines, and thus be less synchronized]. This results in a predisposition of stands to dieback that is spatially differentiated, a process both site and life-stage related.
3. A perturbation that normally upsets growth may become a dieback trigger for stands and subpopulations that have their vitality reduced by this combined predisposition [of site and age]. A regional dieback pattern may then develop into a small-area patchwork related in outline to the spatial extent of cohort stands that are in their terminal life stages. This patchwork is expected to rotate more or less continuously or in pulses throughout the biome.
4. However, the underlying site spectrum can be spatially nonuniform, and the resulting dieback patterns may therefore be spatially unequal. This increases the complexity of a regional dieback pattern. On harsh sites, poorly synchronized cohort establishment patterns may become more synchronized before and during the dieback phase, and this may result in larger-area diebacks. On more favorable sites, poorly synchronized cohort establishment patterns may be maintained, resulting in smaller-area diebacks that are more out of phase with one another.
5. A triggering perturbation can become an aggravating factor in dieback on poorly buffered habitats especially when the perturbation increases in intensity. As such, however, it should become more easily identifiable through its more radical effects on the structure of the impacted stand, e.g., by killing trees in all life stages.
6. Trigger factors may be either biotic agents or abiotic stress factors. In the Hawaiian dieback, biotic agents play only a minor role. They act as weak pathogens and can be regarded as pioneers among the decomposer organisms. However, their function as dieback accelerators may be critical for the regeneration success of a shade-intolerant canopy species. Biotic agents thus can play an important role as determinants of succession.

In this paper Mueller-Dombois' theory is extended by consideration of the mechanisms of senescence and the role of defenses in responses to stress. Loehle (1988) has proposed a model for growth trade-offs that explains differences in longevity between species. This new model provides a mechanism for synchronous stand dieback that is consistent with Mueller-Dombois' theory and may provide a synoptic view of natural and pollution-induced declines.

Defenses and decline

Loehle (1988) suggested that a crucial aspect of tree energetics is allocation to structural defenses (increased wood density for resisting wood pathogens) and chemical defenses to resist insects and other pathogens of leaves and wood. Strong seasonal variation in defensive compound production has been noted (Lorio 1986; Waring 1987) and low production related to increased susceptibility to insect attack (Lorio 1986). Such defenses have a high total energetic cost. Loehle (1988) found that North

American broadleaf tree species with high stemwood defenses (10–13 $J \cdot cm^{-3}$) were two to three times as long-lived as those with low defenses (6–7.5 $J \cdot cm^{-3}$). These facts suggest a strategic trade-off. A tree can grow fast with few defenses or slower with more defenses. But while a high-defense tree will be more resistant to mortality factors, it will be less competitive with respect to neighbors and may be overtopped more easily. This effect is particularly pronounced at the seedling and sapling stages. Thus, we can anticipate that pioneer-type trees (but not those in permanently open habitats) will be lower in defensive investments. Rapid growth may, however, compensate for lack of defenses at certain stages of the life cycle. A vigorous pioneer-type tree may tolerate a substantial load of pathogens by simply outgrowing them (Coley *et al.* 1985); for example, shedding infected leaves for new ones or healing over bark wounds in a few seasons.

As the plant approaches maturity, however, relative growth rate slows due to site limitations of water and nutrients, a decreasing ratio of photosynthesis to respiration, and inherent genetic limits (Waring 1987). Absolute radial increment decreases even faster, because wood must be spread over a greater surface area, giving narrower rings. At this point, rapid growth (particularly rapid trunk growth to heal over wounds) can no longer be used as a defense against pathogens. Then the plant may become increasingly susceptible to pathogen attack. This susceptibility, combined with structural weakness, should make the decline phase for pioneer species quite steep once the peak of vigor has passed. It is possible to summarize Loehle's (1988) theory as a set of hypothesis/prediction pairs:

1. If investment in defenses is crucial to attaining greater longevity, then the resistance to wood decay in particular is significant, especially in warm and wet habitats where decay rates are rapid. Within a particular habitat, increased energy invested in defenses should correlate with increased longevity.
2. If investment in defenses or reproduction slows down the growth rate because of limited energy and/or nutrient supplies, then there should be an inverse relation, within any one habitat, between growth rate and longevity.
3. If decreased growth rate increases mortality rates of juvenile trees, particularly under height-growth competition, then pioneer-type trees should sacrifice defensive investments in order to maximize juvenile growth rates.
4. If pioneer-type trees depend on rapid radial growth rate to resist pathogens, then they should decline rapidly when they approach their maximum size because their growth rate is slower then. This decline should be typified by limb breakage, dead patches on the trunk, increased insect attack, rapid pathogen invasion, and hollowing of the trunk.
5. In contrast, long-lived species should be able to maintain themselves at a low growth rate at maturity without excessive pathogen attack and should have specific adaptations for keeping photosynthesis and respiration in balance. Greater investment in roots and defenses, slower growth at maturity, drought tolerance, cambial retreat (e.g., *Pinus aristata,* LaMarche 1969), and branch-tip dieback are some of the adaptations used.

Analysis of a large data set of attributes of North American tree species strongly supported these predictions (Loehle 1988).

Stand decline

This growth trade-off model has direct bearing on the phenomenon of synchronous stand dieback. The sequence begins with a disturbance that allows invasion by a fast growing, low defense type of species (or several species) over a period of several years. This yields a cohort that is roughly the same age (±10–20 years). These fast growing species will be relatively tolerant of all types of pathogens and stressors as long as they maintain rapid growth. Once they reach full adult size and are growing slowly, the whole stand will be in a marginal position with respect to pathogen or herbivore defenses. When a major stress is applied such as drought or pollution (considered here as a general stressor, possibly acting in various ways), the vigor of all the trees will be reduced below the threshold of adequate defense. As pathogens start to invade, tree vigor will decline further, encouraging even more pathogen invasion.

This is a positive feedback loop of declining vigor (Fig. 1b, 2a) that should result in rapid stand synchronization of individual tree vigor into the dieback stage. Such a feedback loop should be characterized by a variety of opportunistic and nonspecific pathogens such as root and wood decay fungi. The parallels with AIDS patients who have lost their immune system are inescapable; the 'immune system' of such low defense trees results largely from rapid growth and is almost absent under slow-growth conditions. The prediction of varied dieback symptoms and pathogens agrees with Manion's (1981) model of decline. Opportunistic wood pathogens such as the basidiomycete *Armillaria mellea* have been shown to play a secondary role in red spruce decline (Carey *et al.* 1984) and in natural declines in Hawaii (Mueller-Dombois 1986). The prediction that species involved in natural declines should be low in defenses agrees with the occurrence of declines of birch, maple, and ash in the northeastern U.S. during the 1930's to 1960's (Manion 1981). All three of these species are low in defenses and have wood that is not very decay resistant (Loehle 1988).

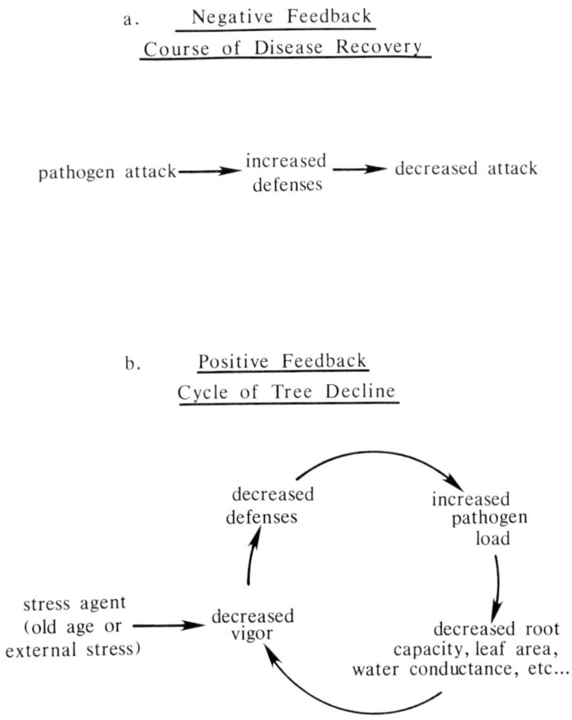

Fig. 1. Comparison of disease recovery with decline syndrome.

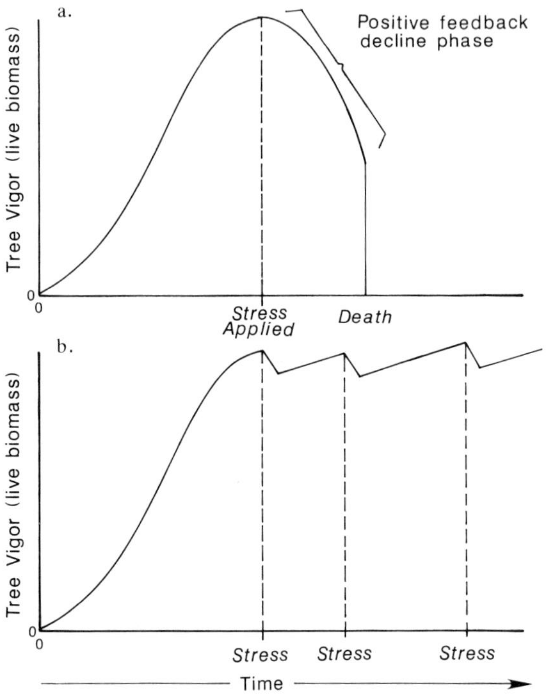

Fig. 2. Hypothetical life-stages for low and high defense tree species. a) Low defense species if stressed at maturity go into a positive feedback loop of declining vigor which encourages pathogen attack, further reducing vigor. Death is relatively rapid. b) High defense species do not go into decline when moderately stressed because past investments in wood defenses reduce pathogen attack even under reduced vigor conditions following stress. Life span is therefore typically greater for high defense types.

In contrast to this scenario is the situation that occurs with high defense types (Fig. 1a, 2b). Bald cypress (*Taxodium distichum*) and redwood (*Sequoia sempervirens*) are species that are shade intolerant and tend to form roughly even-aged stands following disturbance (Fowells 1965). They put a great deal of energy into defenses (in contrast to most pioneer species), producing wood that is highly decay resistant. Their response to a stress is typified by Fig. 2b: a major stress may reduce vigor and even kill part of the crown, but because the wood and roots are so decay resistant and the foliage not very edible, opportunistic pathogens do not invade and the tree keeps growing and gradually recovers. A stand of high defense species may have a dead spire at the top of every tree, yet the trees are neither declining nor dying. Long-lived western conifers are typical examples of this phenomenon. Such species may persist in this condition for centuries with only agents that cause direct mortality (e.g., fire, lightening) limiting the life span. Trees in low-decay habitats (e.g., subalpine) may achieve the same results with lower defensive investments.

Thus, synchronous stand decline should mainly be found in stands made up of species that are low in defensive investments and that use rapid growth as a defense. Such species should be shade intolerant and be characterized by wood that is low density and/or low in defensive chemicals, by low defensive chemicals in foliage, by rapid growth, and by lower belowground/aboveground biomass ratios. These combinations of traits should be found under moist, warm, fertile growing conditions and in colonists of disturbed sites, where juvenile height growth is at a premium (Loehle 1988). Extant data support the contention that high defense types are less sensitive to pollution (D. LeBlanc, pers. comm.), but species testing in this context specifically is needed. What can be said is that for any given level of growth decrease due to pollution, low defense types will suffer more adverse consequences because they depend on rapid growth to overcome insect and pathogen attack. In order to better understand pollution effects, it is first necessary to separate natural from pollution-induced declines and to discern, in high pollution regions, which trees are and which are not affected with respect to age, site, and species.

Application to field studies

It is supposed that there should be a three-way interaction between pollution dose and defensive type, stand age, and site quality. The latter three factors are themselves interrelated.

The hypothesis that low plant defenses (and concomitant susceptibility to pollution) are a key contributor to stand declines suggests the importance of quantifying defenses. It should be possible to rigorously relate defenses to pollution impact in order to predict which species should be susceptible. In addition, the state of tree vigor and pathogen resistance can be related to age and onset of decline symptoms (as discussed below). This will allow prediction of the age at which a species becomes susceptible to decline. Assessment of the status of a stand based on quantification of vigor and defenses may also allow early detection of decline conditions.

Trees have multiple avenues of defense and the most important ones will vary by species and by situation. It is useful to set out the types of defenses and how they might be measured.

Energy budgets

Tree defenses exact a cost in photosynthetic material and nutrients (Waring 1987). The ability to synthesize adequate defensive structures and compounds is related to tree vigor in general (Lorio 1986). The reduced energy budget of low vigor (older or pollution-stressed) trees which results in reduced energy allocation to defenses is in fact a key component of the positive feedback decline model presented in this paper. A general measure of overall vigor is starch reserves, which decrease in tree roots in dieback stands (Manion 1981).

Different compounds have different costs of synthesis. Defensive compounds have a per gram cost of synthesis up to twice that of cellulose (770 to 860 vs. 450 to 470, Bloom *et al.* 1985; Loehle 1988). A vigorous tree attacked by pathogens might shift allocation of energy to defenses (see Lorio 1986; Loehle 1988). In a tree under severe stress, however, we might expect that energy would be allocated to crucial support systems, such as maintaining an adequate water conduction system or minimal leaf area, at the expense of defenses (Lorio 1986). This accords with the description in Waring (1987) and with data of Mueller-Dombois (pers. comm.) that low vigor trees of *Matrocedrus umbellata* in New Zealand showed lower leaf chemical defenses which resulted in defoliation by possums.

Assessing vigor thus contributes to understanding energy available for defense. Understanding how energy allocation patterns shift (Bloom *et al.* 1985; Bazzaz *et al.* 1987) is a key aspect of the problem that requires further work.

Leaf defenses

Leaf defenses have recently received increased attention and their importance is now unequivocal (Bazzaz *et al.* 1987; Bloom *et al.* 1985; Brown 1984; Coley *et al.* 1985; Janzen 1981; Rhoades 1985; Zucker 1983). Several types of defenses exist, including quantitative chemical defenses (e.g. tannins, resins), qualitative chemical defenses (pathogen-specific toxins effective at low doses), leaf toughness, silica (in grass mainly), low mineral nutrient content, and hairs, spines and serrations (Bailey & Deverall 1983; Horsfall & Cowling, 1980). Many of the defensive chemicals are not produced until the plant is attacked.

It has been demonstrated that early successional plants are lower in defensive leaf chemicals (Brown 1984) and therefore suffer greater herbivore damage. Loehle (1988) showed that early successional trees also exhibit lower wood defenses. Reduced defenses may therefore be a general characteristic of pioneer types, but 1) there are exceptions (e.g., *Taxodium* and *Sequoia* mentioned earlier) and 2) it is likely that not all aspects of defense will correlate with one another.

When leaf defenses decline with decreased vigor of older, environmentally-stressed, or severely pollution-stressed trees, then insect attack and disease incidence should rise and the species of leaf pathogens should be relatively general, though single-pest outbreaks are also possible. Leaf defenses are a good variable for assessing pollution impact and tree vigor because leaves are accessible and it should soon be possible to quantify their defenses chemically (e.g., Zucker 1983) or by testing palatability to insects in feeding trials.

Wood defenses

It is argued by Loehle (1988) that wood defenses are crucial for trees that live past a few hundred years. He used a crude measure of level of defenses, volumetric heat content (which combines density and chemical composition) and showed strong correlations of higher heat content with greater longevity. Wood density and decay resistance also correlate positively, and growth rate negatively, with longevity. For studying forest decline, however, more detailed consideration of the components of wood defense is necessary. Two main aspects of wood defense are active and passive defenses, treated separately below.

Shigo has been a pioneer in the area of active wood defense (Shigo & Marx 1977; Shigo *et al.* 1983; Tippett & Shigo 1981). He showed that wood is not a passive decay object, but a dynamic living system with specific strategies of response to pathogens. Upon wounding (including creation of branch stubs), the tree begins to wall off the infection. Cell walls thicken and become impregnated with defensive chemicals. In conifers, massive sap flow seals off the wound externally and impregnates the damaged or diseased wood. Vertical xylem elements become plugged above and below the wound which slows vertical spread of pathogens. These actions effectively compartmentalize the wound, slowing down the rate of pathogen spread. Limited evidence (Loehle 1988) suggests that species that are good compartmentalizers are also high on overall defenses and are longer-lived. The end of pathogen spread does not occur, however, until the wound is healed over. Thus trees with a greater radial increment are well protected because branch stubs and wounds heal over before decay can progress very far.

This discussion shows that there are three primary active wood defenses: barrier zone formation, defensive chemical (e.g., resin) production, and radial increment. All three of these can be quantified. Barrier zone formation can be quantified by experimental wounding and has been used to assess habitat differences in decay rates (Hepting & Shigo 1972). Defensive chemical production should likewise be a quantifiable response to wounding. I assert that the effectiveness of all three active responses will decrease in old or severely stressed trees.

When trees are old, active defenses are likely to be less effective. At this stage passive defenses begin to play a more important role. Wood that is dense and highly impregnated with defensive chemicals will be attacked only quite slowly even in the absence of effective active defenses. This helps explain Loehle's (1988) correlation of high defenses and low decay rates with greater longevity. Chemical defenses are of several kinds and should probably be studied as such. The role of lignin also needs study. A great deal of data exists on physical and chemical properties of wood (Fowells 1965; Hunt & Garratt 1967; Scheffer & Cowling 1966; Taras & Saucier 1967; Wangaard 1950; Wenger 1984; Zucker 1983) which might be related to defensive attributes. Measurement of these properties plus quantification of active defenses could lead to indices of tree susceptibility to pests as influenced by pollution.

Defenses, stand age, vigor, and pollution as factors in declines

If tree defenses are responsive to defensive type, tree vigor, and growth rate, then differences in levels of defenses according to species and age suggest some problems in detecting pollution stress. Stands that consist largely of trees that are more susceptible to pathogens because the species are characteristically low in defenses or because growth rates are slow due to either old age or poor site quality should be more responsive to pollution (given otherwise equal pollution susceptibility). Most documented declines in the United States and Europe involve either higher elevations or infertile sandy or rocky sites where growth rates are slower and climatic stresses are typically greater. Such sites tend to support pioneer type species because of fire histories or general site openness. Such species are the very type postulated to be more susceptible to pollution because of low defenses. In addition, mountain and poor quality sites are the most likely places to find old trees in developed countries because such sites are often made into parks or left alone because the timber is too difficult to harvest or of poor quality. The correspondence of certain soil factors (e.g., low pH, low buffering capacity) with sites containing a preponderance of susceptible trees can also lead to spurious correlations (i.e., the conclusion that trees are dying solely because of the low buffering capacity). Thus there is reason to suspect high occurrences of certain site factors, low defense species, old trees, and slow growth rates on the very sites most often or first showing decline syndrome. This correspondence between pathogen- and general stress-susceptible tree location and decline occurrence seems more than merely coincidental. Thus species type, site quality, and stand age should be measured and controlled for in any geographical study of decline incidence. Otherwise the increased number of old trees in decline stands, if overlooked, can lead to overestimates of pollution effect or, conversely, if casually considered can easily give the impression that declines are 'merely' natural senescence with no role for pollution, when in fact pollution may be a contributing factor. In the next sections, I explore these problems in terms of statistical tests.

Distinguishing natural from pollution-induced declines

Since natural stand declines have been observed (Mueller-Dombois 1986), it is necessary to determine whether current diebacks in Europe and New England (USA) are natural or pollution-induced. It has been shown that there is a likely role for climatic effects (e.g., drought, warming, cold winters following warm summers) in initiating some of the declines in New England (Hamburg & Cogbill 1988; Johnson & Siccama 1983; Johnson *et al.* 1986; Manion 1981; LeBlanc *et al.* 1987a, b, c, d). This does not, of course, demonstrate that observed declines are entirely natural, because pollution could act to make stands more susceptible to climatic stress or to inhibit recovery after a stress by damping the negative feedback cycle of stress recovery (Fig. 1, 2).

If demonstration of correlation of decline initiation with weather events does not prove that a decline is entirely natural, what can be done to separate the two effects? One line of evidence concerns decline characteristics typical of only one type of stress agent (e.g., Hinrichsen 1987; Waring 1987). For example, factors related to water status should be highly correlated with a drought-dominated decline. Mueller-Dombois (1986) showed that where there were sharp boundaries between soil types in Hawaiian forests that differed in their water relations, the adjacent stands often differed completely in their degree of dieback. Similarly, Manion (1981) suggested that trees growing off-site (particularly on sites that are drier than optimal for the species) accounted for a significant number of historical diebacks in New England (see also Johnson *et al.* 1986; LeBlanc *et al.* 1987a, b). Note that defenses are likely to be impaired for such off-site stands. In addition, specific symptoms such as needle burn are more likely to be associated with pollution injury (Hauhs & Wright 1986). Thus, it is likely that information on specific symptoms and decline spatial patterns must be used to assess whether observed climatic and pollution factors played a causative or a contributory role in the onset of decline.

Lagged growth responses

Attempts to correlate pollutant dose with tree response are confounded by the lagged nature of tree growth responses. Pollutant concentrations vary on hourly to monthly time scales, whereas it is difficult to detect growth changes even on an annual time scale. In particular, there is a lag period for recovery after a stress is removed. For severe short term stresses, this recovery period may be quite short. Longer stress periods may, however, result in protracted recovery periods because of reduced starch reserves, unfavorable root/shoot ratios, etc. This can be clearly seen in the response of shade-suppressed seedlings and saplings. Seedlings of many species can withstand a moderate level of suppression. When they are released by removal of some or all of the overstory, they gradually recover and resume normal growth, a positive feedback process of increasing vigor. It may, however, take many years for a suppressed sapling to reach a growth rate equal to that of a non-suppressed sapling of similar size. The longer the period of suppression, the longer the recovery period is likely to be, except in species very tolerant of suppression such as hemlock and red spruce. In extreme cases, suppressed trees are never able to recover, a state we may call the 'point of no return.' For example, suppressed pole-sized loblolly pines (*Pinus taeda*) left after logging will in many cases never fully recover and achieve a large size. McLaughlin *et al.* (1985) showed that, in contrast to younger trees, old sugar maple trees failed to recover after cessation of tent caterpillar infestation. In old trees, active defenses are likely to be less effective. LeBlanc *et al.* (1987c) showed that high vigor trees recovered quickly after drought, whereas low vigor trees responded slowly or not at all to improved conditions. Thus, there is differential recovery based on tree age, vigor, species, and length of suppression or of pollutant stress. For older trees or those chronically stressed, removal of the pollutant may not be sufficient to induce recovery because such trees, once stressed, suffer ongoing increased pest and pathogen loads and may be suffering from physiological damage. Efforts to correlate dose with response could be confounded by such lagged effects. Rather than a model that directly predicts growth suppression from pollutant dose, a lagged variable would be necessary to account for the recovery period. Since a reduction in pollutant dose might not result in recovery of older stands, such stands would be very poor indicators of mitigation effort effectiveness.

Endogenous changes in ring width

There is a natural cycle of ring width (diameter increment) changes during the life of the tree that can also confound analyses of pollution effects. As a tree in an even-aged stand matures, the basal area increment increases up to a point, more or less levels off for a while, and then declines. Three factors contribute to decline in ring width as a tree ages: 1) in larger trees the same annual basal area increment must be spread around an increasingly wider diameter, 2) competition reduces ring width as the canopy closes, and 3) the tree eventually reaches site constraints and/or genetic limits on size. These factors, singly or in combination, can result in spurious correlations between growth decline and the increased pollutant levels of the last few decades. It is informative to examine the effects of geometry alone and then the combined effects of all three factors.

The effect of geometry was assessed by looking at an idealized tree, initially 24 cm diameter, with a constant basal area increment of 23 cm^2. The ring width for such a tree will decrease as diameter increases. Figure 3 depicts this hypothetical tree during the level growth phase, growing unaffected by pollution. Ring width was correlated with pollution dose for two scenarios, 1) a pollution dose increasing linearly over the last 30 years, and 2) no dose until 15 years ago when it increased stepwise by an arbitrary amount (Fig. 3). The linearly increasing dose predicted (with one data point at each year) ring width with $R^2=0.87$ ($P<0.001$) and the step function predicted with $R^2=0.57$ ($P<0.001$). These high correlations are completely spurious and indicate

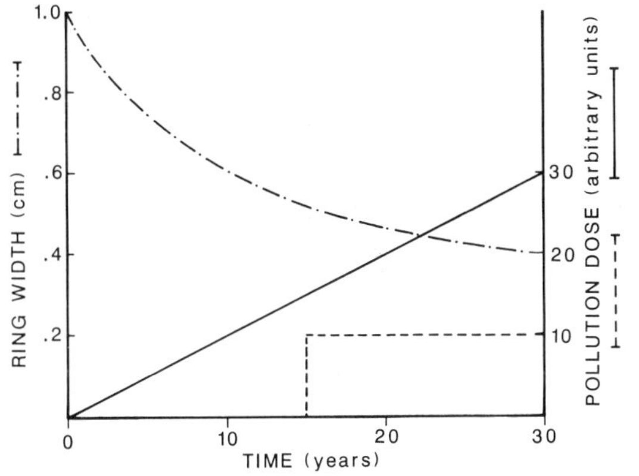

Fig. 3. Ring width for a tree of constant basal area increment (no pollutant effect) versus hypothetical pollution doses. Ring width was predicted from the step-function pollution dose with $R^2=0.57$ ($P<0.001$) and from the monotonically increasing dose with $R^2=0.87$ ($P<0.001$), even though the effect is totally spurious.

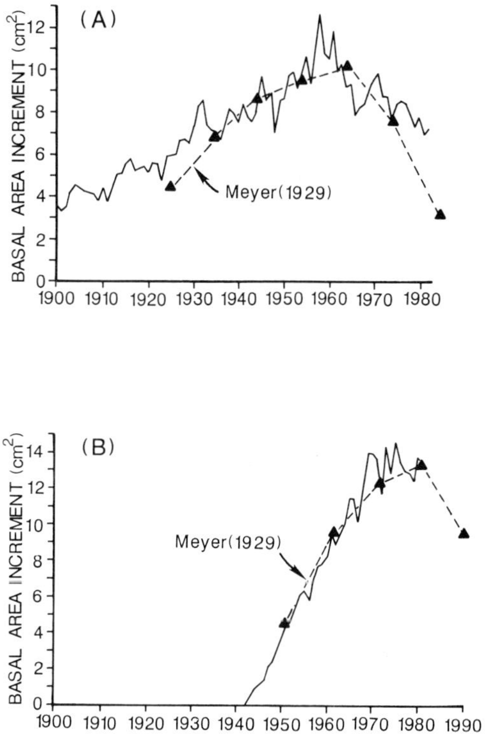

Fig. 4. Annual basal area increment versus time for recent data (solid line) from Hornbeck *et al.* (1986) compared to data from Meyer (1929), showing current stands of red spruce (A) and balsam fir (B) to be growing the same as unpolluted stands in earlier decades. Figure from Hornbeck *et al.* (1986).

the magnitude of the possible effect (in a noise-free case) due to geometry alone.

In a real tree, the effect of geometry combines with competitive effects and tree maturation, as well as random climatic variation. Actual average basal area increments versus time in forest stands are shown in Fig. 4. Data are from Hornbeck *et al.* (1986) for red spruce and balsam fir from northern New England and upstate New York. The decline in basal area increment evident in red spruce after about 1960 (Fig. 4) has often been attributed to pollution or to drought, as mentioned above. However, Hornbeck *et al.* (1986) argue that most red spruce in New England (80% of their large, dispersed sample of 3001 trees) germinated either between 1900–1920 or were suppressed saplings that were released during this period. They attribute this even-aged character of the stands to the combination of extensive logging and budworm activity during these decades. These stands have now passed their peak of growth, and basal area increment is declining. Overlaying data from Meyer (1929) for even-aged second growth stands of red spruce from the turn of the century (Fig. 4) supports this contention. The curves match closely, including the downturn phase, for both spruce and fir. While the overlap of the curves is not hard proof because stand conditions in the two studies are not identical (see McLaughlin & Adams

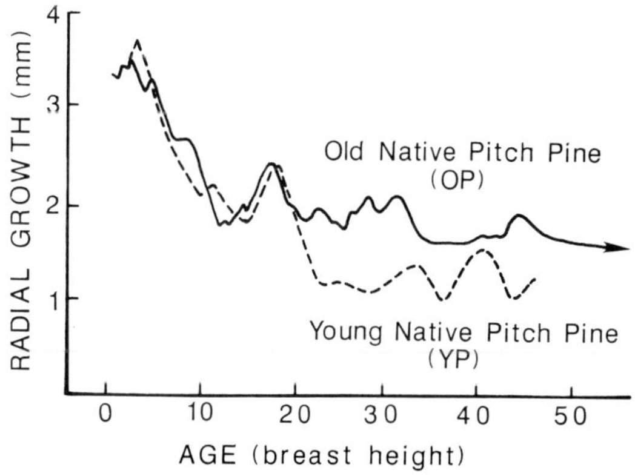

Fig. 5. Comparison of growth of old and young native pitch pine in New Jersey. Young trees grew slowly after age 20 which corresponds to the time of supposed increased air pollution. This same effect could have been due to increased competition at this age. Redrawn from Johnson *et al.* (1981).

1987), it does suggest that endogenous changes in ring width or basal area increment are capable of giving a false impression of recent pollution-induced declines.

While the existence of natural changes in increment would seem evident and dendrochronological texts account for it (e.g., Bonkoungou *et al.* 1983; Fritts 1976; Matalas 1962), many studies have failed to take it into account (e.g., as discussed in LeBlanc *et al.* 1987d; Sheppard & Jacoby 1987). For example, Puckett (1982) dismisses tree age as a factor affecting climate-tree growth relationships, even though his 70-year chronology is long enough and his trees old enough (many over 200 years) relative to species life span that a natural decline in increment is likely, as is also a change in tree response to climate variables (as also noted by Visser 1986). In particular, eastern hemlock, which has a much longer life span than the other species (white pine, pitch pine, and chestnut oak), showed no change in climate-growth response over the study period. The hemlocks in the sample were not as old relative to species typical life span as were the other species. This is not to say that Puckett's (1982) results are incorrect, only that discounting tree age as a factor in this case is difficult to justify.

The use of confounded variables has weakened claimed correlations of tree responses to pollution, leading to the conclusion that no pollution effect has been observed (Burgess 1984; Woodman & Cowling 1987). It is necessary to correct observed tree and stand growth responses for normal dynamics before historical correlations will be valid. Not correcting for natural dynamics may lead to the conclusion that there is no detectable pollution effect (e.g., Cogbill 1977) in stands in which trees are actually in a rapid growth phase, making pollution effects hard to detect. Conversely, it has been concluded that trees are dying from pollution when in fact the effect may be due to natural decline of older stands (e.g., Fig. 4). This problem, of course, is avoided if a definite gradient in pollution dose can be established, as for example near a point source of pollution (Ashby & Fritts 1972; Gemmill *et al.* 1982; Hauhs & Wright 1986; McClenahan & Dochinger 1985; Scheffer & Hedgcock 1955; Stone & Skelly 1974). The confounding variables of stand age, site quality, and tree defenses still need to be considered, however.

Correcting for stand dynamics

The fact that a number of pollution studies have used ring width not corrected for tree sizes as a variable (e.g., Cogbill 1977; Johnson & Siccama 1983; Johnson *et al.* 1981; McLaughlin *et al.* 1987; Smith 1981) suggests that the effect of endogenous changes in tree dimensions may not be universally appreciated. Several methods have been suggested to improve on the use of ring width alone. It has been suggested that basal area increment is better than ring width (radial increment) because it is not affected by stem geometry. However, as is evident from Fig. 4, basal area increment is also subject to reduction as the tree ages because of competition and tree maturation. Height growth has been used (Miller 1977) as a response variable, and while height increment may be less endogenously variable than ring width, it still varies over the life of the tree. Specific volume increment (Leblanc 1987c) has also been used as a response variable.

To overcome these combined problems, some investigators have used dendro-chronological analyses of old trees as controls for young trees that have grown during a polluted period (e.g., Federer & Hornbeck 1987; Hornbeck *et al.* 1986; Johnson *et al.* 1981; Miller 1977). If a tree is now 80 years old, it was 30 years old 50 years ago during a (supposedly) pollution-free period. Its height, diameter, and diameter increment at age 30 can be compared to trees that are now age 30 and that grew to maturity during a largely polluted period. This method very effectively controls for dynamic properties if two conditions are met: 1) the climate (largely precipitation) is similar in the two periods, and 2) stand density is comparable for the two stands at the ages for which they are being compared. Miller (1977), for example, specifically accounted for climate in the two periods he compared. The method commonly used to account for variance associated with climate is response-function analysis (Fritts 1976), in which ring increment is related to key climatic variables by regression. Puckett (1982), for example, looked for a change in the climate-growth relationship over time to show that recent pollution was affecting tree growth-climate relationships. It is difficult to define the magnitude of an effect in this way, however, and furthermore, natural senescence can also affect the relationship (discussed above in relation to Puckett 1982). Indexing can also be used to subtract our natural decline in ring width due to aging (e.g., Fritts 1976 p. 24), but this process may also factor out effects due to pollution. It can be very difficult, on the other hand, to correct for density by establishing the density of a stand 50 years ago, because small dead trees have long since vanished. There is, in fact, historical evidence that many older second growth stands in the United States (ages 60 to 100+) were less dense when young than stands established in the last 40 years. This is because poor logging practices, especially uncontrolled fires that followed the logging crews, resulted in very poor stocking (Eastman 1986). This resulted in a very long period of low competition before crown closure in such stands. Many other stands in the East were established on abandoned farmland or pasture, a process that often results in low stocking, with similar consequences. The early growth of trees that are now fully mature might therefore be significantly greater than young trees today, but when these stands closed up competition could be even more severe than in old-growth stands, giving a spurious apparent decline that correlates in time with current high pollution levels. An ideal solution to this would be to use stands for which historical records of density exist, whether planted at known densities or inventoried as 'fully stocked' when young. Whether there are enough such stands for statistical analysis remains to be determined. In one case, Federer & Hornbeck (1987) were able to derive predictive equations for changes in tree growth due to natural processes based on comparable old data. LeBlanc *et al.* (1987a) were able to obtain historical density data by using planted stands, and to partially correct for competition effects.

An example of the way in which density can confound inference of causation is the study by Johnson *et al.* (1981) of growth decline in the New Jersey pinelands. Plotting radial growth versus year, they believed that they detected a growth decline beginning around 1960. Recognizing that part of the decrease was due to tree aging, they overlaid old native pitch pine and young native pitch pine plots so that ages corresponded (Fig. 5). The growth curves overlapped almost exactly until age 20 (breast height), when the younger, pollution-affected stands dropped sharply and continued to grow at a reduced rate. Since the onset of reduced growth corresponds to the period of increased pollution (specifically pH), it was inferred that the growth reduction was due to pollution. It is worth noting that age 20 breast height is about when we would expect crown closure in young pitch pine (established under fire-suppression policies and therefore denser than old growth stands), at which time competition could reduce ring width considerably. Even dominant trees would be affected by it. Even incomplete crown closure would reduce growth of all trees because of root competition on these dry, nutrient poor soils. In contrast, the old stands were likely established under lower density conditions near the turn of the century, as noted above. While some of the pitch pine sampled were open grown (competition free), Johnson *et al.* (1981) do not say what the percentage was, so the possibility remains that density effects confound their ability to make an inference. Additional data from trees established after 1950 could provide a useful control for the above coincidence between the timing of pollution onset or drought and year of likely crown closure. Such trees should show ring widths less than those of old stands at all ages if they are affected by pollution. An improved analysis would result from consideration of stand growth as a whole, including stand density and basal area.

One final method of controlling for stand dynamics is the use of tree growth models (e.g., Dale & Gardner 1987). Such models can be used to control for the effects of stocking rate, climate, species, and management actions. Dale & Gardner (1987) for example, used a growth model to correct USDA Forest Service data for natural changes in growth as stands age. This correction largely eliminated the supposed growth decline in recent decades evident in the data. An additional modeling project showed that spruce stand dynamics could lead to endogenous declines due to soil nitrogen immobilization (Pastor *et al.* 1987).

A point of caution concerning these growth models

is to be sure that model parameters are for trees unaffected by pollution and not unduly affected by different climatic regimes. Dale (pers. comm.) notes that many of the growth models are based on old data (collected before pollution became severe) and mature trees, so there is no problem of confounding. For growth models based on young stands, such as pulp-rotation pine plantations that are currently 30 years old, model parameters may reflect existing pollution-induced growth suppression. Use of such models to 'correct' inventory data would falsely eliminate pollution-induced growth suppression effects.

It is evident that establishing cause of death is not trivial when dealing with dynamic entities such as trees (as noted by Franklin *et al.* 1987). Nevertheless, the tools do exist to do it right and only proper inclusion of dynamic stand attributes will allow attribution of cause in this important undertaking.

Conclusions

The phenomenon of stand dieback is important and efforts should be made to evaluate the relationship of dieback to possible human impacts. Trees and stands are dynamic entities, however, and simple correlations of pollutant stress with ring increment or tree mortality will not resolve causation. Consideration of stand age and vigor is necessary, as is an understanding of the dynamics of tree defenses and the differences between species and ages in defensive capacity. Tree defenses are clearly multifaceted and should be studied as such.

The recent trend toward recognition of the need to correct tree ring records for natural changes in growth is commendable, but there is a danger of dismissing all stand declines as merely results of natural stand senescence. The model presented in this paper suggests that older trees and species such as red spruce (because of site and age) and fir (because of site, age, and defensive type) should be particularly responsive to pollution damage. Rather than concluding that such hypersensitivity means that we 'don't really' have a problem, I argue that they provide an early warning of potentially serious problems.

Acknowledgements

Research was supported by DOE contract #DE-AC09-76SROO-819 to the University of Georgia's Savannah River Ecology Laboratory. Helpful reviews were provided by John Gladden, James Hornbeck, Robert Jones, Steve Kroeger, Charlie Murphy, Jodi R. Shann, anonymous reviewers and especially David LeBlanc. Pat Davis typed the manuscript. Editorial assistance was provided by Patricia J. West.

References

Ashby, W. C. & Fritts, H. C. 1972. Tree growth, air pollution, and climate near La Porte, Ind. Bull. Am. Meteorol. Soc. 53: 246–251.

Bailey, J. A. & Deverall, B. J. 1983. The dynamics of host defence. Academic Press, New York.

Bazzaz, F. A., Chiarello, N. R., Coley, P. D. & Pitelka, L. F. 1987. Allocation of resources to reproduction and defense. BioScience 37: 58–67.

Bloom, A. J., Chapin III, F. S. & Mooney, H. A. 1985. Resource limitation in plants – An economic analogy. Ann. Rev. Ecol. Syst. 16: 363–392.

Bonkoungou, G. J. E., Raynal, D. J. & Geis, J. W. 1983. Tree population dynamics in relation to climate and forest history in the Oswegatchie Plains, northern New York. Vegetatio 54: 37–59.

Brown, V. K. 1984. Secondary succession: insect-plant relationships. BioScience 34: 710–716.

Burgess, R. L. 1984. Effects of acidic deposition on forest ecosystems in the northeastern United States: an evaluation of current evidence. State University of New York, College of Environmental Science and Forestry (ESF 84-016).

Carey, A. C., Miller, E. A., Geballe, G. T., Wargo, P. M., Smith, W. H. & Siccama, T. G. 1984. *Armillaria mellea* in decline of red spruce *Picea rubens*. Plant Disease 68: 794–795.

Cogbill, C. V. 1977. The effect of acid precipitation on tree growth in eastern North America. Water Air Soil Poll. 8: 89–93.

Coley, P. D., Bryant, J. P. & Chapin III, F. S. 1985. Resource availability and plant anti-herbivore defense. Science 230: 895–899.

Dale, V. H. & Gardner, R. H. 1987. Assessing regional impacts of growth declines using a forest succession model. J. Environ. Manage. 24: 83–93.

Eastman, J. 1986. The ghost forest. Natural History 95: 10–16.

Federer, C. A. & Hornbeck, J. W. 1987. Expected decrease in diameter growth of even-aged red spruce. Can. J. For. Res. 17: 266–269.

Fowells, H. A. 1965. Silvics of forest trees of the United States. USDA For. Ser. Agri. Handbook No. 271, Washington.

Franklin, J. F., Shugart, H. H. & Harmon, M. E. 1987. Tree death as an ecological process. BioScience 37: 550–556.

Fritts, H. C. 1976. Tree rings and climate. Academic Press, New York.

Gemmill, B., McBride, J. R. & Laven, R. D. 1982. Development of tree-ring chronologies in an ozone air pollution-stressed forest in Southern California. Tree-Ring Bull. 42: 23–31.

Hamburg, S. P. & Cogbill, C. V. 1988. Historical decline of red spruce populations and climatic warming. Nature 331: 428–431.

Hauhs, M. & Wright, R. F. 1986. Regional pattern of acid deposition and forest decline along a cross section through Europe. Water Air Soil Poll. 31: 463–474.

Hepting, G. & Shigo, A. L. 1972. Differences in decay rate following fire between oaks in North Carolina and Maine. Plant. Dis. Rep. 56: 406.

Hinrichsen, D. 1987. The forest decline enigma. BioScience 37: 542–546.

Hornbeck, J. W., Smith, R. B. & Federer, C. A. 1986. Growth decline in red spruce and balsam fir relative to natural processes. Water Air Soil Poll. 31: 425–430.

Horsfall, J. G. & Cowling, E. B. 1980. Plant disease V: How plants defend themselves. Academic Press, New York.

Hunt, G. M. & Garratt, G. A. 1967. Wood preservation. McGraw Hill, New York.

Janzen, D. H. 1981. Evolutionary physiology of personal defense. In: Townsend, C. R. & Calow, P. (eds), Physiological ecology. pp. 145–164. Sinauer Assoc., Sunderland, MA.

Johnson, A. H. & Siccama, T. G. 1983. Acid deposition and forest decline. Environ. Sci. Technol. 17: 294A–305A.

Johnson, A. H., Siccama, T. G., Wang, D., Turner, R. S. & Barringer, T. H. 1981. Recent changes in patterns of tree growth rate in the New Jersey pinelands: a possible effect of acid rain. J. Environ. Qual. 10: 427–430.

Johnson, A. H., Friedland, A. J. & Dushoff, J. G. 1986. Recent and historic red spruce mortality: evidence of climatic influence. Water Air Soil Poll. 30: 319–330.

LeBlanc, D. C., Raynal, D. J. & White, E. H. 1987a. Acidic deposition and tree growth: I. The use of stem analysis to study historical growth patterns. J. Environ. Qual. 16: 325–333.

LeBlanc, D. C., Raynal, D. J. & White, E. H. 1987b. Acidic deposition and tree growth: II. Assessing the role of climate in recent growth declines. J. Environ. Qual. 16: 334–340.

LeBlanc, D. C., Raynal, D. J., White, E. H. & Ketchledge, E. H. 1987c. Characterization of historical growth patterns in declining red spruce trees. In: Jacoby, G. C. (ed.), Proc., International Symposium on Ecological Aspects of Tree-Ring Analysis, Aug. 17–21, 1986. pp. 360–372. Tarrytown, New York, NTIS # Conf-8608144.

LeBlanc, D. C., Raynal, D. J. & White, E. H. 1987d. Dendroecological analysis of acidic deposition effects on forest productivity. In: Hutchinson, T. C. & Meema, K. M. (eds), Effects of Atmospheric Pollutants on Forests, Wetlands and Agricultural Ecosystems. pp. 291–306. NATO ASI series, Vol. G16, Springer-Verlag, Berlin.

LaMarche Jr, V. C. 1969. Environment in relation to age of bristlecone pines. Ecology 50: 53–59.

Loehle, C. 1988. Tree life history strategies: the role of defenses. Can. J. For. Res. 18: 209–222.

Lorio Jr, P. L. 1986. Growth-differentiation balance: a basis for understanding southern pine beetle-tree interactions. For. Ecol. Manage. 14: 259–273.

Manion, P. D. 1981. Tree-disease concepts. Prentice-Hall, Englewood Cliffs, NJ.

Matalas, N. C. 1962. Statistical properties of tree ring data. Bull. Int. Assoc. Sci. Hydrol. 7: 39–47.

McClenahan, J. R. & Dochinger, L. S. 1985. Tree ring response of white oak to climate and air pollution near the Ohio River Valley. J. Environ. Qual. 14: 274–280.

McLaughlin, D. L., Linzon, S. N., Dimma, D. E. & McIlveen, W. D. 1985. Sugar maple decline in Ontario. Ontario Ministry of the Environment, Toronto, Canada.

McLaughlin, S. B., Downing, D. J., Blasing, T. J., Cook, E. R. & Adams, H. S. 1987. An analysis of climate and competition as contributors to decline of red spruce in high elevation Appalachian forests of the Eastern United States. Oecologia (Berl.) 72: 487–501.

McLaughlin, S. B. & Adams, H. S. 1987. Growth declines in red spruce (correspondence). J. For. 85: 50–51.

Meyer, W. H. 1929. Yields of second-growth spruce and fir in the northeast. USDA For. Serv. Tech. Bull. 142, 52 pp.

Miller, P. R. 1977. Photochemical oxidant air pollutant effects on a mixed conifer forest ecosystem. US EPA Publ. EPA-600/3-77-104, Corvallis, OR.

Mueller-Dombois, D. 1986. Perspectives for an etiology of stand-level dieback. Ann. Rev. Ecol. Syst. 17: 221–243.

Pastor, J., Gardner, R. H., Dale, V. H. & Post, W. M. 1987. Successional changes in nitrogen availability as a potential factor contributing to spruce declines in boreal North America. Can. J. For. Res. 17: 1394–1400.

Puckett, L. J. 1982. Acid rain, air pollution, and tree growth in southeastern New York. J. Environ. Qual. 11: 376–381.

Rhoades, D. F. 1985. Offensive-defensive interactions between herbivores and plants: their relevance in herbivore population dynamics and ecological theory. Am. Nat. 125: 205–238.

Scheffer, T. C. & Hedgcock, G. C. 1955. Injury to Northwestern forest trees by sulfur dioxide from smelters. USDA For. Serv. Tech. Bull. No. 1117. Washington.

Scheffer, T. C. & Cowling, E. B. 1966. Natural resistance of wood to microbial deterioration. Ann. Rev. Phytopathol. 4: 147–174.

Sheppard, P. R. & Jacoby, G. C. 1987. Tree-ring data (Correspondence). J. For. 85: 50–51.

Shigo, A. L. & Marx, H. G. 1977. Compartmentalization of decay in trees (CODIT). USDA Inf. Bull. 405, Washington.

Shigo, A. L., Dorn, D. & Lee, H. C. 1983. Selections of maple and birch trees with high resistance to spread of decay associated with wounds. pp. 110–117. In: Proc., 28th Northeastern Forest Tree Improvement Conference, Durham, NH, 1982.

Smith, W. H. 1981. Air pollution and forests. Springer, Berlin.

Stone, L. L. & Skelly, J. M. 1974. The growth of two forest tree species adjacent to a periodic source of air pollution. Phytopathology 64: 773–778.

Taras, M. A. & Saucier, J. R. 1967. Influence of extractives on specific gravity of southern pine. For. Prod. J. 17: 97–99.

Tippett, J. T. & Shigo, A. L. 1981. Barrier zone formation: a mechanism of tree defense against vascular pathogens. IAWA Bull. 2: 163–168.

Visser, H. 1986. Analysis of tree ring data using the Kalman filter technique. IAWA Bull. N.S. 7: 289–297.

Wangaard, F. F. 1950. The mechanical properties of wood. John Wiley & Sons, New York.

Waring, R. H. 1987. Characteristics of trees predisposed to die. BioScience 37: 569–574.

Wenger, K. F. 1984. Forestry handbook. John Wiley & Sons, New York.

Woodman, J. N. & Cowling, E. B. 1987. Airborne chemicals and forest health. Environ. Sci. Technol. 21: 120–126.

Zucker, W. V. 1983. Tannins: Does structure determine function? An ecological perspective. Am. Nat. 121: 335–365.

Vegetatio 77: 79–85, 1988
© Kluwer Academic Publishers, Dordrecht – Printed in the Netherlands

Factors in *Pinus ponderosa* and *Calocedrus decurrens* mortality in Yosemite Valley, USA

Robert J. Sherman & R. Keith Warren
Biology Department, Sonoma State University, Rohnert Park, CA 94928 USA

Accepted 27.6.1988

Keywords: Bark beetle, Drought, Fire exclusion, *Heterobasidion annosus*, Root rot, Yosemite National Park

Abstract

Management practices over the past 100 years, especially fire exclusion and suppression, in Yosemite Valley, Yosemite National Park, CA resulted in forest stands with basal areas in excess of 110 m^2 ha^{-1}. Accelerated attack by several species of bark beetles (Scolytidae) followed a severe drought in 1976–77. *Pinus ponderosa*, already under attack by root rot (*Heterobasidion annosus*) was affected severely, and mortality exceeded 70%. *Calocedrus decurrens* did not suffer comparable mortality and has replaced *P. ponderosa* as the dominant species. Size class distributions as well as comparisons of relative density and relative basal area from 1961, 1976 and 1983–85 also demonstrate a major change in forest composition. This change is most pronounced on alluvial landforms.

Nomenclature of the vascular plants follows: Munz, P. A. & Keck, D. D. 1973. A California Flora and Supplement. Univ. California Press, Berkeley.

Introduction

The nineteenth century Sierran forests were described by Muir (1894) as open in nature and subject to periodic fires. Others including Reynolds (1959) and Vankat (1970) drew similar conclusions. The situation was summarized by van Wagtendonk (1974) who stated: There is general agreement that the forests were quite open compared to present conditions and that fire was important in maintaining the openness.

Gibbens & Heady (1964) conducted a study in 1961 using historical records and photographs to document vegetational changes in Yosemite Valley over the last century. They concluded that *Pinus ponderosa* and *Calocedrus decurrens* had overtopped open stands of old growth *Quercus kelloggii* and that highly shade tolerant species, especially *Calocedrus*, would become codominant with the less tolerant *Pinus*. Heady & Zinke (1978) concluded that, barring catastrophe or managerial impact, the forest stands in the Valley would become dominated increasingly by *Calocedrus*, *Abies concolor*, *Pseudotsuga menziesii*, and *Q. chrysolepis*. These four species, however, would not replace *Q. kelloggii* and *Pinus*, but would combine with them to form a dense, mixed conifer-oak forest.

These studies suggest that, prior to 1976, succession in Yosemite Valley was gradual. Precipitation in western North America, however, is especially variable from year to year and droughts may result in stress and rapid deterioration of healthy trees. In 1976–77 an extreme drought affected the western United States. Precipitation at Yosemite National

Park Headquarters was only 39% of average (U.S. Weather Bureau, 1976–77) and in Yosemite Valley, mean flow of the Merced River for September 1977 was only 8.2% of the 1971–81 September average (U.S. Dept. Interior 1971–81). This drought was indicated as responsible for inducing tree mortality as early as June 1977 in the national forests of northern California (Smith *et al.* 1983).

The interactions of specific factors that provoke the attack of drought stressed trees by phytophagous insects and fungi are complex. Manion (1981) provided a comprehensive model which illustrated the way in which many factors eventually contribute to mortality. Larsson *et al.* (1983) demonstrated that reduced vigor led to bark beetle attack and mortality for *P. ponderosa* stands with basal areas >34 m^2 ha^{-1}. *P. ponderosa* stands with basal areas from 26 to 172 m^2 ha^{-1} (average = 46.49) were reported for the southern Sierra Nevada in California by Vankat (1970). Many of these stands may suffer reduced vigor and have high potential of bark beetle attack.

There is often a strong association between the occurrence of root pathogens and infestation of *Pinus* by bark beetles (Cobb *et al.* 1974). Smith *et al.* (1983), in an inventory of conifer mortality in California, stated: . . . a substantial portion, usually a majority, of dead trees have both insects and diseases contributing to their deaths. The presence of *Heterobasidion annosus* (Fr.) Bref. in Yosemite Valley was reported in 1966 (Bega & Smith) and by 1974, 144 disease centers were documented (Felix *et al.* 1974).

The present study, undertaken in 1976, documents the changes in forest composition in Yosemite Valley since 1961. The drought of 1976–77 provided the opportunity to examine the differential mortality of the major tree species under stressful conditions and to demonstrate that major successional changes in forest ecosystems may be abrupt rather than gradual.

Study sites

Site selection

In May of 1976 five stands (C, D, H, 6 and 7) in Yosemite Valley (37°43′N, 119°37′W) previously studied by Gibbens & Heady in 1961 were resampled for tree species composition. The 1976 work was repeated over a two year period (1983–1985) and expanded to include three additional numbered Gibbens & Heady sites (1, 2 and 3). Two other numbered sites (4 and 5) were not resampled because they are now occupied by, or are in close proximity to, campgrounds.

Landforms

Heady & Zinke (1978) classified the Valley using the system illustrated in Fig. 1. Differences in landforms and soil types have great influence on both surface runoff and ground water availability which, in turn, influence plant moisture balance. To emphasize these site differences all figures and tables in this study are arranged to conform to Heady & Zinke's system and sites are designated A1–A3, EA, F, and C1–C3 to more clearly reflect the soil/landform type.

Prescribed fire

Yosemite Valley has been treated with prescribed fire since 1970. Many of the sites sampled by Gibbens & Heady in 1961 have since been affected by this program. Density of smaller understory trees may have been reduced by as much as 50%, especially *Calocedrus* in the 1–3 m height class (van Wagtendonk 1974), leading to an unquantified reduction in the 5–9.9 cm diameter breast height (DBH) size class on some sites. Because all sites were not treated at

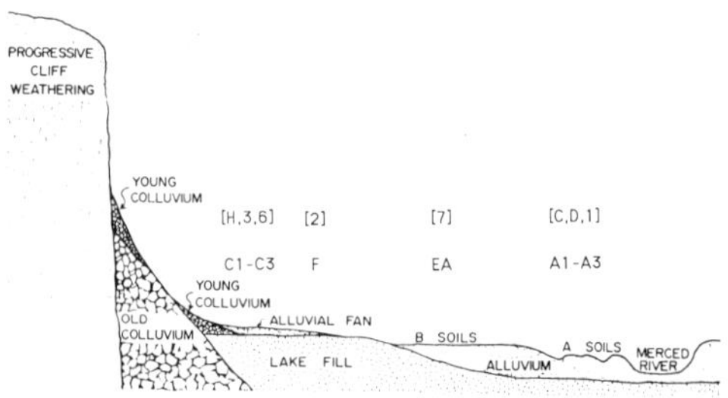

Fig. 1. Cross section of Yosemite Valley showing site designations and relationships of colluvial (C), fan (F), elevated alluvium (EA), and alluvial (A) deposits. Site designations used by Gibbens & Heady in brackets. Revised from Heady & Zinke with permission.

the same time or by fires of the same intensity, differences in the 5–9.9 cm DBH size class from the 1961 study and the present study are complicated.

Methods

Gibbens & Heady (1964) utilized two sampling approaches. In some stands they used a standard point-centered-quarter technique of Cottam & Curtis (1956) measuring trees greater than two inches (5 cm) DBH. In other stands they used a modified step-point method of sampling, mixing two different life forms, trees and herbaceous plants.

We sampled five of their numbered (point-centered-quarter method) and three of their lettered (step-point method) sites using the point-centered-quarter method exclusively. All values in Table 1, except those in parentheses, are based on point-centered-quarter samples. Values in parentheses were derived by excluding the herbaceous species from the data gathered by Gibbens & Heady using the step-point method. Points were laid out at 25 m intervals along five parallel transects. In each stand live trees in two size classes; 5–9.9 cm and ≥10 cm DBH were sampled at each of 25 points. In the summer of 1985 all eight sites were resampled for dead trees ≥10 cm DBH only. Sampling of the smaller size class of dead trees was not appropriate because of the recent treatments with prescribed fire. On some sites the density of dead trees was low, resulting in missing data in some quarters. The correction factor table developed by Warde & Petranka (1981) was used to compensate when dead trees were not found within 12 m of the sample point. No corrections were necessary for any of the live tree samples. Adequacy of sampling was determined using performance curves of mean values and standard error of the mean/mean ratios (SEM/MEAN) (Mueller-Dombois & Ellenberg 1974). In all cases the cumulative means became insensitive to variations in the raw data at the 5% level with ≤20 points and SEM/MEAN values ranged from 0.06–0.11 (average = 0.07).

Mortality was calculated for each species using only the ≥10 cm size classes. The value (density or basal area) of the dead trees of a given species was divided by the combined value of live and dead trees of that species. Because it is likely that mortality was the result of a number of complex interactions, determination of specific causes of mortality was not attempted.

Results

Density and basal area

Nine species of trees were encountered: *P. ponderosa, C. decurrens, Q. kelloggii, Q. chrysolepis, A. concolor, P. menziesii, Cornus nuttallii, Acer macrophyllum,* and *Populus trichocarpa.* Data from two size classes of live trees (5–9.9 cm and ≥10 cm) and one size class of dead trees (≥10 cm) are reported. Relative density and relative basal area data for live trees for 1961, 1976, and 1983 85 are presented in Table 1. In addition, total absolute density and total absolute basal area for live and live plus dead trees are reported for the 1983–85 sampling period. Because absolute density and basal area values were not reported by Gibbens & Heady from the 1961 data, comparisons are limited to relative density and relative basal area.

Size class distributions

Relative densities by size class based only on live trees from the 1983–85 sample for the ecologically dominant species, *Pinus* and *Calocedrus,* are compared in Fig. 2. In all cases, even with the recent reductions of the 5–9.9 cm size class by prescribed fire, the relative density of *Calocedrus* is greater in the smaller size class than in the larger size class. For *Pinus* the situation is reversed. With the exception of Site C1, the relative density of the smaller size class is less than that of the larger size class.

Mortality

Most mortality, both in terms of reductions in density and basal area, was of large *Pinus.* The reductions in density for *Pinus* and *Calocedrus* are illustrated

Table 1. Relative density and relative basal area for nine species of live trees on eight forest sites for the years 1961, 1976 and 1983 – 85. Total site absolute density and absolute basal area for live and live plus dead trees for 1983 – 85 only. Relative density values for 1961 in parentheses are not strictly comparable to other values. See text for details. Entries of '– –' in 1961 data indicate basal area not sampled and in 1976 data indicate site not sampled. Entries of 't' indicate a trace with a relative value <0.5.

Relative Density – Trees ≥5 cm DBH

Species	*Pinus*			*Calocedrus*			*Q. kelloggii*			*Q. chrysolepus*			*Abies*			*Pseudotsuga*			*Cornus*			*Acer*			*Populus*			1983 – 85 Total Density (#/ha)	
Year	61	76	83 – 85	61	76	83 – 85	61	76	83 – 85	61	76	83 – 85	61	76	83 – 85	61	76	83 – 85	61	76	83 – 85	61	76	83 – 85	61	76	83 – 85	Live	Live + Dead (± 1 SE)
Site																													
A1	(28)	33	16	(30)	55	73	(0)	6	1	(0)	0	2	(15)	2	6	(0)	1	1	(0)	0	1	(28)	0	t	(0)	0	0	387	503 (± 53)
A2	(14)	18	13	(53)	61	79	(8)	2	1	(0)	0	0	(23)	12	5	(2)	0	0	(0)	0	0	(0)	0	0	(0)	0	1	415	579 (± 51)
A3	40	--	14	53	--	75	4	--	1	0	--	1	3	--	8	0	--	2	0	--	0	0	--	0	0	--	0	502	727 (± 65)
EA	12	25	23	20	48	59	7	10	6	2	t	t	54	5	7	4	9	4	1	3	t	0	0	0	0	0	0	601	655 (± 61)
F	45	--	25	39	--	63	11	--	5	1	--	0	4	--	6	0	--	1	0	--	0	0	--	0	0	--	0	472	540 (± 56)
C1	(48)	64	53	(0)	8	9	(26)	19	30	(26)	3	5	(0)	0	2	(0)	0	1	(0)	0	0	(0)	0	0	(0)	0	0	638	749 (± 48)
C2	54	--	38	29	--	28	14	--	10	2	--	4	1	--	4	0	--	16	0	--	0	0	--	0	0	--	0	556	573 (± 149)
C3	59	30	26	28	63	57	8	5	9	0	0	2	5	0	1	0	0	5	0	0	1	0	0	0	0	0	0	700	762 (± 77)

Relative Basal Area – Trees ≥5 cm DBH

Site	*Pinus* 61	76	83 – 85	*Calocedrus* 61	76	83 – 85	*Q. kelloggii* 61	76	83 – 85	*Q. chrysolepus* 61	76	83 – 85	*Abies* 61	76	83 – 85	*Pseudotsuga* 61	76	83 – 85	*Cornus* 61	76	83 – 85	*Acer* 61	76	83 – 85	*Populus* 61	76	83 – 85	Total Basal Area (m^2/ha) Live	Live + Dead (± 1 SE)
A1	--	45	46	--	53	48	--	2	t	--	0	t	--	t	4	--	t	1	--	0	t	--	0	t	--	0	0	55	83 (± 11)
A2	--	33	21	--	44	77	--	t	0	--	0	0	--	13	3	--	9	0	--	0	0	--	0	0	--	0	t	53	117 (± 14)
A3	64	--	16	29	--	79	7	--	t	0	--	t	t	--	3	0	--	1	0	--	0	0	--	0	0	--	0	45	88 (± 10)
EA	33	39	53	18	29	23	9	23	11	1	t	t	31	3	6	7	6	7	t	1	t	0	0	0	0	0	0	63	72 (± 10)
F	68	--	60	20	--	29	9	--	7	t	--	0	3	--	3	0	--	t	0	--	0	0	--	0	0	--	0	100	113 (± 16)
C1	--	59	56	--	1	3	--	33	38	--	7	1	--	0	t	--	0	t	--	0	0	--	0	0	--	0	0	61	90 (± 10)
C2	54	--	69	29	--	14	16	--	5	t	--	1	t	--	2	0	--	9	0	--	0	0	--	0	0	--	0	55	61 (± 14)
C3	87	52	37	11	40	55	2	9	5	0	0	t	1	0	t	0	0	2	0	0	t	0	0	0	0	0	0	80	96 (± 14)

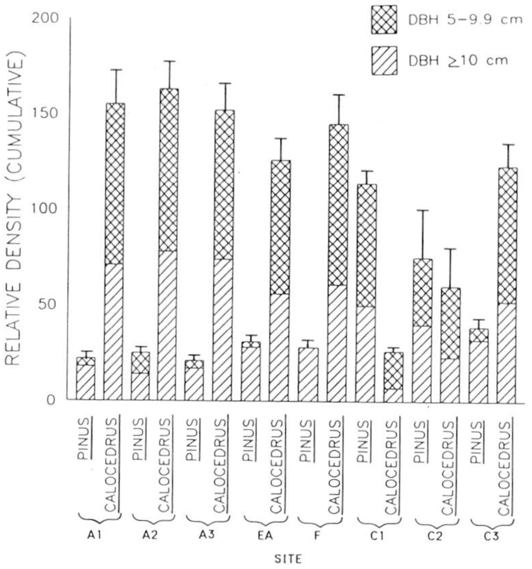

Fig. 2. Relative density of *Pinus* and *Calocedrus* in two size classes (5–9.9 cm and ≥10 cm DBH) on eight sites sampled in 1983–85. Values for the two size classes calculated separately. Error bars for cumulative total +1 SE.

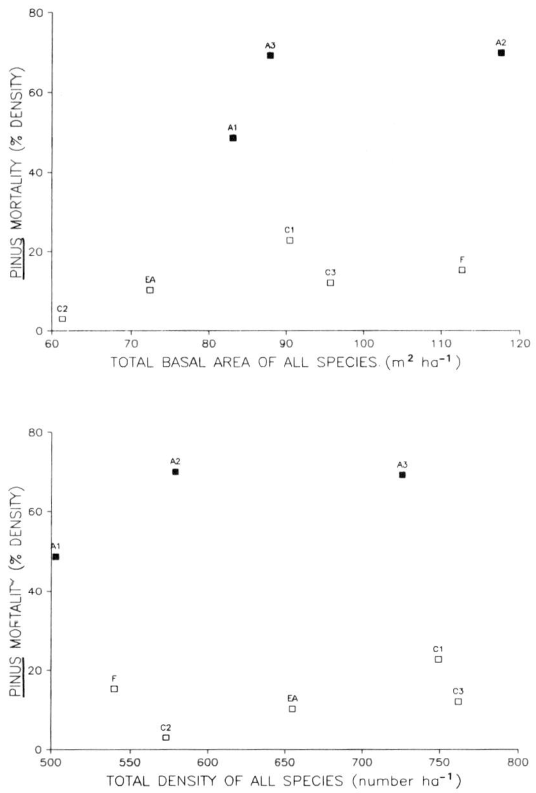

Fig. 4. Relationships of total density and total basal area of all species to mortality of *Pinus* on eight sites sampled in 1985. Alluvial sites (■) are significantly different (*t* test, $P<0.01$) from elevated alluvial or non-alluvial sites (□).

in Fig. 3. The relationships of *Pinus* mortality to total density of all species and to total basal area of all species are illustrated in Fig. 4. The three sites on alluvium (A1, A2 and A3) experienced significantly greater *Pinus* mortality, as determined by a simple *t*-test ($P<0.01$), than the remaining five sites on more elevated and generally more coarsely textured soils.

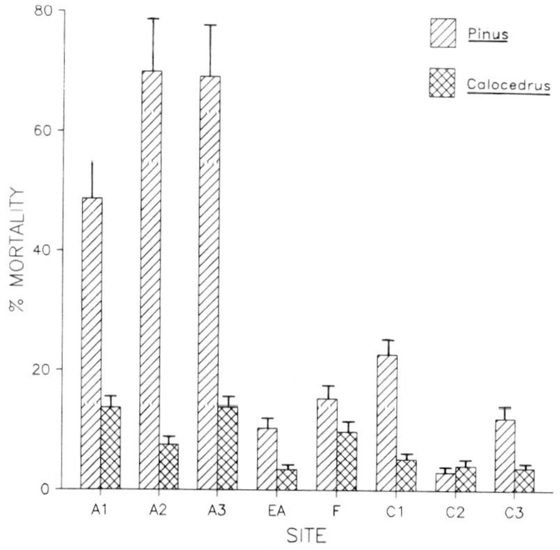

Fig. 3. Mortality as a % reduction in mean density (error bars indicate $\bar{X}$ + 1 SE) of *Pinus* and *Calocedrus* on eight sites sampled in 1985. Only trees ≥10 cm DBH considered.

Discussion

Mortality patterns, taken in consideration with size class distributions, present a picture in which mature *Pinus* have died in large numbers and at the same time smaller size classes have failed to become well established. This mortality, following the drought of 1976–77, is attributed to the combined attack of bark beetles and root rot fungi on drought-stressed trees. The exceptionally high basal areas on all of the sites, which far exceeded the 34 m^2 ha^{-1} levels suggested by Larsson *et al.* (1983) may have served as a predisposing factor in this decline. In no case was the mortality of the larger trees attributable to prescribed fire. In contrast, *Calocedrus* has experienced low mortality in the larger size class and is well represented in the smaller size class.

The relative density for *Pinus* increased between 1961 and 1976 in four of five sites (Table 1). This trend was reversed between 1976 and 1983–85 with *Pinus* decreasing in relative density on all five of the sites. The three sites which were not sampled in 1976 also showed declines from 1961 to 1983–85. *Calocedrus* followed a different pattern. It has continually increased since 1961 on all landforms except those with colluvium (C1, C2 and C3). On Site C1 *Calocedrus* has failed to become well established and on Site C2, although well established, it has not expanded. *Calocedrus* on Site C3 expanded substantially between 1961 and 1976 but has decreased slightly since that time.

Conclusions

There have been major changes in the forest structure in the Yosemite Valley since 1961. By 1976, basal areas, and perhaps densities, had increased to levels that may be considered to exceed those appropriate to promote tree vigor. The increases in density and basal area were not shared equally by all species. The dramatic increase in relative density of *Calocedrus* and the decline in relative density of *Pinus*, especially on alluvium, represents the most obvious change. Much of this growth and change was the product of over 100 years of management that attempted to exclude fire (and other causes of tree mortality) from the Valley ecosystem.

Eventually, in 1976–77, a drought placed many of the stands in a stressed condition. Reductions of *Pinus* by as much as 85% basal area and 70% density on alluvial sites are attributed to the expansion of infestations of bark beetles and root rot. Since *Pinus* was already in a less competitive position with regard to reproduction, *Calocedrus* quickly became the dominant tree species in many areas of the Valley.

The recent reintroduction of fire into the Valley forests may arrest this trend. On alluvial landforms, however, the transition from *Pinus* to *Calocedrus* is nearly complete. Fire may only serve to hold future reproduction of *Calocedrus* in check with *Pinus* continuing to decline due to its failure to reproduce under what has become a *Calocedrus* overstory. The effect of the continued spread of *H. annosus* is not clear at this time. Research in progress by Parmeter (1987 pers. comm.) suggests that nearly all of the dense stands of *Pinus* and *Calocedrus*, especially on alluvium, will sustain high mortality from this pathogen over the next several decades and that the Valley may revert to a patchwork of mixed conifers and open oak woodland, similar to what existed prior to 1850. The implications of these trends for forest and park management throughout the western United States are considerable.

Acknowledgements

The authors thank the members of the Sonoma State University plant ecology classes of 1976, 1983 and 1985 for their assistance in the collection of data. We are especially grateful to Richard Zook who resampled all of the sites to gather the dead tree data. We also wish to thank Jan van Wagtendonk for his assistance in gathering fire history records and in reviewing the early draft of the manuscript.

References

Bega, R. V. & Smith, R. S. 1966. Distribution of *Fomes annosus* in natural forests of California. Plant Dis. Reptr. 50: 832–836.

Cobb Jr, F. W., Parmeter Jr, J. R., Wood, D. L. & Stark, R. W. 1974. Proceedings of the Fourth International Conference on *Fomes annosus*. Athens, GA. Sept. 17–22, 1973. IUFRO. U.S.D.A. Forest Service. pp. 8–15.

Cottam, G. & Curtis, J. T. 1956. The use of distance measurements in phytosociological sampling. Ecology 37: 451–460.

Felix, L. S., Parmeter Jr, J. R. & Uhrenholdt, B. 1974. Proceedings of the Fourth International Conference on *Fomes annosus*. Athens, GA. Sept. 17–22, 1973. IUFRO. U.S.D.A. Forest Service. pp. 2–7.

Gibbens, R. P. & Heady, H. F. 1964. The influence of modern man on the vegetation of Yosemite Valley. California Agr. Expt. Station Manual 36.

Heady, H. F. & Zinke, P. J. 1978. Vegetational changes in Yosemite Valley. National Park Service Occasional Paper Nr 5.

Larsson, S., Oren, R., Waring, R. H. & Barnett, J. W. 1983. Attacks of mountain pine beetles as related to tree vigor of ponderosa pine. Forest Sci. 29: 395–402.

Manion, P. D. 1981. Tree disease concepts. Prentice-Hall, Inc. Inglewood Cliffs, NJ.

Mueller-Dombois, D. & Ellenberg, H. 1974. Aims and methods of vegetation ecology. Wiley & Sons, NY.

Muir, John. 1894. The mountains of California. Doubleday Inc., Garden City, NY.

Reynolds, R. D. 1959. Effect of natural fires and aboriginal burning upon the forests of the central Sierra Nevada. M. A. thesis. Univ. California, Berkeley.

Smith Jr, R., Byler, J., Wood, R. E., Caylor, J., Hoskins, W. & Scharpnack, N. 1983. Pest damage inventory - A method for measuring pest-caused tree mortality in California. In: Forest insect and disease survey methods manual, Section 3.3.1 pp. 1–15. Methods Applications Group, Forest Pest Management, U.S.D.A. Forest Service, Ft. Collins, CO.

U. S. Department of Interior, United States Geological Survey, Water Resources Division. 1971–1981. Water resources data for California. Water-Data Reports CA-71 through CA-81.

U. S. Weather Bureau, N.O.A.A. 1976–1977. Climatological data, California. Monthly reports from Jan. 1976–Dec. 1977.

Vankat, J. L. 1970. Vegetation change in Sequoia National Park, California. Ph.D. dissertation, Univ. Calif., Davis.

van Wagtendonk, J. 1974. Refined burning prescriptions for Yosemite National Park. National Park Service Occasional Paper Nr 2.

Warde, W. & Petranka, J. W. 1981. A correction factor table for missing point-center quarter data. Ecology 62: 491–494.

Vegetatio 77: 87–93, 1988
© Kluwer Academic Publishers, Dordrecht – Printed in the Netherlands

Stand-level dieback and regeneration of forests in the Galápagos Islands

Jonas Erik Lawesson
Botanical Institute, University of Aarhus, Nordlandsvej 68, DK-8240 Risskov, Denmark

Accepted 28.6.1988

Keywords: Cohort, Disturbance, Oceanic island, Pioneer tree, *Scalesia*, Senescence

Nomenclature: Wiggins, I. & Porter, D. M. 1971. Flora of the Galápagos Islands. Stanford University Press, Stanford, and Eliasson, U. 1974. Studies in Galápagos plants XIV. The genus *Scalesia* Arn. Opera Botanica 36: 1–117.

Abstract

Stand-level dieback and regeneration of forests in Galápagos are discussed, and related to natural- and man-made disturbances. The dieback of *Scalesia pedunculata* (Asteraceae) in Galápagos seems to follow the etiology and patterns seen in other pacific island groups. The currently large and synchronized cohorts of this early-successional species, found on Santa Cruz Island, may be explained by severe disturbances such as the natural el Niño phenomenon and human caused fires. It is suggested, that slow-growing species such as *Zanthoxylum fagara, Psidium galapageium* and *Acnistus ellipticus* have been suppressed by frequent fires in the highlands of Santa Cruz. The population dynamics of *Scalesia pedunculata* are suggested to follow a cyclus of 10–20 years duration, caused by senescence in the *Scalesia* stand, and promoted by extreme stress. Stand-level dieback is also reported from *Scalesia cordata, Erythrina velutina* and *Miconia robinsoniana.*

Introduction

The succession and dynamics of vegetation in the Galápagos Islands have been reported on in a number of publications (e.g. de Vries & Tupiza in press; Eliasson 1984; Hamann 1979, 1981, 1985; Nowak *et al.* in press). However, there is little evidence on natural stand-level dieback and regeneration of large populations of species in Galápagos. Kastdalen (1982) mentioned the rapid dynamics of the endemic composite tree *Scalesia pedunculata* in the highlands of Santa Cruz as a naturally occurring successional event. He found almost complete dieback of the *Scalesia* trees in extremely wet years. In subsequent dry years regeneration was noticed. An important contribution to understanding the population dynamics of *Scalesia pedunculata* was given by Hamann (1979). He found a very high initial mortality, and a growth rate like in pioneer trees. Most of the studied individuals died within 10 years of the germination, but a few seed-producing trees remained alive. *Scalesia* appeared to maintain dominance as a persistent monolayer tree, always being in the juvenile stage. Hamann suggested, that the life history of *Scalesia pedunculata* is that of an early successional species, with a rapid turnover. Eliasson (1984) described communities of *S. pedunculata* on Santa Cruz as climax forest. Van der Werff (1978) found that the *Scalesia* forests on Santa Cruz showed several characteristics traditionally found in secondary forest, such as rapid growth, soft wood, and heliophilous germination. However, he found no indications of this forest being secondary, and suggested that its appearance was because of lack of late-successional tree species, which are found in continental forest systems.

During the extremely wet 'Niño' years of 1982–1983 (Hamann 1985) massive stand-level dieback of several square kilometers were observed in the *Scalesia pedunculata* forests on four of the Galápagos Islands. Dieback was reported also in the *Miconia robinsoniana* shrub vegetation on Santa Cruz between December 1984 and November 1985, (de Vries manuscr, Adsersen, pers. comm.) presumably caused by severe drought following the Niño period. Similar cases of extensive stand-level dieback were observed in other types of vegetation (Nowak *et al.* in press; de Vries & Tupiza in press) involving *Scalesia cordata* and *Erythrina velutina*. Thorough knowledge of population dynamics is important to understand this phenomenon (Mueller-Dombois 1983, 1987).

The purpose of this paper is to review stand-level dieback and in particular the development of *Scalesia* forests in Galápagos. Preliminary results from a study on dynamics and succession in the *Scalesia* forests of Santa Cruz are included.

Previous evidence of stand-level dieback in Galápagos

Published evidence of diebacks is very scanty. Kastdalen (1982) is probably the first to report on the dynamics of *Scalesia pedunculata* on Santa Cruz. He commented: 'during the first five or six years of our stay it seemed that *Scalesia pedunculata* was disappearing... there were only mature trees and no young ones were coming up to replace the old ones which were dying. Then in 1944 or 1945 there were prolonged droughts of about four months duration in each year... most of the underbrush died, and the damage was further increased by the wild cattle and donkeys. The succeeding garua season showed a simply fantastic recovery of the *Scalesia* forest. Where formerly there had been only scattered or single trees, there were now forests of young *Scalesias'*. Like Kastdalen, Hamann (1979, 1985) found that *Scalesia* could recover very rapidly under certain conditions, especially after prolonged droughts, or other disturbances. However, it was assumed, that the presence of other tree species, like *Zanthoxylum* and *Pisonia* could slow down the regeneration process of a new cohort, as the ground was covered with more trees, shrubs and vines. At other places an almost pure *Scalesia* forest was found. Hamann concluded that *Scalesia* rarely becomes older than 15 yr.

Disturbances in the highlands of Santa Cruz

The highlands of Santa Cruz have been influenced considerably for the last 50 yr. Man-made fires seem to have occurred frequently (Eckhardt 1972; de Roy pers. comm.). The precise locations of the fires are not known, but their extent may have been considerable in dry years. It has generally been assumed, that fires only affected the tree-less pampa vegetation of Santa Cruz. However, new data from Isabela (Nowak *et al.*in press) show that fire in humid *Scalesia* vegetation may occur after a drought. Photographic evidence for this is present in the files of the Charles Darwin Research Station (CDRS), with photographs from a fire in the most humid *Scalesia* forest on Santa Cruz. Kastdalen (1982) also reported how domestic cattle and donkeys ran wild since the 1930s and had a considerable effect on the vegetation.

Figure 1 shows the relation between the precipitation data from CDRS, completed with indications

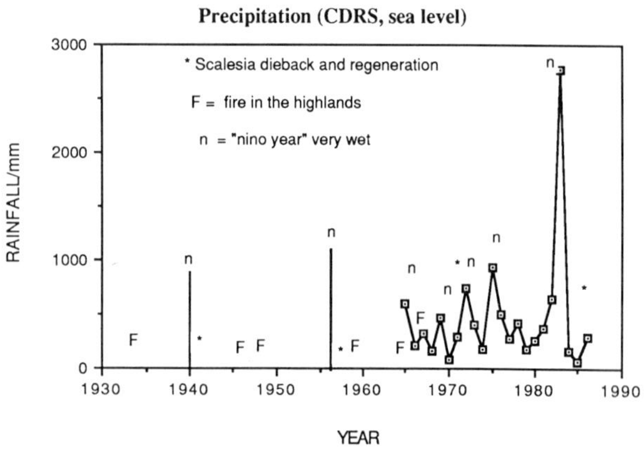

Fig. 1. Precipitation as measured at the Charles Darwin Research Station, Santa Cruz, sea-level, with indication of Niño years, reported fire in the highlands, and occurrence of *Scalesia pedunculata* dieback.

on wet years from Kastdalen (1982) and Naranjo (1985), information on major diebacks and confirmed fires in the highlands. Diebacks seem to have occurred in connection with extremely rainy years, and a recovery of new cohorts in subsequent dry years.

Study area

Most of the Galápagos Islands are dominated by one or more typical shield volcanoes. The lowlands and northern slopes are largely covered with lava fields, whereas the southern slopes and the higher parts above 200 m generally receive more precipitation and therefore have better developed soil profiles.

The climate is determined by sea currents around the archipelago. The cool season occurs between July and December, due to predominance of cold water around the Archipelago. In December, warm water currents from Central America often start to dominate and sea-temperatures rise. This causes a more or less pronounced wet period. In some years an unusual upwarming of the surface waters occurs in the entire eastern equatorial Pacific, connected with a large scale oceanwide air-sea oscillation, called El Niño. (Perry 1984; Robinson & del Pino 1985). The average precipitation in the highlands is much higher than at the coast, but extreme drought or wetness may still occur there as well.

There are only a few common tree species in the highlands (Table 1). Some could be called biological or heliophilous nomads (Hallé *et al.* 1978) or weed trees (*Scalesia* spp., *Trema micrantha, Solanum erianthum*), while others are slow-growing and shade-tolerant sciaphilous species (Hallé *et al.* 1978) of late successional stages, (*Psidium galapageium, Zanthoxylum fagara, Acnistus ellipticus*). The fast growing pioneer trees are favoured by disturbances and will soon become dominant in the vegetation. As a stand of a pioneer species dies from senescence, slow growing shade-tolerant species may emerge, unless the area again is subject to external disturbance or extreme stress. Such successional relations have been studied in an area on the rim of volcano Fernandina, which was subject to a tephra eruption in 1968 (Hendrix & Smith 1986). *Scalesia microcephala* and *Solanum erianthum* rapidly colonized the area, whereas *Zanthoxylum fagara* invaded very slowly. A similar pattern has been observed in a large area on Isabela which burnt in 1985 (Nowak *et al.* in press).

Methods

Populations of *Scalesia pedunculata* were studied in two permanent quadrats in homogeneous vegetation on Santa Cruz Island. The quadrat size was 10 × 10 m. One quadrat was established close to Cerro Crocker, at its northern slope, 700 m a.s.l., another in the Caseta area at 170 m a.s.l. (established by T. de Vries). All *Scalesia* individuals were marked with metal tags and monitored from 1985 to 1987, measuring height and DBH. Data on structure and composition of the plant communities were collected as well.

Results and discussion

Before the dieback the *Scalesia* population included almost a pure stand of old trees, presumably in a stable phase. Small seedlings or saplings were rare or absent (Eliasson 1984). As shown by Eliasson (1984), the competition among these adult trees of sufficient vigour was so strong, that newly opened treefall patches quickly regenerated through self-replacement by meristem growth (but not by seeds) and thereby were covered by the neighboring crowns.

Table 1. Common tree species in natural highland forests in Galápagos, and their characteristics. (Arborescent shrubs not included).

Species	Type of growth	Type of wood	Importance
Scalesia cordata	rapid	soft	locally dominant
S. pedunculata	rapid	soft	very dominant
S. microcephala	unknown	soft	not dominant
Zanthoxylum fagara	slow	hard	locally dominant
Psidium galapageium	slow	hard	locally dominant
Trema micrantha	rapid	soft	dominant
Acnistus ellipticus	slow	unknown	not dominant
Solanum erianthum	rapid	soft	locally dominant

This would happen provided the disturbance was only small, and the forest in a vigorous state. However, in December 1980–January 1981, when a large area was cleared, *Scalesia* achenes rapidly germinated as a consequence of the increased light. This represented a rejuvenation phase, as no meristem growth was involved but only the intrastructural seed reserve. This may be seen in the data from before the Niño period. Densities of tall trees in the Caseta quadrat were 52, 121, 43 and 38 trees per 100 sq. m, in 1970, 1972, 1975 and 1977, respectively (Hamann 1979). The last record before the Niño phenomenon is from March 1982, where 17 trees were still alive in one quadrat (Hamann 1985). These latter died before August 1983, except four of which two were dying. The remaining two old trees stayed alive through the Niño rain.

In 1985 a remarkable rejuvenation of *Scalesia* was observed. Estimates in the Caseta quadrat show 5500 saplings per 100 sq. m. In Table 2, the appearance of the 1985 cohort is expressed by the height classes encountered in the Caseta quadrat. Similar figures are shown in Table 3 for the Mount Crocker area. *Scalesia pedunculata* is clearly divided into an old cohort and the new one. Mortality in the initial phase is very large, which agrees with Hamann's observations. In the Caseta quadrat ca. 5500 new saplings came up, but only 61 of these were still alive in July 1987, wich gives a mortality of 99%. The mortality in Mount Crocker was apparently lower, but here no counts of saplings were made.

Table 2. Appearance of the 1985 cohort of *Scalesia pedunculata* on Santa Cruz as shown by numbers of saplings and old trees in the Caseta quadrat.

Height classes/m	Aug. 1985	Nov. 1985	Jan. 1986	Mar. 1986	June 1986	Oct. 1986	July 1987
0.5	5500	17	7	0	0	0	0
1	0	135	71	25	4	0	0
1.5	0	95	89	45	19	2	1
2	0	21	29	27	35	7	0
2.5	0	0	0	2	24	13	0
3	0	0	0	0	5	15	0
3.5	0	0	0	0	0	19	2
4	0	0	0	0	0	13	9
4.5	0	0	0	0	0	2	12
5	0	0	0	0	0	0	8
5.5	1	1	1	1	1	1	8
6	1	1	1	1	1	1	11
6.5	0	0	0	0	0	0	9
7	0	0	0	0	0	0	3
Number of trees	5502	270	198	101	89	73	63

Table 3. Scalesia regeneration Crocker 1985–1987.
Appearance of the 1985 cohort of *Scalesia pedunculata* on Santa Cruz, as shown by numbers of saplings and old trees in the 'Cerro Crocker' quadrat.

Height classes/m	Nov. 1985	Jan. 1986	Mar. 1986	June 1986	Oct. 1986	July 1987
0.5	24	13	5	3	0	0
1	40	54	24	4	2	0
1.5	4	22	43	31	3	0
2	0	5	33	26	17	0
2.5	0	0	5	26	19	2
3	0	0	1	18	15	4
3.5	0	0	0	2	23	12
4	0	0	0	0	3	11
4.5	0	0	0	0	0	16
5	0	0	0	0	0	15
5.5	0	0	0	0	0	1
12.5	3	3	3	3	2	2
Number of trees	71	97	114	113	84	63

During 1986 and 1987, no new saplings emerged. The growing *Scalesia* trees shaded out most other species, except for shade tolerant ferns and shrubs. Synchronized dieback and recovery of *Scalesia pedunculata* on Santa Cruz may be due to repeated fires and grazing by domestic animals. Relative dominance of *S. pedunculata* as compared to other tree species may be explained by such disturbances, which favour rapidly growing early successional heliophilous species, and disfavour slower growing fire-sensitive species. The stands of *Scalesia* may therefore have become more and more uniform due to repeated catastrophic events such as fires and severe browsing. Regarding the etiology of the nearly complete stand-level dieback, it is probably important, that before the Niño phenomenon 1982–1983 the *Scalesia* forest was an adult or old stand with large trees, forming a closed canopy (Hamann 1981; Eliasson 1984). The expected maximum life-span for *Scalesia pedunculata* in the Caseta area may be around 10–15 yr (Hamann 1979). It is therefore probable that natural cohort senescence of the species was important for the dieback encountered, in the same manner as has been reported for forests on Hawaii and New Zealand (Mueller-Dombois 1987). Dieback could therefore be explained by assuming that the mature forests of *Scalesia* were reaching a state of reduced vigour or partial senescence, and hence they were highly susceptible to disturbance. Together with the extreme Niño conditions (strong winds, water saturated soil and high air-temperatures) the intrinsic characteristics of the *Scalesia* may have resulted in the dieback of large stands. As stated by Mueller-Dombois (1987), during the terminal phase of its cyclus, a cohort may break down, either suddenly or gradually. If the disturbance is small and patchy, a quick regeneration will occur and the cohort survives. If the disturbance is large this is not possible, and a seed-dependent rejuvenation is necessary. Environmental extremes may occur during the whole life-span of a cohort stand. As long as the cohort is in a normal state of vigour, it will survive such extremes. However, when a senescence stage is reached, a certain extreme, such as climatic stress, may act as a trigger for a general collapse of the cohort. With the expected life-span of *Scalesia pedunculata* cohorts as well as the climatic setting, it is probable that the combination of uniform cohorts, life-span and population structure will cause repeated stand level dieback, as indeed has been observed. It seems also probable that dieback is a cyclic event, with a rejuvenation phase, an adult phase and a senescent phase. If new disturbance occurs in the latter phase, a dieback phase will follow. If no disturbance occurs, the cyclus may continue with the formation of a more diverse forest, where slow-growing trees become more important. This proposed life history has been summarized in Table 4.

After the ‘Niño’ years, the accumulated seed bank of *Scalesia pedunculata* quickly germinated. It was probably further promoted by the strong drought, which killed or mimimized much of the undergrowth. This agrees with Kastdalen’s (1982) observations. A few trees of the old cohort survived the extreme conditions, and were still alive in 1987. This may be due to genetic variation within the cohort.

Table 4. Suggested model for *Scalesia pedunculata* forest dynamics on Santa Cruz, Galápagos. 1. Germination phase; 2. Rejuvenation phase, where *Scalesia* saplings grow from seedling to 5 – 10 m tall trees, depending on local environmental conditions; 3. First optimal (adult) phase, which is a steady state; 4. First senescence phase of the *Scalesia* population; 5. Partial break down of the cohort structure; 6. Second optimal phase, where a combination of new *Scalesia* trees and slow growing trees is formed; 7. Second senescence phase, where the second cohort of the species is further broken down, and replaced by other species.

Succession phase:	Duration (yr)	Dominant species:
1. Germination	1 – 2	Herbs & *Scalesia*
2. Rejuvenation	2 – 5	*Scalesia*
3. Adult	5 – 10	*Scalesia* & Shrubs
4. First *Scalesia* senescence	1 – 5	*Scalesia*
5. Partial rejuvenation and displacement	5 – 15	*Scalesia*, some other trees
6. Adult diverse forest	until disturbance	*Scalesia*, some other trees
7. Second *Scalesia* senescence	10 – 15	Various tree species

More likely, the former forest, although of uniform appearance, contained different cohorts, in different phases, viz. adult and senescent phases, differently susceptible to abiotic stress-factors.

General dieback of large stands of trees or shrubs is not restricted to *Scalesia pedunculata*. *S. cordata* in the Sierra Negra area died over large areas after the Niño, and rejuvenated in 1985. However, most of this new generation died during 1987 because of severe attacks from endemic insects, while surviving plants became covered by a dense growth of *Ipomoea triloba* and *I. alba*.

The uniformity of *Miconia robinsoniana* in the highlands of Santa Cruz may be explained in a similar way. Several fires have occurred in the area, the latest one in 1967. In 1974 *Miconia* was restricted to a small area around Media Luna and its vicinity, probably delimitated by the 1967 fire. *Miconia* now covers a large area extending to Cerro Crocker. Young *Miconia* were severely affected by the Niño, with many individuals apparently dying during 1982–1983, while the old populations first survived the Niño, but then started to die during 1985.

A related observation concerns the apparent uniformity of some of the lowland tree cohorts, notably of *Bursera graveolens* which is dominant in large areas of dry coastal steppe forest. A synchronized population structure and dynamic pattern, including large stand-level dieback is to be expected in such areas (Mueller-Dombois 1987). Stands of *Bursera graveolens, Piscidia carthagensis, Erythrina velutina* and *Parkinsonia aculeata* seem to be very uniform in large areas. According to Grant (pers. comm.) rejuvenation of *Erythrina* is extremely difficult, although this tree may produce seeds annually. Similar observations of stands with a few old trees of this species and no saplings, have been made by the present author.

A succession theory based on observations on Hawaiian *Metrosideros* forests (Mueller-Dombois 1983, 1987) explains temporally recurring dieback as a pattern and process sequence in primary succession, where large initial cohorts have been formed due to catastrophic events. Canopy dieback of these large cohorts during their senescent stage gives rise to successively smaller and more patchy cohorts at each dieback cycle. The canopy dieback of *Scalesia pedunculata* on Santa Cruz island seems to fit this theory, and the uniformity of the cohorts may be explained with repeated man-made catastrophes such as fires and 'El Niño' events.

It is here suggested that the occurrence of dieback and uniform cohorts of *Scalesia pedunculata* is a result of combined natural and man-made disturbances. As to the natural ones, the climatic setting forces the biotic environment to change constantly, as the climate is an unpredictable factor. One year may be extremely dry with desert conditions, whereas the subsequent year may show extremely wet conditions. This everlasting fluctuation in the physical conditions must cause some rather dramatic changes in the Galápagos plant and wildlife, including provoked stand-level dieback of whole populations of plants and also animals (Robinson & del Pino 1985).

The rapid dynamics of *Scalesia* also ensure, that the former tall forest, known from the highlands of Santa Cruz, will soon again appear, but then within another 10–20 years, a new stand level dieback is to be expected. It is of scientific and conservation interest to continue long-term studies of the highland Galápagos forests. They may reveal whether the pure *Scalesia* forests are true natural forests, as stated by several authors, or, if left alone, will be displaced in part by other slow-growing species, leading to a more diverse forest.

Acknowledgements

This study was supported by The Charles Darwin Research Station, The Galápagos National Park Service and the Danish Natural Science Research Council, grants nrs 11-5471, 11-5663, 11-6288 and 81-6050. Thanks are due especially to Yolanda Carvajal, Bosko Nowak, Luis Ortiz and Felix Burgos for field-assistance and good company. H. Adsersen, H. Balslev and O. Hamann are thanked for valuable comments on the manuscript. This study is contribution nr 125 of the AAU Ecuador project and Contribution nr 423 from the Charles Darwin Foundation for the Galápagos Isles.

References

de Vries, T. & Tupiza, A. in press. The dynamics of six stands of *Scalesia cordata* in permanent plots in different habitats in Sierra Negra, Isabela, Galápagos. Monogr. Syst. Bot.

Eckhardt, R. C. 1972. Introduced plants and animals in the Galápagos Islands. Bioscience 22: 585–598.

Eliasson, U. 1984. Native climax forests. In: Perry, R. (ed.), Key environments. Galápagos, pp. 101–114. Pergamon Press, Oxford.

Hallé, F., Oldeman, R. A. A. & Tomlinson, P. B. 1978. Tropical trees and forests. An architectural analysis. Springer Verlag, Berlin.

Hamann, O. 1979. Dynamics of a stand of *Scalesia pedunculata* Hooker fil., Santa Cruz Island, Galápagos. J. Linn. Soc. Bot. 78: 67–84.

Hamann, O. 1981. Plant communities of the Galápagos Islands. Dansk Bot. Ark., 34: 1–163.

Hamann, O. 1985. The el Niño influence on the Galápagos vegetation. In: Robinson, G. & del Pino, E. M. (eds), The Niño in the Galápagos Islands: the 1982–1983 event: 299–330. Charles Darwin Foundation, Quito, Ecuador.

Hendrix, L. B. & Smith, S. D. 1986. Post-eruption Revegetation of Isla Fernandina, Galápagos: II. Nat. Geogr. Res. 2(1): 6–16.

Kastdalen, A. 1982. Changes in the biology of Santa Cruz Island between 1935 and 1965. Not. Galápagos 35: 7–12.

Mueller-Dombois, D. 1983. Canopy dieback and successional

processes in pacific forests. Pacific Sci. 37: 317–325.

Mueller-Dombois, D. 1987. Natural dieback in forests. Bioscience 37: 575–583.

Naranjo, P. 1985. El fenomeno el Niño y sus efectos en el clima del Ecuador. In: G. Robinson & E. M. del Pino (eds), El Niño in the Galápagos Islands: The 1982–1983 Event. Charles Darwin Foundation, Quito, Ecuador.

Nowak, J. B., Lawesson, J. E., Adsersen, H. & de Vries, T. in press. The post fire vegetation dynamics on southern Isabela, Galápagos Islands. Monogr. Syst. Bot.

Perry, R. 1984 (ed.). Key environments. Galápagos. Pergamon Press, Oxford.

Robinson, G. & del Pino, E. M. 1985 (eds). El Niño in the Galápagos Islands: the 1982–1983 event. Charles Darwin Foundation, Quito, Ecuador.

van der Werff, H. H. 1978. The vegetation of the Galápagos Islands. Thesis Utrecht.

Vegetatio 77: 95–101, 1988
© Kluwer Academic Publishers, Dordrecht – Printed in the Netherlands

Mechanisms of succession on fallow lands*

Reinhard Bornkamm
Institute of Ecology, Technical University Berlin, Rothenburgstr. 12, D 1000 Berlin 41, Federal Republic of Germany

Keywords: Coexistence, Invasion, Oscillation, Trend

Accepted 20.7.1988

Abstract

Vegetation changes during succession can be regarded as plant-to-plant replacement processes. For deeper insight into the nature of these process we need to investigate the mechanisms involved. Therefore two experiments with herbaceous plant communities were analyzed. The data confirm the view that all three types of interaction: competition, coexistence by niche separation, and cooperation, act together. Likewise it can be concluded that the three models of succession proposed by Connell & Slatyer (1977) are not exclusive but describe mechanisms acting together in the same succession process. Evidence is given that seasonal events can act like a switch and influence the successional trend at a later time. It is, therefore, not meaningful to differentiate between fluctuations in the sense of yearly oscillations and the successional trend itself.

Introduction

Vegetation succession is the successive occurrence of phytocoena at a given site (van der Maarel & Werger 1980). The successive occurrence is brought about by changes. Succession is a pattern of changes in specific composition of a community after a radical disturbance or the opening of a new patch in the physical environment for colonization by plants and animals (Horn 1974). Changes usually are separated into two different phenomena: fluctuations and trends. Fluctuations are short time changes which reflect the influence of annual changes of climate or biological events on the performance of a given species. They are reversible (Miles 1979) and oscillate around the trend. The trend is the average change over a longer period, leading from one stage of succession to the next one. Stages are characterized by typical plant communities following each other in time at the same site (Knapp 1974). The predictability of this process increases with the time span involved: In a region with forest climax a sere of stages like annuals – herbaceous perennials – colonizing woody plants – late successional trees is highly predictable (Hayashi 1977; Bornkamm 1985), whereas fluctuations are hardly predictable at all.

Formally, we can describe succession in models as a plant-by-plant replacement process (Horn 1974) or as a replacement of vegetation types (Debussche *et al.* 1977; Lippe *et al.* 1985). There is some evidence that this replacement is a non-random process (Usher 1979, 1980; Lippe *et al.* 1985; van Dorp *et al.* 1985; van Hulst 1987). Already Egler (1954) recognized the initial floristic composition as an important factor in old-field succession. But the existence of a determinated sequence of plant species remains an open question.

In order to obtain a better insight into this problem it has to be investigated by which mechanisms plant populations replace each other (Connell & Slatyer 1977; Bornkamm 1985). Looking closer at these mechanisms reveals the fact that succession is

*Dedicated to Prof. Dr Dr h.c. Heinrich Walter on the occasion of the 90th anniversary of his birthday.

a continuous process in which not only changes from year to year but also seasonal events play a role (Bornkamm 1986). Some gap models consequently use very small time steps (van Tongeren & Prentice 1986). This makes the differentiation between trends and fluctuations of succession doubtful. In the present paper, therefore, two examples will be analyzed under both viewpoints, the replacement mechanisms and the seasonal events.

Methods and results

Example I – short-term experiment

The first experiment included 9 subplots of 1 m^2 and was designed for estimating the phytomass of a ruderal vegetation on sand during the first two years of succession (for details see Bornkamm 1986, 1987). The plots were harvested at different seasons between 18.6.1969 and 1.4.1971. After harvesting the plots were allowed to start a new succession. Vegetation composition was recorded for 9 more years. It turned out that the colonizing species had an effect on the species composition in the later years, although they died out after 2 yr. The subplots started in the summer (June–August) of the first year were dominated by a legume (*Trifolium arvense*), whereas the subplots started in other seasons were dominated by species like *Conyza canadensis* or *Bromus hordeaceus* (Table 1). In later years *Solidago canadensis* and *Sarothamnus scoparius* were less frequent in the summer subplots (= former *Trifolium* subplots), whereas some other species were more frequent. The summer subplots were not inaccessible for *Solidago* and *Sarothamnus*, but these species invaded later (Table 2).

Table 1. Short-term experiment: Cover value of important species in two successional periods and their correlation to the initial cover values of *Trifolium arvense*.

	Years 4–6			Years 7–9		
	A	B	C	A	B	C
Solidago canadensis	0.8	12.0	+	3.3	19.9	ns
Sarothamnus scoparius	–	11.9	ns	6.7	20.5	ns
Solidago + Sarothamnus	0.8	23.9	+	10.0	40.4	+
Agrostis tenuis, Pinus sylvestris (juvenile) + Taraxacum officinale	4.8	1.1	ns	20.2	0.1	+

A = cover within plots started in summer (June–August, $n = 4$) in %; mean cover of *Trifolium arvense* after 1 year = 53%; B = cover within plots started in other seasons (October–May, $n = 5$), in %; mean cover of *Trifolium arvense* after 1 year = 2%; C = Significance (5% confidence limit) of the correlation between the cover values in A or B with the initial cover values of *Trifolium arvense* (log/log transformation).

Example II – long-term experiment

The second experiment was set up in 1953 in a dry calcareous pasture near Göttingen (Lower Saxony) which was abandoned about 40 yr ago. The plot measured 4 m^2. The vegetation of half of this plot (B) was completely removed, whereas in the other half (A) the two most dominant grasses (*Bromus erectus* and *Brachypodium pinnatum*) were allowed to continue their growth (for details see Bornkamm 1974). After a few years the grass sward recovered in both parts, and looking superficially one could not recognize large differences within the last 30 yr. In the very first years a method of direct mapping was applied. Later on, when the structure became more complex and two strata occurred, signs on the maps were used to indicate the size of individual plants. From 1974 on, only the frequency was recorded, using a 10 × 10 cm grid scale (= 200 microplots in each subplot) and all earlier records were transformed into this frequency scale. If we look at the number of species per dm^2 (Fig. 1) we see that the

Table 2. Average year of invasion. Group A = plots started in summer (June – August, $n = 4$); Group B = plots started in other seasons (October – May, $n = 5$).

	A	B	Signif. (5%)
Conyza canadensis	1.3	1.0	ns
Bromus hordeaceus	1.0	1.2	ns
Vicia angustifolia	3.0	1.8	ns
Daucus carota	3.8	3.0	ns
Solidago canadensis	6.5	2.4	+
Sarothamnus scoparius	6.8	5.2	ns

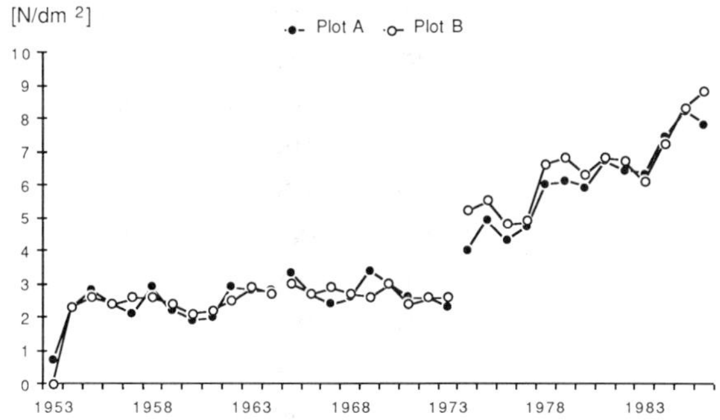

Fig. 1. Number of species (*N*) per microplot (dm^2)
A = subplot starting with buried-seed bank and with plants of *Bromus erectus* and *Brachypodium pinnatum*
B = subplot starting with buried-seed bank only. One subplot comprises 200 microplots.

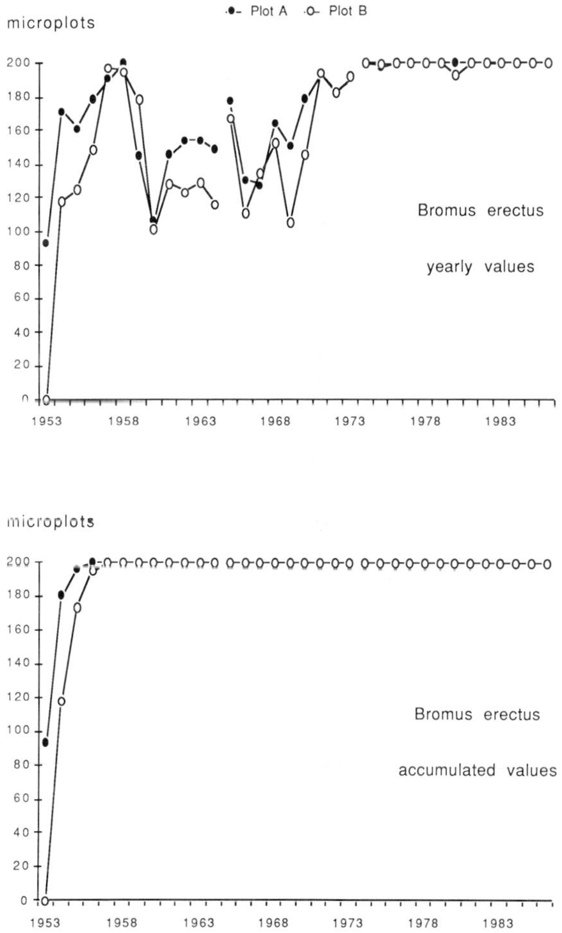

Fig. 2. Bromus erectus. A, B as in Fig. 1.
Above: frequency (200 microplots = 100%)
Below: microplots reached.

plant species were gradually mixing up, in other words the niche formation became more and more fine-grained. We also recognize that the mapping method was adequate for the first period, the frequency method for the last one, whereas the method for the central periods was not as reliable.

Now the behaviour of some important species will be discussed. In the very first years *Bromus erectus* (Fig. 2) spread rapidly, but faster in subplot A than in B, because in A it had not been removed. There it very soon became the absolutely dominating species forming longliving tussocks. After ca 20 yr the stand was closed. After only 5 yr *Bromus* had arrived at all microplots at least once.

Brachypodium pinnatum (Fig. 3) spread much slower, but again faster in A than in B. Its rhizomatous structure seemed to make it rather mobile and a large number of settlements was abandoned after a short time. Although the frequency did not exceed 50% all microplots had been reached at least once after only 10 yr.

Whereas the first two species mentioned were

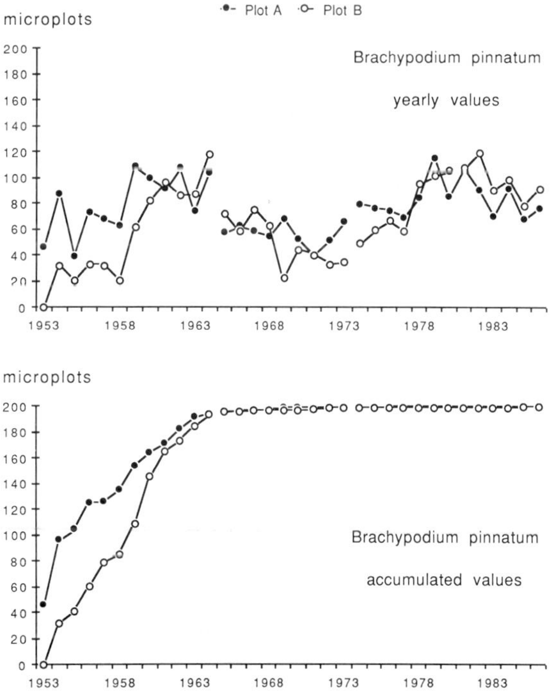

Fig. 3. Brachypodium pinnatum. A, B as in Fig. 1.
Above: frequency (200 microplots = 100%)
Below: microplots reached.

present from the beginning, *Ononis spinosa* (Fig. 4) was an early invader. It also showed a high mobility: most primary settlements were left in the following year. As a pioneer species it showed a clear maximum of frequency after 5 yr, then it died and invaded again. All microplots had been reached after only six yr, but here – with the invading species – the spreading was faster in B than in A.

Festuca rupicola (Fig. 5) started after 3 yr with very few settlements, colonizing further sites later on, but spreading very slowly. After 20 yr finally, when the *Bromus* canopy was closed, it formed a second grass layer as a strong codominant (in the intermediate years it was severely underestimated). It spread much faster in B than in A, but it reached all microplots only after 20 yr.

Rhinanthus minor (Fig. 6) invaded after 28 yr and spread very rapidly like an infestation. No differentiation between subplots A and B could be detected and after six years nearly all microplots had been reached at least once.

The species mentioned so far, were frequent spe-

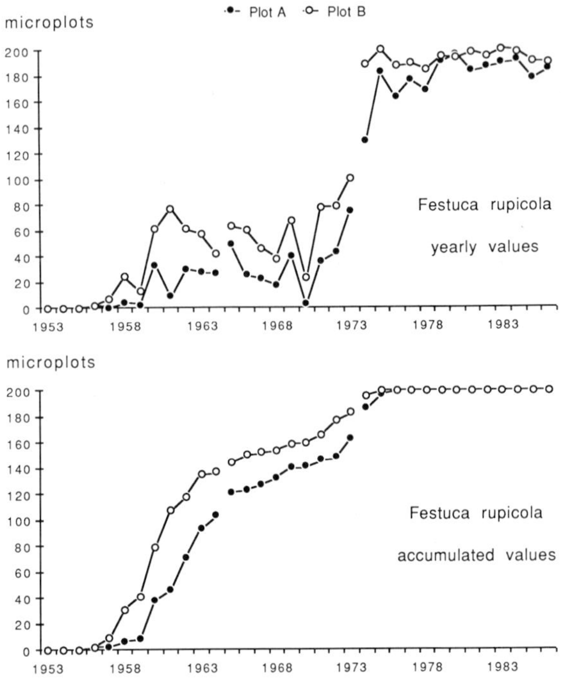

Fig. 5. Festuca rupicola. A, B as in Fig. 1.
Above: frequency (200 microplots = 100%)
Below: microplots reached.

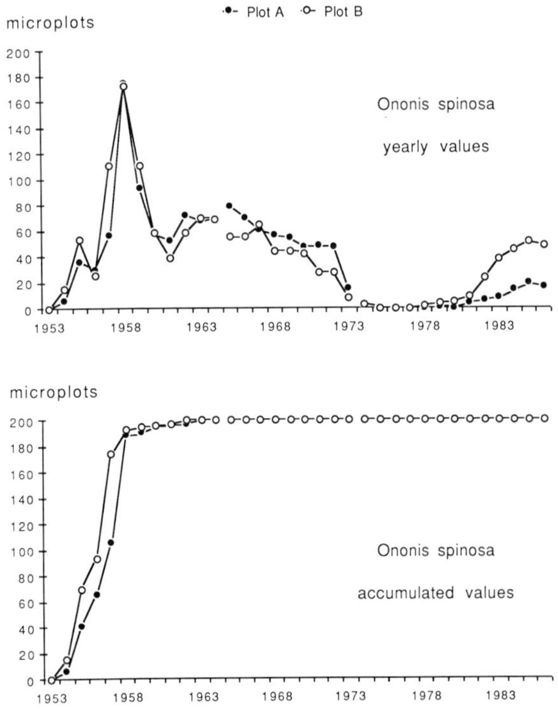

Fig. 4. Ononis spinosa. A, B as in Fig. 1.
Above: frequency (200 microplots = 100%)
Below: microplots reached.

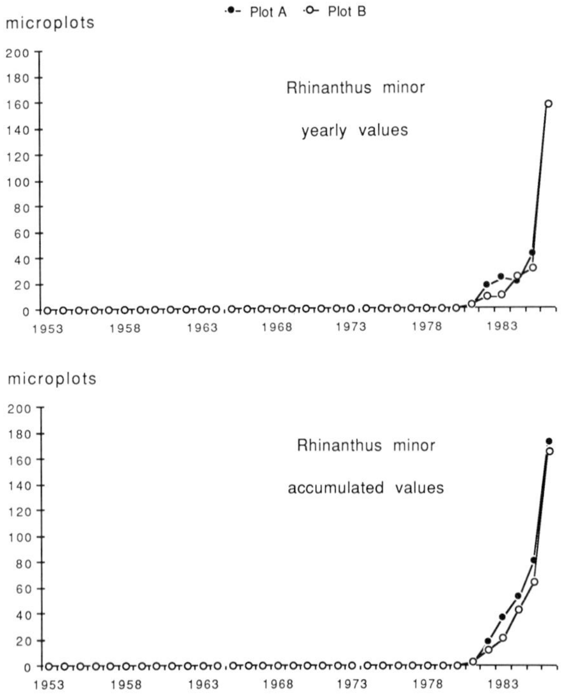

Fig. 6. Rhinanthus minor. A, B as in Fig. 1.
Above: frequency (200 microplots = 100%)
Below: microplots reached.

cies with a relatively rapid population growth. In Fig. 7 two species of only local distribution are shown. *Plantago media* invaded subplot A after 9 yr. After several attempts it became established in one locality only where it slowly increased in stand size. Using a linear model with data from 1973 onwards it can be predicted that this species will have frequented all microplots after ca one century. *Hippocrepis comosa* invaded subplot B after 27 yr and became established. Due to its creeping shoots it increased its stand size faster than *Plantago* and will have touched all microplots after half a century.

In the first experiment one of the mechanisms observed was a retarded invasion of some later successional species (Table 2). If we compare the years of invasion in the present experiment for all relevant species (Table 3) no striking differences can be detected except the late arrival of *Scabiosa columbaria* in subplot A. Looking closer at this species we see, that the early invasion of subplot B was not successful (Fig. 8). In the early sixties new populations

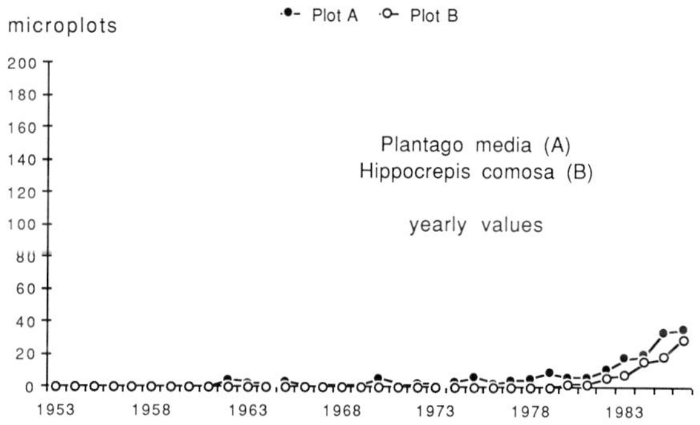

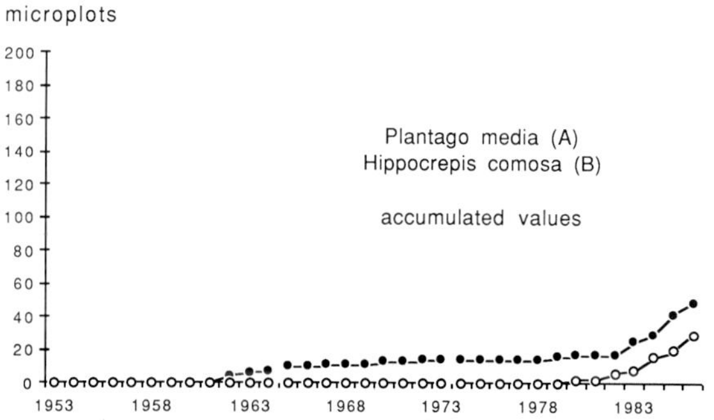

Fig. 7. Plantago media (only in plot A) and *Hippocrepis comosa* (only in plot B). A, B as in Fig. 1.
Above: frequency (200 microplots = 100%)
Below: microplots reached.

Table 3. First observation of species in a grassland experiment started in 1953. A: all vegetation removed except *Bromus erectus* and *Brachypodium pinnatum*; B: all vegetation removed.

	A	B
Bromus erectus	1953	1954
Brachypodium pinnatum	1953	1954
Sanguisorba minor	1954	1954
Convolvulus arvensis	1954	1954
Medicago lupulina	1954	1954
Achillea millefolium	1954	1954
Ononis spinosa	1954	1954
Daucus carota	1954	1954
Pimpinella saxifraga	1954	1954
Lotus corniculatus	1954	1954
Koeleria pyramidata	1954	1954
Anthyllis vulneraria	1954	1955
Scabiosa columbaria	1962	1954
Festuca ovina	1956	1956
Dactylis glomerata	1957	1957
Arrhenatherum elatius	1958	1958
Poa angustifolia	1959	1958
Potentilla verna	1959	1959
Carex flacca	1961	1960
Knautia arvensis	1959	1963
Plantago media	1962	–
Agrimonia eupatoria	1968	1968
Carex caryophyllea	1971	1974
Hippocrepis comosa	–	1980
Rhinanthus minor	1981	1981

started in both subplots and spread slowly but continuously. This latter species, too, appears rather mobile and will have touched all microplots in the near future.

Discussion and conclusions

First of all the results of both experiments stress the importance of invasions. Invasions take place continuously throughout the succession observed so far, although not every invasion is successful and not every successful invasion results in a consequent spreading: *Festuca rupicola* invaded in 1956 but did not spread considerably until 20 yr later.

Looking at different forms of interaction (competition, coexistence and cooperation) the second experiment stresses the importance of coexistence by

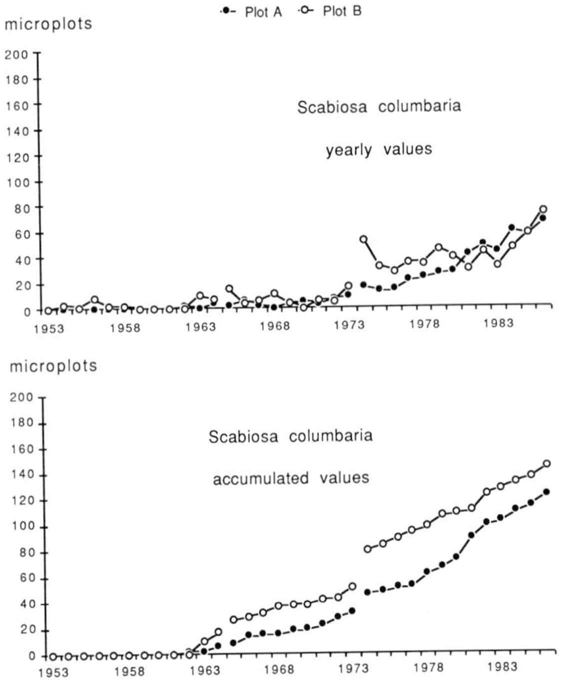

Fig. 8. Scabiosa columbaria. A, B as in Fig. 1.
Above: frequency (200 microplots = 100%)
Below: microplots reached.

niche separation. The number of species per dm^2 increases four-fold, whereas the total number of species increased only by 65% (from 26 to 43 spec/4 m^2). This means a closer and closer growing together of the species. It is not known to what extent closing of the *Bromus erectus* canopy gives a hint in this direction. Moreover it can be assumed that competition for space is present as an overall process.

The results bear characters of all three succession models proposed by Connell & Slatyer (1977). As already mentioned the behaviour of *Festuca rupicola* fits the facilitation model. The dominant *Bromus erectus* and the codominant *Brachypodium pinnatum* are founders in subplot A and early invaders of subplot B. They showed higher frequency in A than in B for 4 (*Bromus*) resp. 7 years (*Brachypodium*). Other early invaders showed higher frequency in B (where open space was available), than in A, where *Bromus* and *Brachypodium* modified the environment as early occupants making it less suitable for the recruitment of new species (*Ononis spinosa, Festuca rupicola, Scabiosa columbaria*). A late invading species like *Rhinanthus minor* does not differentiate between A and B. This fits the tolerance model (model 2) proposed by Connell & Slatyer 1977). The retarded invasion of later successional species (*Solidago canadensis* in experiment I) by an early occupant (*Trifolium arvense*) finally fits the inhibition model (model 3). In agreement with Hils & Vankat (1982) the conclusion can be drawn that the models of Connell & Slatyer (1977) are not exclusive but describe succession mechanisms which can act together in the same succession process.

The differences in frequency of the early invaders of subplot A and B can be explained by the differences in open space, which disappear after a few years. In slower penetrating species like *Festuca rupicola* and *Scabiosa columbaria*, however, the differences in frequency can persist much longer than the factor (open space) causing them. The same indication of a deterministic structure of the succession process can be seen in experiment I, where the performance of later successional species, several years after extinction of a colonizing species, is still influenced by the performance of this colonizing species. We need more and better data for this interesting fact.

The last conclusion concerns the nature of fluctuation. We have seen that seasonal events influence successional events later on. Likewise invasions depend on the season in which open space is available (Bornkamm 1986). This means that fluctuations in the sense of yearly oscillations are not completely reversible and do not show other mechanisms than the trend itself. The trend of succession acts through its oscillations. Since it is evident that short term events oscillate around trends I propose to abandon the term 'fluctuation' for this special feature, and to use it in such cases only where the character of fluctuation can be clearly defined e.g. as fluctuation of dominant species (Rabotnov 1966) or as overlapping rhythms of different time scale (Cramer & Hytteborn 1987).

References

Bornkamm, R. 1962. Über die Rolle der Durchdringungsgeschwindigkeit bei Kleinsukzessionen. Veröff. Geobot. Inst. Rübel 37: 16–26.

Bornkamm, R. 1974. Zur Konkurrenzkraft von *Bromus erectus* II. Ein zwanzigjähriger Dauerversuch. Bot. Jahrb. Syst. 94: 391–412.

Bornkamm, R. 1985. Vegetation changes in herbaceous communities. Handb. Veget. Sci. 3: 89–109.

Bornkamm, R. 1986. Ruderal succession starting at different seasons. Acta Soc. Botan. Pol. 55: 403–419.

Bornkamm, R. 1987. Veränderungen der Phytomasse in den ersten zwei Jahren nach einer Sukzession auf unterschiedlichen Böden. Flora 179: 179–192.

Connell, J. H. & Slatyer, R. O. 1977. Mechanisms of succession in natural communities and their role in community stability and organization. Am. Nat. 111: 1119–1144.

Cramer, W. & Hytteborn, H. 1987. The separation of fluctuation and long-term change in vegetation dynamics of a rising seashore. Vegetatio 69: 157–167.

Debussche, M., Godron, M., Lepart, J. & Romane, F. 1977. An account of the use of a transition matrix. Agro-Ecosystems 3: 81–92.

Egler, F. 1954. Vegetation science concepts. I. Initial floristic composition, a factor in old-field vegetation development. Vegetatio 4: 412–417.

Hayashi, I. 1977. Secondary succession of herbaceous communities in Japan. Jap. J. Ecol. 27: 191–200.

Hils, M. H. & Vankat, J. L. 1982. Species removals from a first-year old-field plant community. Ecology 63: 705–711.

Horn, H. S. 1974. The ecology of secondary succession. Ann. Rev. Ecol. Syst. 5: 25–37.

Knapp, R. 1974. Some principles of classification and of terminology in successions. Handb. Veget. Sci. 8: 169–177.

Lippe, E., de Smidt, J. T. & Glenn-Lewin, D. C. 1985. Markov models and succession: A test from a heathland in the Netherlands. J. Ecol. 73: 775–791.

Miles, J. 1979. Vegetation dynamics. Chapman & Hall, London.

Rabotnov, T. A. 1966. Peculiarities of the structure of polydominant meadow communities. Vegetatio 13: 109–116.

Usher, M. B. 1979. Markovian approaches to ecological succession. J. Animal Ecol. 48: 413–426.

Usher, M. B. 1980. Modelling ecological succession with particular reference to Markovian models. Vegetatio 46: 11–18.

van Dorp, D., Boot, R. & van der Maarel, E. 1985. Vegetation succession on the dunes near Oostvoorne, The Netherlands, since 1934, interpreted from air photographs and vegetation maps. Vegetatio 58: 125–136.

van Hulst, R. 1987. Invasion models of vegetation dynamics. Vegetatio 69: 123–131.

van der Maarel, E. & Werger, M. J.A. 1980. On the treatment of succession data. Phytocoenosis (Warszawa–Białowieza) 7: 257–278.

van Tongeren, O. & Prentice, I. C. 1986. A spatial model for vegetation dynamics. Vegetatio 65: 163–173.

Vegetatio 77: 103–114, 1988
© Kluwer Academic Publishers, Dordrecht – Printed in the Netherlands

An experimental study of old-field succession in relation to different environmental factors

W. Schmidt
Systematisch-Geobotanisches Institut der Universität Göttingen, Untere Karspüle 2, D-3400 Göttingen, Federal Republic of Germany

Accepted 25.5.1988

Keywords: Dispersal, Fertilization, Invasion, Nutrient, Phytomass, Water level

Abstract

From 1984 to 1986, old-field succession on sterilized sand and loam was studied under different water- and nutrient regimes. Within one month, moss and phanerogam species appeared on all experimental plots but further succession was rather varied. *Salix* species established quickly on loam and formed within 3 years a shrub layer up to 3 m in height. On sand, woody plant species were observed only at a high ground-water level. On loam, the well-known old-field succession from short-living therophytes to long-living phanerophytes of clearings and woodlands proceeded very quickly. In contrast, on sand, therophytes, hemicryptophytes and herbaceous chamaephytes of ruderal- and grassland communities were still dominant after three years. A high ground-water level as well as mineral fertilization had sometimes positive, sometimes negative effects on this succession. Periodic estimates of cover, made during the succession were supplemented at the end of the experiment by the measurements of phytomass and bioelement storage. The highest amount of biomass was measured on the three loamy soils where shrub layers were well developed. In comparison with data published elsewhere, the above-ground biomass of 2.2–2.8 kg dry matter m^{-2} and the below-ground biomass up to 7.2 kg dry matter m^{-2} were both extraordinarily high. Over the three years, the vegetation on sandy soils accumulated between 1.2 and 5.1 g N m^{-2} yr^{-1} and on loamy soils between 17.1 and 24.7 g N m^{-2} yr^{-1}.

Nomenclature: Ehrendorfer, F. 1973. Liste der Gefäßpflanzen Mitteleuropas. Fischer, Stuttgart.

Introduction

The aim of succession research is to describe changes in ecosystems in time as well as to understand the causes of change. The most conspicuous results of successions are changes in community composition. The most reliable method of defining such changes exactly, is by the direct observation of permanent plots (Braun-Blanquet 1964; Schmidt 1981a, 1983a).

It usually takes many years or even decades to produce useful results from permanent plots. Moreover, it is subsequently often difficult to explain the courses of a turnover in any particular sere. This is also true for the now well-known succession of old-fields and ruderal communities. Besides the comparison of plots of different ages on the same sites, many permanent plot studies exist. The results of these showed that the starting conditions influenced the succession for a long time, especially on old-fields which were formerly ploughed. The persistence of competitive early colonists is particularly important for the woody plant species of the climax

(Hard 1972; Schmidt 1983a) but also for a number of perennials such as *Solidago canadensis, Urtica dioica* and *Agropyron repens* which established quasi-stable permanent ruderal communities (Hahn *et al.* 1978; Wittig 1978; Schmidt 1981b, 1983b; Werner 1983).

Most of the old-field successions described above, started on plots with homogeneous site conditions which were observed many years. Often, part of the plot was treated experimentally during the succession. Moreover Schmidt (1981a, 1983a) compared two old-field successions on the same site that began in two different years. The successions differed considerably, which was explained mainly by climatic differences, but also by differences in the diaspore supply from year to year.

The vegetation development of ruderal plant communities on five different soil types with different nitrogen supply was investigated by Bornkamm & Hennig (1982) and Bornkamm (1984) during ten years. The soil types as well as the nutrient supply altered the composition of the vegetation, the duration of the successional phases, biomass development and nitrogen- and ash concentrations in the biomass. Differences in competitive behaviour, niche separation, co-evolution and/or genetically-caused environmental tolerance, can be responsible for the establishment of different species during particular successional phases (Parrish & Bazzaz 1976, 1979; Harper 1977; Ernst 1978; Bornkamm 1984). Therefore Pickett & Bazzaz (1976, 1978) and Parrish & Bazzaz (1976, 1979, 1982a, 1982b) tested a number of species of early and late successional communities under controlled water- and nutrient supply. They concluded that in early stages, niche overlap is common and therefore competition is more important than in late successional stages where competition avoidance is supposed to be more dominant.

These investigations as well as ecological research on competition behaviour and vegetation dynamics of *Solidago canadensis* and *Urtica dioica* (Schmidt 1981b, 1983b, 1986b) were the reason for this study which followed early succession in two sterilized soil types with two levels of water and nutrient supply. The study area bordered directly on a competition experiment with *Solidago canadensis* and *Urtica dioica* in order to get detailed information on the establishment of both species on bare old-fields with different site conditions.

Experimental design

The experiment took place in the new botanical garden of Göttingen University. Six concrete lysimeter basins (5 m long, 1.2 m wide, 2 m deep) were arranged in a west-east direction (Figs 1, 2). In April 1984 the basins (each with 6 m^2 of colonizable surface) were filled with soils sterilized by heating: three with washed sand (origin: Northeim, gravel pit), three with alluvial loam (origin: Göttingen, Leine flood plain). Details on the physical and chemical composition of both soils were described by Schmidt (1981b). Because of its high silt content the loam had a good water and nutrient supply. The calcium carbonate, humus and nitrogen content of the loam were markedly higher than of the sand. Only the lactate-soluble phosphorus and potassium concentrations were higher in the sand.

From 1984 to 1986 the water level as well as the fertilization varied in two ways. The following fertilization variants were chosen:
N-rich (+ N) (basin 1 and 2): 30 g N m^{-2} yr^{-1}
N-poor (– N) (basin 3 to 6): without any fertilization.
At the beginning of each vegetation period, N-fertilizer was given as a commercial, complete granular fertilizer. Thus unbalanced nutrition was prevented. The basins got some nutrients from rain

Fig. 1. Start of the experiment in basin number 1 to 6 – April 15th, 1984. For explanation see text.

Fig. 2. Two years later dense *Salix*-stands occurred in basins number 1, 3 and 5 – May 12th, 1986.

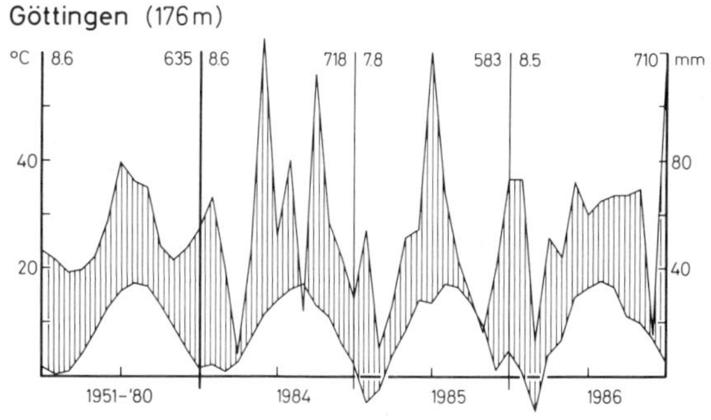

Fig. 3. Climatic diagrams of Göttingen. Left: Long-term mean (1930–1980). Right: Investigation period 1984–1986.

as well as the soil-borne nutrients and the mineral fertilization. The nitrogen in rain water collected nearby was 13.1 g N m^{-2} from the start of the experiment in April 1984 to the end of the investigations in August 1986.

During the whole period no rain water was stored in a ground water table within the 2 m depth of basins 1 to 4 ('dry' conditions), while under 'wet' conditions (basins 5 and 6) the soils were watered by aqua dest so that the ground water level normally ranged from 10 to 30 cm.

This experimental design gave the following combinations of treatments:

Basin number	4	2	6	3	1	5
Soil	sand	sand	sand	loam	loam	loam
Water supply	dry	dry	wet	dry	dry	wet
Fertilization	−N	+N	−N	−N	+N	−N

During the investigation period the weather conditions were also important for the water supply as well (Fig. 3). Compared to the long-term average, the vegetation periods 1985 and 1986 were humid-cold and the winters 1984/85 and 1985/86 rather chilly. Short dry periods happened only during late summer 1984 and autumn 1985.

Methods

Since any disturbance of vegetation during succession affects its subsequent development, harvesting of unreplicated plots to describe this process was precluded. In 1984 to 1986 the height and cover of the shrub-, herb- and moss layers were estimated monthly during the growing season. Cover of living biomass of all phanerogams was specified directly as the percentage of the soil surface covered by each species.

The arithmetic means from all the relevés from each growing season and from each basin were calculated. These are the basic data used in this paper. The sampling design did not incorporate information about the time of first arrival of particular plant species or of the phenology of the plant communities. However, the focus in this study is on the community or the population-community interface rather than on specific demographic and phenological dynamics or responses.

The vegetational evaluation of the succession was concentrated on the number of species, the height and cover of the vegetation layers, life forms and phytosociological and ecological behaviour. Mean annual cover values of the species (Schmidt 1981a) were used to calculate indicator values (Ellenberg 1979).

Early in September 1986, the phytomass of all basins was harvested. The above-ground biomass and litter were taken from the total basin surface. The plant material was separated into species and dried at 105 °C to constant weight. The below-ground biomass was estimated from three soil monoliths (50 × 50 cm surface) in each basin to 60 cm depth by the method of Werner (1983). Below 60 cm depth the biomass of roots was negligible. During washing, the roots were separated into large and fine roots of woody plants (> 2 mm and < 2 mm in diameter, respectively), roots of grasses and herbs. All phytomass values were specified in kg dry matter m^{-2}. After pulverizing a mixed sample, the nitrogen, phosphorus and potassium content of each phytomass component and of soil samples were analyzed following Schmidt (1981a).

The experimental design with its limited numbers of plots without any replication restricted strongly the statistical evaluation. To test the influence of soil, water supply and fertilization the statistical sign test of Dixon and Mood was used. This test of difference of the central tendency is well qualified for paired samples which must not belong to an equal sample unit (Sachs 1978). The chosen tolerance limit was $P < 5\%$.

Results

Structure of the vegetation layers (height and cover)

Within one month after starting the experiment, phanerogams were noticed on all six experimental plots. Of course their height and cover were quite different among treatments (Table 1). Until August/September 1984 the herb layer on all loamy basins covered more than 50% of the soil surface. Under this layer many *Salix*-seedlings (mainly *Salix caprea*) grew up rapidly under the favourable natural nutrient- and water supply of the loam, forming a shrub layer up to 100 cm tall in the first year of succession. In 1985 the shrub layer was so closely

Table 1. Development of the structure of the vegetation layers (cover and height), the number of phanerogam species and the moisture and nitrogen values (sensu Ellenberg 1979) from 1984 to 1986.

Year	1984						1985						1986					
Soil (S: sand; L: loam)	S	S	S	L	L	L	S	S	S	L	L	L	S	S	S	L	L	L
Water supply (D: dry; W: wet)	D	D	W	D	D	W	D	D	W	D	D	W	D	D	W	D	D	W
Fertilization (−N: none; +N: plus 30 gN/m²)	−N	+N	−N	−N	+N	−N	−N	+N	−N	−N	+N	−N	−N	+N	−N	−N	+N	−N
Mean annual cover of the shrub layer (%)	–	–	–	1.2	3.8	11.3	–	–	2.5	30.5	59.6	22.3	–	–	3.1	52.4	41.1	44.6
Mean annual cover of the herb layer (%)	3.0	2.4	9.4	34.0	36.1	58.7	14.1	65.8	18.7	55.8	20.2	70.8	3.1	56.4	26.4	22.1	12.6	43.6
Mean annual cover of the moss layer (%)	–	0.3	0.3	6.7	8.2	1.9	4.7	0.3	7.7	1.5	2.5	0.5	10.0	0.6	25.0	–	1.1	0.6
Mean height of the shrub layer (cm)	–	–	–	52	51	56	–	–	59	178	198	159	–	–	100	266	283	230
Maximum height of the shrub layer (cm)	–	–	–	100	100	100	–	–	100	230	260	195	–	–	120	300	300	270
Mean height of the herb layer (cm)	17	37	40	125	97	82	85	129	85	150	136	120	62	59	51	116	114	112
Maximum height of the herb layer (cm)	25	60	80	185	170	140	120	200	110	210	170	165	100	100	65	180	160	170
Number of species	5	9	19	17	18	21	16	14	29	19	20	19	20	22	25	15	16	14
Moisture (F−) values (Ellenberg 1979)	4.4	5.0	5.3	5.2	5.8	6.1	4.6	4.1	5.2	5.8	6.0	6.0	4.8	5.7	5.4	5.6	5.8	6.4
Nitrogen (N−) values (Ellenberg 1979)	6.5	5.4	6.0	6.9	7.0	5.9	6.2	6.9	5.6	6.8	6.9	6.3	4.9	7.3	5.6	6.8	6.9	6.6

packed, especially under the conditions of additional fertilization, that the herb layer was greatly reduced, presumably because of the lack of light and nutrients. In 1986 the cover of the shrub layer on the fertilized loam decreased, while it increased on the unfertilized soil. On the other hand, the shrubs increased in height continuously. At the end of the third year (1986) the *Salix*-species had reached 270 to 300 cm in height.

In comparison with a nearby succession experiment on calcareous fertile level loam, studied on permanent plots by Schmidt (1981a, 1983a) since 1969, the woody plant species on the loamy basins developed much more quickly. The main reason was high seed production of the *Salix*-species in the spring of 1984 followed by good germination conditions so that the willows grew up quickly before meeting competition from tall herbs and grasses. In addition in the fenced basins no browsing by hares, roe deer or other herbivores took place during winter, a time when woody plants in many old-fields are often severely cut back by browsing.

In contrast to the loamy basins, on sand, a closed herb layer occurred only with nitrogen fertilization (Table 1). To a certain degree the poor nutrient status of the unfertilized sand was compensated by a high ground water level which favoured mainly mosses besides higher plant species. Dense carpets of in particular *Bryum argenteum* were observed on wet sand. On loam the mosses were only important during the first year. Later on they decreased rapidly as the light intensity was reduced.

Generally, soil quality had the strongest influence on the vegetation structure. On loam, both the height of shrub- and herb layers were significantly higher than on sand. Under wet conditions the herb layer covered more of the soil surface than on dry soils, but its height was determined more by the soil type. On wet sand, woody plant species grew faster than on dry sand, while on dry loam the willows grew better than under a high water level. The response to fertilization was non-uniform. During the first stage of succession the additional nutrient supply favoured markedly the shrub layer but later on the cover and height of the herb layer was limited by insufficient light.

Life form spectrum

Just as on abandoned fields (Schmidt 1981a; Pickett 1982; Armesto & Pickett 1986), colonization of the soils was characterized by a rapid change of life form spectra (Fig. 4). Generally, therophytes were a main component of the plant communities in the first two years. In the third year they played an important role only on the dry sand; elsewhere they formed less than 5% of the total cover. The decrease of the annuals was compensated quite differently according to the environmental conditions. On dry sand, herbaceous chamaephytes and hemicryptophytes dominated in

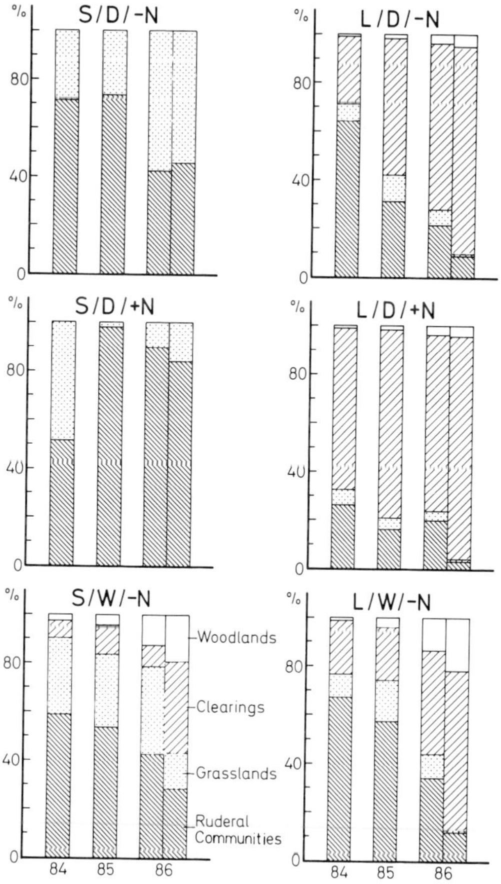

Fig. 4. Life form spectra from 1984 to 1986 on the basis of mean annual cover percentage (=100%). For 1986, the left column represents the cover values while the right column gives the results on the basis of the aboveground biomass in September 1986. S: sand, L: loam, D: dry, W: wet, –N: without fertilization, +N: fertilization.

1986. On wet sand, phanerophytes appeared. In the third year, woody plants contributed more than 50% of the total cover on all loamy soils.

Comparing the results of the three years study, the portion of therophytes, hemicryptophytes and herbaceous chamaephytes was higher on sand than on loam. On the other hand, geophytes (mostly *Tussilago farfara* with low cover values) and phanerophytes developed better on loam than on sand. A wet soil was advantageous for the growth of geophytes and hemicryptophytes. In contrast, therophytes were more successful on dry soils. The phanerophytes reacted differently according to the soil type. On dry loam, tree and shrub species grew better than on wet loam. Dry sand however, hampered the establishment of woody plants, while on wet sand they composed about one fourth of the plant communities at the end of 1986. The reasons for these differences were the desiccation of saplings during the germination phase on sand, and the rise in water level and consequent oxygen deficiency on loam, respectively. Both situations inhibit the growth of woody plant species. Contrasting nutrient supply had no influence on the proportion of therophytes and geophytes. Hemicryptophytes and herbaceous chamaephytes were strongly suppressed by mineral fertilization. On loam, the phanerophytes clearly profited by their decrease.

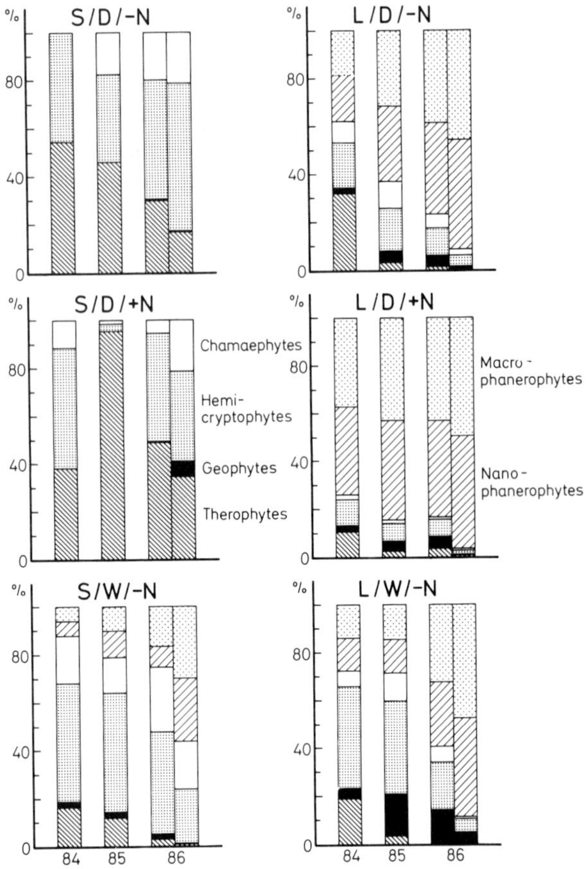

Fig. 5. Spectra of the phytosociological behaviour (sensu Ellenberg 1979) from 1984 to 1986 on the basis of mean annual cover percentage (=100%). For 1986, the left column represents the cover values while the right column gives the results on the basis of the aboveground biomass in September 1986. S: sand, L: loam, D: dry, W: wet, –N: without fertilization, +N: fertilization.

Phytosociological spectrum

Figure 5 presents a phytosociological spectrum based on the percentage cover of species groups characterizing phytosociological units (Ellenberg 1979). Only four groups of classes were distinguished because classes and lower units were often represented by only one species. At the beginning of the succession, species from ruderal, disturbed places (mainly character species of the classes *Chenopodietea, Secalietea, Artemisietea* and *Plantaginetea*) generally occurred with the highest proportion. They decreased significantly to the end of the experiment, except on dry fertilized sand. On unfertilized sand, the class character species of regularly used meadows and pastures (*Molinio-Arrhenatheretea*) increased. From the beginning on wet sand, woody plant species of clearings (*Epilobietea*) and willow-riparian woodlands (*Salicetea purpureae*) participated also in the plant communities. Members of these classes dominated clearly on the loamy soils. Firstly on dry loam, the clearing species *Salix caprea* increased rapidly, while on wet loam *Salix fragilis, Salix triandra* and *Salix viminalis,* as character species of the *Salicion albae,* also increased during the course of succession. Therefore the well-known grassland stage, which has been described for many old-field successions between the 3rd and 10th year after abandonment (Büring 1970; von Borstel 1974; Schmidt 1981a), was missed here. It was also remarkable that on dry sand no clearing- and woodland species were noticed. Therefore, besides bare soil, a well-balanced water supply was also important for the development of woody plant species. If a dense vegetation of herbs and grasses had established at first, tree and shrub species could not have competed successfully against them even with a good soil surface water supply (Hard 1972; Schmidt 1981a, 1983a).

Species richness

The experiment can be subdivided obviously into two groups by the number of species (Table 1). During the first year on dry sand, less than 10 phanerogamic species were found. Later, the number of species increased significantly. Already during the first year on wet sand and on all loamy soils species numbers were markedly higher than on dry sand. The numbers of species reached a peak during the second year and later decreased significantly. In 1986 all loamy soils had lower species numbers than all sandy soils. The persistence of competitive early colonists in the moist fertilized treatments is not the reason for low species richness. Mineral fertilization had positive effects on the number of species but altogether the absolute increase of species number caused by fertilization was rather low.

Surprisingly, the six plots differed little in the total number of species from 1984 to 1986. In general 24 to 26 higher plant species were encountered per 6 m^2. Only the wet sand had, with 36 species, a significantly higher diversity and was also, with 25 species, the most species rich community in 1986.

The ratio of the total number of species to mean annual number of species showed a remarkable turnover of species on all plots during the succession. The proportion of species which occurred during all three years, out of the total number of species from 1984 to 1986, was significantly lower on sand (17–36%) than on loam (39–46%). This characterizes the slower colonization of sandy soils by new species. It also indicates that under good soil conditions, as on loam, competitive species soon hampered the further invasion of non-competitive species.

F- and N-values

The F (moisture)-values of Ellenberg (1979), calculated on the basis of cover values, extend from 4.1 to 6.4 (Table 1). Under the given climatic conditions the plant communities thus indicate a medium water supply. During the three years under investigation a small increase of the F-values appeared. Whether this was the result of the establishment of a closed stand with a well-balanced microclimate or of the cool and wet summers of 1985 and 1986 (Fig. 3), is unknown. In any case, distinct correlations between soil type, water level, fertilization and F-value exist. On loamy soils, the F-values were higher than on sand, just as they were increased by a high ground water level and additional mineral fertilization. The well-known experience of farmers that 'nitrogen replaces water' was confirmed again by this experiment.

The N (nitrogen)-values of Ellenberg (1979) (Table 1) showed that the converse of this rule is not necessarily valid. The N-values ranged from 4.9 to 7.3, a similar amplitude to that of the F-values, but at a higher level. Clear changes from year to year were only observed on dry sand. Without mineral fertilization the N-value here decreased significantly from 1984 to 1986 but increased markedly with the use of fertilizer. Also, the small decrease of the N-value on wet sand indicates a decrease in nutrient supply during vegetation development. Compared with this change, the N-value of the wet loam increased slightly from 1984 to 1986. On the other hand, on dry loam there were no changes in the N-values, with or without fertilization.

Generally the plant communities on loamy soils and on fertilized sand were characterized by a higher nitrogen indication value than on unfertilized sand. Also on dry soils a better nutrient supply was indicated by higher N-values. With a high ground water level, leaching of nutrients from the soil was also possible, involving nitrogen losses by denitrification. Therefore the increasing number of indicators for a decreased nitrogen status of the soils was reasonable.

Phytomass

Phytomass was determined by harvesting, but only at the end of the experiment in September 1986. Figures 4 and 5 compare the results of the cover values in 1986 and the aboveground biomass. In general, the use of cover values overestimated the proportion of therophytes and geophytes, and partly also of hemicryptophytes and herbaceous chamaephytes, in comparison with their biomass. On the other hand, phanerophytes of clearings and woodlands were underestimated. This is because long-living trees and shrubs produce a considerable amount of non-photosynthesizing aboveground biomass over many years while herbs and grasses consist only of leaves and shoots which are renewed every year. Moreover, in this comparison it was also important that a number of species like *Tussilago farfara* and *Taraxacum officinale* had long passed their annual peak of biomass when harvested.

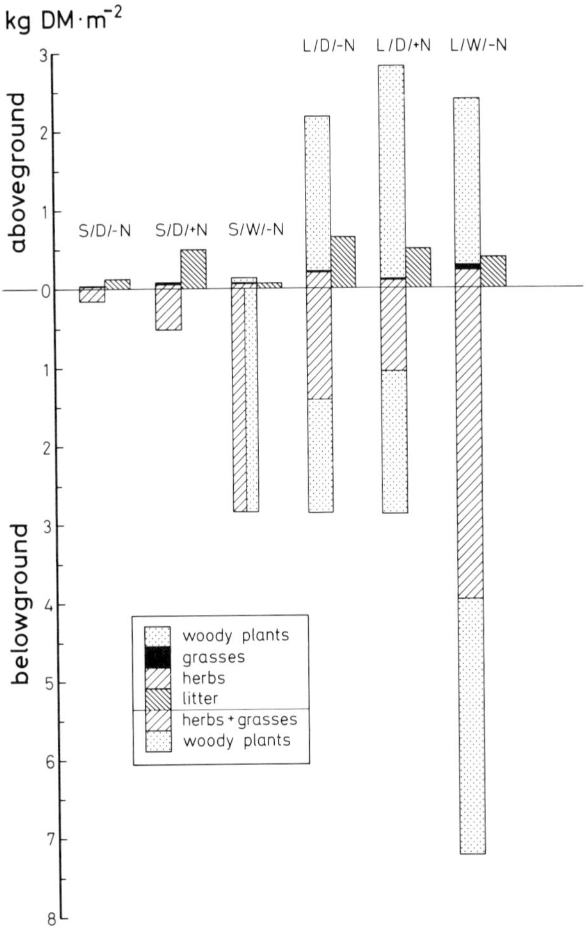

Fig. 6. Above- and belowground phytomass (dry matter) harvested in September 1986. S: sand, L: loam, D: dry, W: wet, −N: without fertilization, +N: fertilization. For the basin S/W/−N it was not possible to separate the roots of herbs, grasses and woody plants.

Figure 6 presents data on aboveground and belowground biomass. The comparison of the plots highlights differences in the share of woody plant species with their long-living root and shoot systems. In contrast, the biomass of herbs and grasses was rather low. In this connection, it must be stressed that a calculation of the net primary production would obviously change this ratio in favour of the non-woody plant species. The highest biomass was measured in the three loamy basins with a well-developed shrub layer. In contrast to reports in many well-known syntheses (e.g. Ellenberg 1986) the shoot-root ratio did not shift with increasing dryness in favour of the root biomass. The root biomass was higher on both wet sand and on wet loam than on the dry soils. This may be a further indication that there was no severe water stress during any stage of the succession. But it was not clear why such a high ground-water level produced such an intensive growth of the root system. The aboveground biomass on sand increased more than on loam with mineral fertilization. For the belowground biomass, the same effect was noticed only on sandy soils.

Nitrogen, phosphorus and potassium

The bioelement storage of the phytomass (Fig. 7) is related to the dry weight. The nitrogen concentration was only of minor significance. Thus the shoot and root system of the woody plants had a high biomass but a rather low nitrogen content (as well as phosphorus and potassium), whereas the herbs with low biomass often showed high bioelement concentrations. Mineral fertilization increased nitrogen and phosphorus concentrations mainly in the belowground biomass, while potassium concentrations were particularly higher in the aboveground parts. On the other hand, a high ground-water level lo-

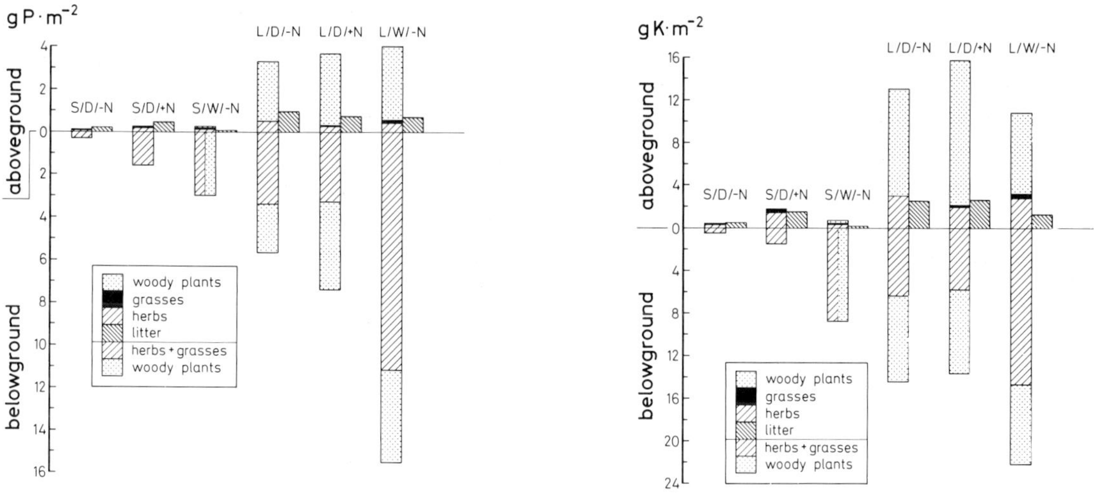

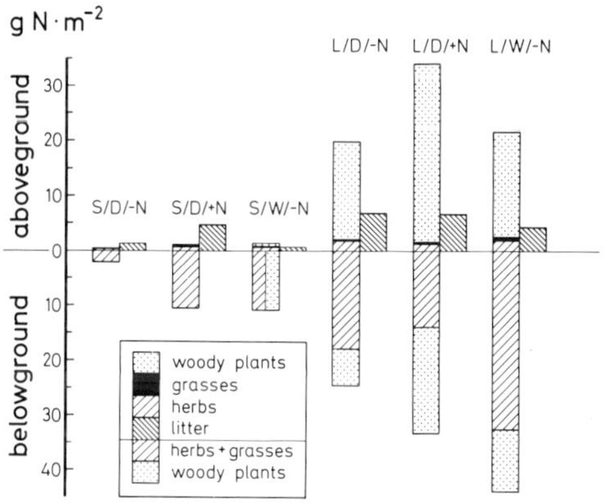

Fig. 7. Nitrogen, phosphorus and potassium storage of the phytomass harvested in September 1986. S: sand, L: loam, D: dry, W: wet, –N: without fertilization, +N: fertilization. For the basin S/W/–N it was not possible to separate the roots of herbs, grasses and woody plants.

wered the nutrient concentrations. Altogether, Fig. 7 clearly shows the well-known contrast of loam and sand, as well as of dry and wet soil conditions.

Table 2 summarizes nitrogen input and output of all six basins. The nitrogen storage of the phytomass was supplied on sand and loam in a rather different way. On unfertilized sand, the low N storage could in theory have been supplied just by wet deposition. Since there was a simultaneous increase of the nitrogen store of the soil, while microbial N_2-fixation presumably took place as well. The sand fertilized with 90 g N m^{-2} produced significant increases in

Table 2. Calculation of the nitrogen budget (g N m^{-2}).

Basin number Site conditions	4 S/D/ – N	2 S/D/ + N	6 S/W/ – N	3 L/D/ – N	1 L/D/ + N	5 L/W/ – N
Input (1984 – 1986)						
rain (wet deposition)	13.1	13.1	13.1	13.1	13.1	13.1
fertilization	–	90.0	–	–	90.0	–
Storage (September 1986)						
phytomass	3.7	16.5	13.0	51.3	74.2	69.9
soil (0 – 30 cm) (difference from April 1984)	+ 13.1	+ 28.7	+ 4.9	– 58.5	– 48.1	– 53.8

nitrogen in the phytomass and in the soil. Together with the wet deposition, the total nitrogen input could not be fixed by the vegetation and the top soil, and substantial nitrogen losses by leaching must have happened on the porous sand. At the end of the first year, the nitrogen store of the phytomass on loamy soils was 51–74 g N m^{-2}, more than the atmospheric input. At the same time, the nitrogen store of the top soil decreased by 48–59 g N m^{-2} by mineralisation from decomposing organic material after sterilization of the alluvial loam. Thus nearly all the nitrogen demand of the developing vegetation could be covered by precipitation input and mineralisation from soil organic matter. Mineral fertilization influenced the nitrogen uptake by plants on loam less than on sand. Nevertheless, the nitrogen losses of the soil were mostly limited on fertilized soils which had the largest uptake by plants. Water level had little influence on this nitrogen balance, which does not include leaching, herbivory, losses or gains by wind. On wet soils there were no obvious nitrogen losses by denitrification.

Discussion

The aim of these investigations was to compare succession on soils of different nutrient and water status. In comparison with the known secondary succession of old-fields (Schmidt 1981a, 1983a) and ruderal communities on deposit soils (Bornkamm & Henning 1982; Bornkamm 1984) two results were particularly interesting.

First, the different site conditions caused different patterns of colonization on the open sterilized soils in the early stages of succession. Assuming that all plots received the same supply of diaspores, it was surprising that only a small number of species occurred as early colonizers in more than half of all basins. In 1984, *Epilobium tetragonum, Plantago intermedia, Salix caprea, Solidago canadensis, Taraxacum officinale, Trifolium repens* and *Urtica dioica* were noticed on four, and *Cerastium holosteoides, Chenopodium album* and *Leucanthemum vulgare* on five of the six plots. Only *Poa annua* grew in all six basins during the first year. The other 25 species of 1984 were either linked to special site conditions or their distribution was caused by chance. Therefore the great diversity of many old-field – and ruderal communities (Schmidt 1981a) can be explained by differences in environmental factors over short distances. Further, because varying climatic conditions from year to year mean that these site factors themselves are not constant, this may explain the coexistence of many species on old-fields (Pickett & Bazzaz 1978; Schmidt 1983a).

Secondary successions are strongly influenced in their development by the initial plant stand and soil seed bank. Accordingly, abandoned grasslands can keep their original species composition for many years. They change mainly in their dominance structure (Schiefer 1981; Schreiber & Schiefer 1985). Annuals very rarely invade such grasslands. Most new species are woody plants which become important in the final successional stages. On old-fields and ruderal sites, the more classic succession takes place. After the annual stage during the first one to three years, a grassland stage with perennial grasses and herbs follows for several years before shrubs and trees become dominant (Schmidt 1981a).

It is often assumed that most diaspores colonizing open soils come from the surrounding area. In contrast, von Borstel (1974) showed that the seed bank of abandoned fields can be extremely high, and seed bank species can produce the bulk of the vegetation present during the early first years of succession (Schmidt 1981a). Bornkamm & Henning (1982) also concluded that most plant species recorded in their experiments originated from the soil seed bank. In the experiment described here, there was no soil seed bank. A typical therophyte stage was noticed only on dry sand and lasted for only two years. Among the other experimental treatments, perennial species dominated within the first two years. Therefore the classic succession can be observed only when the soil seed bank is rich in annuals from the previous vegetation or when site conditions are unfavourable for the rapid growth of perennials.

The second point of particular interest was the fast development of woody plant species on all loamy soils. Subsequently the phytomass and its content of nitrogen, phosphorus and potassium was outstandingly high. Clearly woody plants can dominate early in succession if conditions allow quick germination after a high input of seeds. Because of their fast growth in height and biomass, the *Salix* species allowed only a very few species to exist in the understorey. This contrasts with the results of Bornkamm & Henning (1982), who observed a quick development of shrubs on sand while on all loamy soils perennial herbs, mainly *Solidago canadensis,* dominated for many years. Although *Solidago canadensis* is common in the neighbourhood of the experimental field and dominates many old-fields around Göttingen, it played only a minor role in this experiment. The main reason is probably that *Solidago canadensis,* which needs many years to build up a complete root system (Werner 1983; Bornkamm 1984) was preempted as a colonizer by *Salix* species.

As the experiment was run for only three years, the importance of niche differentiation and competition during succession cannot be answered easily. The species number, structure and biomass development on the loamy soils with their favourable site conditions support the final conclusions of Bornkamm & Hennig (1982) that the development of communities rich in species to stages dominated by only a few plant species represent an increase of competitive interactions and of interference on community organization. Even after three years on dry sands, the soil surfaces were partly open and therefore new species could potentially have colonized until the end of the experiment.

Acknowledgements

The relevés were accurately made by Mrs S. Hermann-Borchert. I thank Dr John Miles (Banchory), Prof. Eddy van der Maarel (Uppsala) and Dr S. T. A. Pickett (Millbrook, N.Y.) for many critical and helpful comments and for the correction of the English.

References

Armesto, J. J. & Pickett, S. T. A. 1986. Removal experiments to test mechanisms of plant succession in oldfields. Vegetatio 66: 85–93.

Bornkamm, R. 1984. Experimentell-ökologische Untersuchungen zur Sukzession von ruderalen Pflanzengesellschaften. II. Quantität und Qualität der Phytomasse. Flora 175: 45–74.

Bornkamm, R. & Henning, U. 1982. Experimentell-ökologische Untersuchungen zur Sukzession von ruderalen Pflanzengesellschaften auf unterschiedlichen Böden. Flora 172: 267-316.

Braun-Blanquet, J. 1964. Pflanzensoziologie, 3rd ed. Springer, Wien, New York.

Büring, H. 1970. Sozialbrache auf Äckern und Wiesen in pflanzensoziologischer und ökologischer Sicht. Thesis Univ. Gießen.

Ellenberg, H. 1979. Zeigerwerte der Gefäßpflanzen Mitteleuropas. Scripta Geobot. 9.

Ellenberg, H. 1986. Vegetation Mitteleuropas mit den Alpen. 4th ed. Ulmer, Stuttgart.

Ernst, W. 1978. Discrepancy between ecological and physiological optima of plant species. A re-interpretation. Oecol. Plant. 13: 175–188.

Hahn, W., Wolf, A. & Schmidt, W. 1979. Untersuchungen zum Stickstoff-Umsatz von *Tussilago farfara* und *Agropyron repens*-Beständen. Verhandl. Ges. Ökol. 7: 369–380.

Hard, G. 1972. Wald gegen Driesch. Das Vorrücken des Waldes auf Flächen junger 'Sozialbrache'. Ber. Dtsch. Landesk. 46: 49–80.

Harper, J. L. 1977. Population biology of plants. Academic Press, London, New York, San Francisco.

Parrish, J. A. D. & Bazzaz, F. A. 1976. Underground niche separation in successional plants. Ecology 57: 1281–1288.

Parrish, J. A. D. & Bazzaz, F. A. 1979. Difference in pollination niche relationships in early and late successional plant communities. Ecology 60: 597–610.

Parrish, J. A. D. & Bazzaz, F. A. 1982a. Competitive interactions in plant communities of different successional ages. Ecology 63: 314-320.

Parrish, J. A. D. & Bazzaz, F. A. 1982b. Responses of plants from three successional communities to a nutrient gradient. J. Ecol. 70: 233–248.

Pickett, S. T. A. 1982. Population patterns through twenty years of oldfield succession. Vegetatio 49: 45–59.

Pickett, S. T. A. & Bazzaz, F. A. 1976. Divergence of two co-occurring successional annuals on a soil moisture gradient. Ecology 57: 169–176.

Pickett, S. T. A. & Bazzaz, F. A. 1978. Organisation of an assemblage of early successional species on a soil moisture gradient. Ecology 59: 1248–1255.

Sachs, L. 1978. Angewandte Statistik. 5th ed. Springer, Berlin, Heidelberg, New York.

Schiefer, J. 1981. Bracheversuche in Baden-Württemberg – Vegetations- und Standortsentwicklung auf 16 verschiedenen Versuchsflächen mit unterschiedlichen Behandlungen (Beweidung, Mulchen, kontrolliertes Brennen, ungestörte Sukzession). Veröff. Naturschutz Landschaftspfl. Baden-Würtemberg Beih. 22: 1-325.

Schmidt, W. 1981a. Ungestörte und gelenkte Sukzession auf Brachäckern. Scripta Geobot. 15.

Schmidt, W. 1981b. Über das Konkurrenzverhalten von *Solidago canadensis* und *Urtica dioica*. Verhandl. Ges. Ökol. 9: 173–188.

Schmidt, W. 1983a. Experimentelle Syndynamik - Neuere Wege zu einer exakten Sukzessionsforschung, dargestellt am Beispiel der Gehölzentwicklung auf Ackerbrachen. Ber. Dtsch. Bot. Ges. 96: 511-533.

Schmidt, W. 1983b. Über das Konkurrenzverhalten von *Solidago canadensis* und *Urtica dioica*. II. Biomasse und Streu. Verhandl. Ges. Ökol. 11: 373-384.

Schmidt, W. 1985. Mahd ohne Düngung - Vegetationskundliche und ökologische Ergebnisse aus Dauerflächenuntersuchungen zur Pflege von Brachflächen. Münstersche Geogr. Arb. 20: 81-99.

Schmidt, W. 1986a. Über die Dynamik der Vegetation auf bodenbearbeiteten Flächen. Tuexenia 6: 53-74.

Schmidt, W. 1986b. Über das Konkurrenzverhalten von *Solidago canadensis* und *Urtica dioica*. III. Stickstoff- und Phosphorhaushalt. Verhandl. Ges. Ökol. 14: 537-550.

Schreiber, K.-F. & Schiefer, J. 1985. Vegetations- und Stoffdynamik in Gründlandbrachen - 10 Jahre Bracheversuche in Baden-Württemberg. Münstersche Geogr. Arb. 20: 111-153.

Stüssi, B. 1970. Naturbedingte Entwicklung subalpiner Weiderasen auf Alp La Schera im Schweizer Nationalpark während der Reservatsperiode 1939-1965. Ergeb. Wiss. Unters. Schweiz. Nationalpark 13.

von Borstel, U. O. 1974. Untersuchungen zur Vegetationsentwicklung auf ökologische verschiedenen Gründland- und Ackerbrachen hessischer Mittelgebirge (Westerwald, Rhön, Vogelsberg). Thesis Univ. Gießen.

Werner, W. 1983. Untersuchungen zum Stickstoffhaushalt einiger Pflanzenbestände. Scripta Geobot. 16.

Wittig, R. 1978. Zur pflanzensoziologischen und ökologischen Stellung ruderaler Bestände von *Solidago canadensis* und *Solidago gigantea* AIT. (Asteraceae) innerhalb der Klasse *Artemisietea*. Decheniana 131: 33-38.

Vegetatio 77: 115–128, 1988
© Kluwer Academic Publishers, Dordrecht – Printed in the Netherlands

Succession, regeneration and fluctuation in the Białowieża Forest (NE Poland)*

Janusz Bogdan Faliński
Geobotanical Station, Warsaw University, PL – 17-230 Białowieża, Poland

Accepted 1.6.1988

Keywords: Degeneration, Dynamic process, Primary forest, Primary succession, Secondary succession, Vegetation dynamics

Abstract

A general description is given of the most important processes occurring in forest communities and of the spatial range and duration of the cycles of particular processes.

In the initiation and course of nearly all processes involved in non-periodical vegetation dynamics a particular role is played by two groups of biotic-biocenotic factors: 1. Behaviour of main forest edificators (oak, hornbeam, spruce and pine); 2. Primary environmental preferences of the big herbivorous animals (bison, red-deer, roe-deer, elk, wild boar).

The range of the particular processes is practically limited by the size of the potential forest phytocoenose, and the duration of a single cycle is determined by the properties of the habitat and the stability of the climate.

Introduction

This paper will present general conclusions from a long-term research of the fundamental ecological processes involved in temperate forest dynamics. In addition to the extensive monograph on this research (Faliński 1986a) the present paper provides a wider generalisation and more information on the spatial range and duration of the various processes. The paper also attempts a formal description of forest communities undergoing dynamical processes.

The surviving fragments of primaeval forest in Białowieża – in view of its geographical location, state of preservation and conditions favouring permanent ecological research – may be considered as a model object for understanding the dynamics of forest communities and ecosystems.

Vegetation dynamics concentrate on phytocenoses and plant populations as integral components of ecosystems, subjected to the laws of their development and the principles of their functioning. Vegetation dynamics include both autonomous endogenous transformations and changes due to exogenous (anthropogenic) forces.

Contemporary vegetation dynamics aim at following the phenomena and processes occurring in phytocenoses and in the ecotones between them. Hence the great significance of permanent areas for research on vegetation dynamics. The role of reconstructive methods, indispensable in the study of vegetation history and the reconstruction of the dynamics of now extinct phytocenoses, is less important here.

Vegetation dynamics can be specified as follows: 1. Two distinct directional processes: succession (both primary and secondary) in the strict sence of the word and regression as the reverse process; 2. Three processes with variable duration: fluctua-

*This work was supported by grant CPBP 04.10.07 (Warsaw Agricultural University)

tion, degeneration and regeneration;

3. The strictly cyclic, periodical seasonal process.

All these processes depend strongly on time. As a measure of intensity of the process, a conventional dynamics index (DI) can be adopted (Faliński 1986a). This index may represent the state of biomass, the relation of primary production to respiration, the leaf area index, some index of species diversity, cover plant corm volume, etc. (see Faliński 1986a, and further Knapp 1974; van der Maarel & Werger 1978; van der Maarel 1988).

Study area

The Białowieża Forest occupies an area of 1 250 km^2 on the boundary between Poland and the USSR and it represents the type of forests proper to the boreal-nemoral zone. The prevailing substrates are Riss glaciation formations. The relatively short vegetation period (185 days) and the long duration of snow cover (92 days per year) favour boreal elements in the plant cover and fauna. The transitional biogeographical character of this forest complex is reflected in the characteristics of the forest: species-rich, especially geophytes, typically central-European mixed deciduous (*Tilio-Carpinetum* communities, alliance Carpinion betuli mixed with coniferous forest communities of the alliances Dicrano-Pinion and Vaccinio-Piceion). A specific regional feature of the Białowieża forests is the presence of thermophilous oak forest (*Potentillo albae-Quercetum* of the order Quercetalia pubescentis) together with spruce forests on peatland (*Sphagno girgensohnii-Piceetum* of the alliance Vaccinio-Piceion). *Picea abies* occurs in nearly all forest communities; it easily takes possession of all habitats under favourable circumstances and it influences thus the course of the major ecological processes. The boreal element in the geographical spectrum is mainly represented in spruce forest communities, where it meets suboceanic and central-European elements of deciduous forest communities (Faliński & Matuszkiewicz 1963; Faliński 1986a).

The best preserved primaeval forest communities have found protection in the Białowieża National Park since 1921 (47 km^2). They then exhibited a spontaneous primary vegetation with many species and layers and different-aged stands (Fig. 1). The canopy, especially *Quercus, Carpinus* and *Tilia*, was over 200 yr old, and many single trees over 300 or even 400 yr. They occurred almost uninterruptedly, the only gaps being some pasture clearings and overgrown belt fellings of 1910. Some stands show traces of plundering fellings, filled by agglomerations of anemochoric trees such as *Betula, Populus tremula* and *Salix*. They also included numerous trees with traces of ancient primitive beehives, charcoal pits from the 18th and 19th century, and extensive burying grounds with tumuli of the 10th and 13th century. The road network consisted of old hunting paths and newer ones leading along the compartment lines. It has hardly changed and at present the density of the roads maintained is 0.75 km/km^2.

The most conspicuous consequence of human activity was the damage caused by game, owing to its artificial multiplication beyond the available nutritional resources. During the period 1892–1915 the Białowieża Forest actually acted as a game-breeding park of the Russian tzar. Damage included: 1. Complete destruction of the lower forest layer, especially the undergrowth and up-growth; 2. Debarking of young deciduous trees; 3. Inhibition of natural rejuvenation of most broadleaved trees. As a result, *Picea* spread under the canopy of deciduous trees in nearly all habitats. In the course of 50–70 yr, spruce became the absolute dominant although earlier scientific and literary descriptions mentioned it as subordinate to oak and pine.

Between 1948 and 1952 and also from 1980 to 1986 meadows in the river valleys of the Narewka and Hwoźna were included in the National Park and subsequently abandoned. This meant a good protection of the last fragments of primary forests and its also enabled studies on secondary succession of forest sites which had been deforested 200 yr ago.

Material and methods

The material and methodical basis of my research has been described in detail in a earlier monograph (Faliński 1986a), and may be summarized as follows:

Fig. 1. Primary forest stand of the *Tilio-Carpinetum* in Białowieża National Park. Big *Tilia cordata* in the foreground. Photo J.B. Faliński.

1. Observations on vegetation dynamics on permanent plots in successive series 1952–1962 and 1963–1988 (continued with repeatable photographic registration) (cf. Faliński 1977, 1986a);
2. Observations on forest stand dynamics in the Białowieża National Park performed in the period 1936–1973 by Włoczewski and collaborators (see Włoczewski 1954, 1972; Kowalski 1972, 1975, 1982);
3. Investigations of windfall, particularly its role in forest community dynamics and organic matter balance (Faliński 1976, 1978);
4. Studies on the influence of wild boar and other herbivores on the structure and dynamics;
5. Observations and experiments on permanent study plots on the course of secondary succession on abandoned farmland 1970–1986 (Faliński 1980a, 1980b, 1986a, 1986b);
6. Comparison of herb layer structure, leaf-fall and soil properties in the primary and secondary stands of the *Tilio-Carpinetum* (Faliński, Canullo & Biały in prep.);
7. Dating of the beginning and course of secondary succession on abandoned farmland by analysis of the age of individual trees of the main woody species forming pioneer populations: (*Juniperus communis, Populus tremula, Salix caprea, S. aurita, S.*

repens ssp. rosmarinifolia); dating of older spruce individuals subject to self-thinning; elaboration of a system of developmental phases for secondary succession, degeneration and regeneration;

8. Interpretation of panchromatic aerophotos from the point of view of revealing changes in forest structure and texture and elaboration of a map of the dynamic tendencies in the vegetation of the Białowieża National Park; 9. Establishment of a system of development phases for the main processes of secondary succession, degeneration and regeneration;

10. Reconstruction and modelling of the course of some selected processes on the basis of the abovementioned observation data, experiments and historical sources (cf. Faliński 1986a, 1986b).

Results

Fluctuation

Fluctuation is the dominant process in the Białowieża National Park communities (Fig. 2), particularly in the moist forms of the *Tilio-Carpinetum*, in flood-plain forest (*Circaeo-Alnetum*), in bog alder forest (*Carici elongatae-Alnetum*). Fluctuation is most typical in the well-preserved primary community, with different-aged, multi-species and multi-layer stands, in which changes caused by game, and subsequent spruce invasion, were not important.

FLUCTUATION:
Tilio-Carpinetum
Circaeo-Alnetum
Carici elongatae-Alnetum
Pino-Quercetum
Querco-Piceetum

REGRESSION:
Pino-Quercetum (fragm.)
Potentillo albae-Quercetum (out BNP)

REGENERATION:
Pino-Quercetum
Tilio-Carpinetum (fragm.)
Peucedano-Pinetum (fragm.)

SECONDARY SUCCESSION:
Tilio-Carpinetum (fragm.)
Circaeo-Alnetum (fragm.)
Salicetum pentandro-cinereae
Carici elongatae-Alnetum

PRIMARY SUCCESSION:
Sphagnetum medio-rubelli
Vaccinio uliginosi-Pinetum

REGRESSION → SECONDARY SUCCESSION:
Potentillo albae-Quercetum → Tilio-Carpinetum (out BNP)
Pino-Quercetum, Tilio-Carpinetum } ecotone → Carici elongatae-Quercetum

Fig. 2. Dominant ecological processes in forest communities in the Białowieza National Park.

Fluctuation is manifested in most communities by a gradual exchange of components. Single older trees, disappearing from the stand are substituted by individuals of the same species or other long-lived

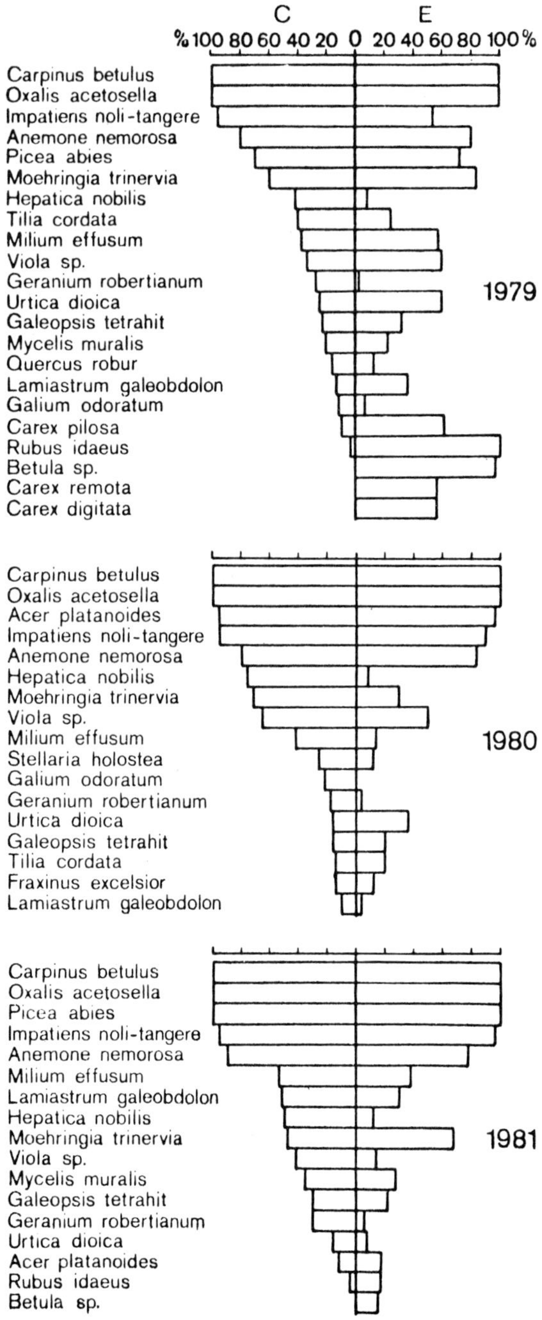

Fig. 3. Species composition of seedlings in the herb layer of a *Tilio-Carpinetum typicum* community. C – control data, E – experimental data (after uprooting of mature plants). After E. Piroznikow (mscr.).

species, for instance: *Alnus* and *Fraxinus* in the *Circaeo-Alnetum, Tilia, Acer* and *Carpinus* in the *Tilio-Carpinetum*. At present, no regeneration of *Quercus* and *Pinus* is observed in deciduous and mixed forests, neither is there any new spruce invasion.

In the undergrowth and herb layers there is permanent up-growth and seedlings of the main tree species are permanently present. The seed bank in the soil (Piroznikow mscr) contains seeds of all permanent components of the communities, as well as seeds of ephemeral species which take advantage of the transitional changes in the forest (anemochoric trees and shrubs, e.g. *Betula, Rubus idaeus*, and therophytes; Fig. 3). The herb layer has a constant structure, which varies only locally through the vegetative spread of the main components, rooting of wild boars and uprooted trees. The curve of the total number of species for three subassociations of the *Tilio-Carpinetum* shows only small variations from year to year (Fig. 4b, c, d). Changes under the influence of the above-mentioned factors do not cause penetration of foreign (non-forest) species, but only concern fluctuations in abundance of therophytes, especially *Impatiens nolitangere*. On the other hand, Fig. 4a shows a pattern of change for a community in regeneration from a previous disturbance (see below).

Degeneration and regeneration

Degeneration and regeneration are coupled processes. Degeneration means the process of deformation of the community structure, and disturbance of its functioning, either in the whole phytocenosis or in

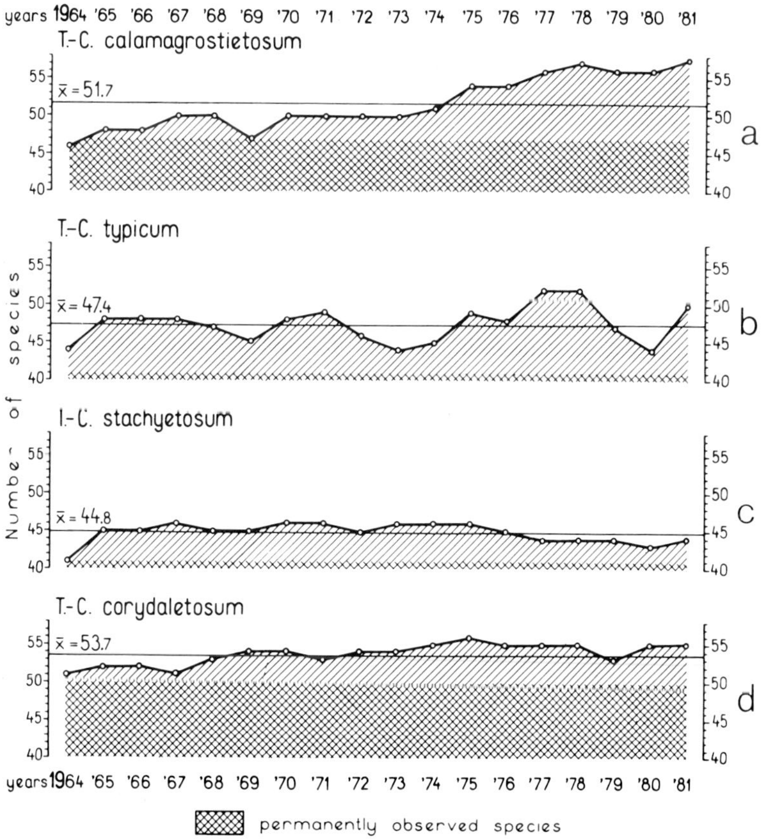

Fig. 4. Changes in the total number of herb layer species in the main subassociations of the *Tilio-Carpinetum* in the course of 18 years. a – community in regeneration, b–d – stable communities under fluctuation. (After Faliński 1986a).

a part. It is caused chiefly by the action of external factors, whether natural or anthropogenic. The main contemporary factors responsible for degeneration of forest communities outside the National Park are timber exploitation, road building, tourist activities and game rearing; in the past, they were cattle grazing, charcoal burning and potash production (Faliński 1966, 1986a). Apart from the advance of invasive trees such as spruce, birch and aspen, the result of degeneration is the penetration of alien species which partly establish themselves permanently and then acquire the status of local neophyte (e.g. Faliński 1966, 1968, 1986a).

Within the limits of the National Park the main degeneration factor in the past (1892–1915) was the excess of game, including species alien to the local fauna (dapple deer) and cattle grazing.

The presence of big herbivores as such may be considered as a permanent internal biotic-biocenotic factor within the forest ecosystem. It is only when the numbers of game become excessive in any one area that game becomes a degeneration factor.

Regeneration means the process of reconstruction of degenerated communities by internal forces. In the National Park, this occurred after removal of excessive game during World War I. Spruce invasion after destruction of the up-growth appeared to be only a short-term process relative to the scale of forest history, since as early as after 50–60 yr self-thinning started (Fig. 5). Under the spruce canopy a new generation of deciduous trees developed. Together with old oaks and lindens saved in earlier fellings, they form a characteristic two-layer structure. On air photographs these stands show a fine-grained structure because of the small tree crowns. The intensive impact of animals in the past is indicated by the absence in these stands of intermediate age classes of trees (70–90 yr).

The regeneration process occurs relatively faster in fertile, moderately moist habitats with *Tilio-Carpinetum stachyetosum* and *T.-C. typicum*. In poorer and drier habitats of the *Tilio-Carpinetum calamagrostietosum* and *Pino-Quercetum* the process of regeneration is slowed down by the lack of oak regeneration and the low accessibility of these habitats to hornbeam and linden. Other factors

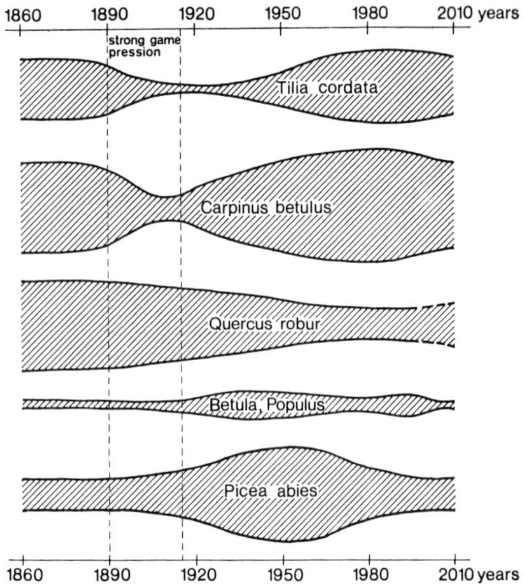

Fig. 5. Changes in species extension in a deciduous forest stand of the *Tilio-Carpinetum*. Note the temporary increase of spruce due to strong game impact and the prognosis of stand regeneration under strict reserve conditions with moderate game impact. (After J. B. Faliński 1986a).

delaying regeneration are the big dimension of invaded aspens and birches and the slow decomposition of their leaf litter.

Communities undergoing regeneration show a simplified vertical structure. Species exchange in the herb layer results in an increase of species number from year to year (Fig. 4a).

Secondary succession

Secondary succession means the process of forest reconstruction on sites where forest used to grow, but was destroyed by human activity. Such a succession is called secondary-recreative, and not regenerative, to distinguish it from proper regeneration as a process within the phytocenosis. We deal with three cases of secondary succession:

1. Moist meadows in deforested river valleys; cessation of their exploitation leads to the renewal of *Circaeo-Alnetum, Ficario-Ulmetum* and *Carici elongatae-Alnetum* forest communities;
2. Permanent clearings for pasture and agriculture,

which had been abandoned when the National Park was established. In this habitat spontaneous development occurred towards the primary type of the *Tilio-Carpinetum.* In the course of this secondary succession the range of direct seed supply from the neighbouring forest is very important. Beside the less numerous anemochoric tree species *Populus tremula, Betula pubescens, B. pendula,* and *Salix caprea*, fruit trees (*Malus domestica, M. sylvestris)* grew wild and still persist after more than 70 years and so do relicts of ruderal communities (*Rumex obtusifolius*) in the herb layer of compact brushwood; 3. Belt fellings within the *Tilio-Carpinetum* after the forest felling in 1910, that is before the establishment of the National Park. This case was documented by Paczoski in 1930. Because of the significance of this object for understanding the nature and consequences of succession we will devote some attention to it.

A system of three parallel belts is found, each of 1 076 m (1 verst) length and ca 107 m width, separated by bands of primary forest of the same width (Fig. 6). The belt system with stands of different origin runs in a south-north direction, following the general direction of terrain depression. Nearly the whole of this system lies within the boundaries of a large *Tilio-Carpinetum* phytocoenosis. In relation to changes in terrain relief and primary soil properties the *Tilio-Carpinetum* forest is differentiated into three basic units: *Tilio-Carpinetum calamagrostietosum, T.-C. typicum* and *T.-C. stachyetosum.* This primary differentiation is combined with a secondary differentiation of dynamic character, that is corresponding to stands of different origin and age. The primary *Tilio-Carpinetum* has a considerable participation of spruce and a highly differentiated vertical structure with maximal ages of trees, 150–250 yr. At the site of the former belt felling there appeared a birch-aspen stand with a large participation of *Salix caprea.* At present self-thinning is observed in this stand, in the first place of willow, next birch and finally aspen. A younger maple-linden-hornbeam understory has developed in time together with spruce which gradually substitutes the anemochoric tree stand. The presence of the latter is still clear from the composition of leaf-fall and litter (Fig. 7).

Fig. 6. Alternation of primary and secondary tree stands of a *Tilio-Carpinetum* in Białowieża National Park. Result of forest exploitation in 1910, before the establishing of the park. Fragment of aerophotos from 1971. After aerophotos by PPG-K, Warsaw.

The diversity in the structure of both stands and their parallel variability due to habitat and topographic variation is clearly visible on air photographs (Fig. 6).

The undergrowth and herb layer in the primary stand are better developed and richer in species. The herb layer has a homogeneous spatial structure, whereas, in the secondary stands, it has a mosaic pattern with numerous spots completely deprived of vascular plants, owing to the aspen litter. The herb layer in the secondary stand is characterized above all by the absence of several mesohygrophilous- and sciophilous species and an increased proportion of xero-mesophilous- and heliophilous species (Table 1). Ecological interpretation of the floristic composition of the herb layer of the secondary stand suggests that it is influenced by soil moisture deficit and enhanced illumination, as compared with the herb layer under the primary stand.

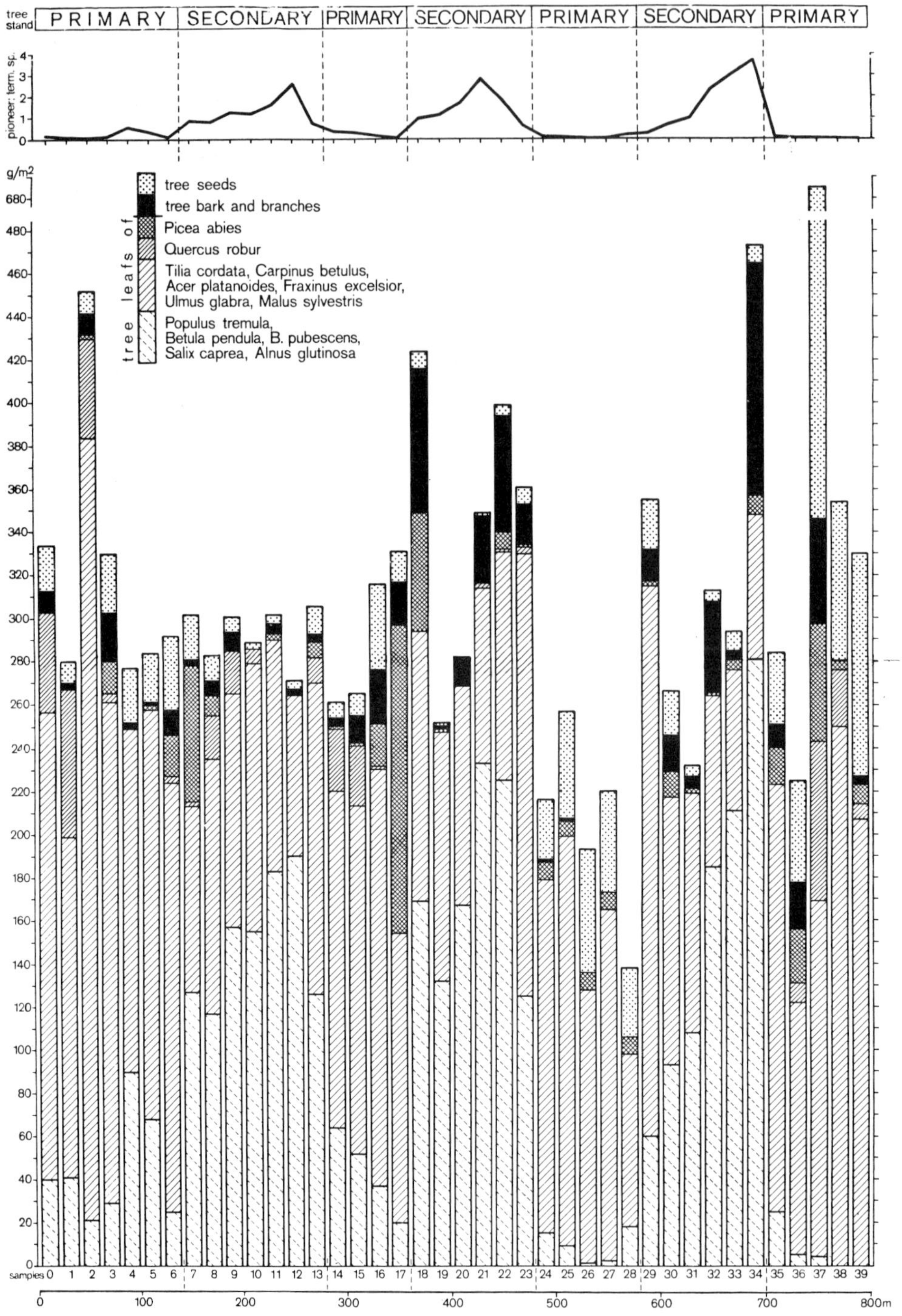

Fig. 7. Changes in tree litter composition under primary and secondary tree stands in a *Tilio-Carpinetum* forest community. Data from a transect perpendicular to forest belts of different origin.

Table 1. Comparison of herb layer species composition in primary and secondary tree stands in a *Tilio-Carpinetum* community (after J. B. Faliński, R. Canullo, K. Biały, in prep.)

Herb layer under	Primary stand			Secondary stand			Primary regenerated stand		
	Frequency	Cover % $\bar{x}$	Cover % min–max	Frequency	Cover % $\bar{x}$	Cover % min–max	Frequency	Cover % $\bar{x}$	Cover % min–max
Oxalis acetosella	100.0	12.23	1–37.5	100.0	23.00	5–37.5	96.7	19.03	0–37.5
Lamiastrum galeobdolon	93.3	11.67	0–37.5	36.7	3.80	0–12.5	100.0	16.67	5–37.5
Stellaria holostea	56.7	5.17	0–37.5	93.3	13.07	0–37.5	76.7	8.13	0–37.5
Aegopodium podagraria	83.3	3.87	0–12.5	36.7	1.03	0– 5	46.7	1.40	0– 5
Carpinus betulus juv.	56.7	1.10	0– 5	43.3	0.43	0– 1	56.7	0.83	0– 5
Adoxa moschatellina	30.0	0.57	0– 5	63.3	0.90	0– 5	56.7	1.10	0– 5
Anemone nemorosa	56.7	0.70	0– 5	73.3	0.73	0– 1	20.0	0.20	0– 1
Geranium robertianum	66.7	2.40	0– 5	10.0	0.10	0– 1	66.7	2.50	0–12.5
Impatiens noli-tangere	73.3	8.27	0–37.5	23.3	0.88	0–12.5	40.0	1.72	0–12.5
Maianthemum bifolium	40.0	0.67	0– 5	73.3	3.27	0– 5	20.0	0.47	0– 5
Galium odoratum	40.0	1.20	0– 5	33.3	1.40	0– 5	60.0	2.32	0–12.5
Acer platanoides juv.	40.0	0.67	0– 5	36.7	1.37	0– 5	33.3	0.60	0– 5
Viola riviniana	30.0	0.70	0– 5	33.3	0.87	0– 5	56.7	2.03	0– 5
Tilia cordata juv.	66.7	0.80	0– 5	6.7	0.07	0– 1	26.7	0.27	0– 1
Populus tremula juv.	26.7	0.40	0– 5	53.3	0.53	0– 1	10.0	0.10	0– 1
Hepatica nobilis	23.3	0.90	0– 5	3.3	0.03	0– 1	56.7	1.67	0– 5
Viola reichenbachiana	26.7	1.18	0–12.5	30.0	0.83	0– 5	20.0	0.73	0– 5
Urtica dioica	36.7	1.40	0–12.5	3.3	0.03	0– 1	30.0	0.30	0– 1
Ranunculus lanuginosus	[illegible]	1.33	0– 5	6.7	0.20	0– 5	20.0	0.47	0– 5
Fraxinus excelsior juv.	[illegible]	0.30	0– 1	10.0	0.10	0– 1	40.0	0.53	0– 5
Dentaria bulbifera	20.0	0.47	0– 5	6.7	0.20	0– 5	23.3	0.90	0– 5
Sanicula europaea	6.7	0.07	0– 1	6.7	0.20	0– 5	20.0	0.73	0– 5
Lathyrus vernus	13.3	0.27	0– 5	13.3	0.40	0– 5	6.7	0.20	0– 5
Glechoma hirsuta	56.7	3.62	0–37.5	3.3	0.17	0– 5	.	.	.
Equisetum pratense	33.3	0.33	0– 1	3.3	0.03	0– 1	.	.	.
Asarum europaeum	3.3	0.03	0– 1	3.3	0.52	0–12.5	.	.	.
Polygonatum multiflorum	10.0	0.23	0– 5	3.3	0.17	0– 5	.	.	.
Ranunculus cassubicus	3.3	0.03	0– 1	3.3	0.03	0– 1	.	.	.
Chrysosplenium alternifolium	33.3	0.33	0– 1	.	.	.	13.3	0.27	0– 5
Milium effusum	6.7	0.20	0– 5	.	.	.	13.3	0.27	0– 5
Stellaria nemorum	13.3	0.40	0– 5	.	.	.	6.7	0.07	0– 1
Moehringia trinervia	13.3	0.27	0– 5	.	.	.	6.7	0.07	0– 1
Picea abies juv.	6.7	0.07	0– 1	.	.	.	10.0	0.10	0– 1
Circaea lutetiana	13.3	0.67	0– 5	.	.	.	.	.	.
Mercurialis perennis	10.0	0.50	0– 5	.	.	.	.	.	.
Paris quadrifolia	10.0	0.23	0– 5	.	.	.	.	.	.
Cardamine parviflora	6.7	0.07	0– 1	.	.	.	.	.	.
Athyrium filix-femina	3.3	0.42	0–12.5	.	.	.	.	.	.
Geum urbanum	3.3	0.17	0– 5	.	.	.	.	.	.
Cardamine amara	3.3	0.03	0– 1	.	.	.	.	.	.
Rumex sanguineus	3.3	0.03	0– 1	.	.	.	.	.	.
Scrophularia nodosa	3.3	0.03	0– 1	.	.	.	.	.	.
Stachys sylvatica	3.3	0.03	0– 1	.	.	.	.	.	.
Quercus robur juv.	.	.	.	10.0	0.10	0– 1	3.3	0.03	0– 1
Ajuga reptans	.	.	.	6.7	0.07	0– 1	3.3	0.03	0– 1
Carex pilosa	.	.	.	13.3	0.53	0– 5	.	.	.
Galeopsis tetrahit	.	.	.	6.7	0.07	0– 1	.	.	.
Vicia sylvatica	.	.	.	6.7	0.07	0– 1	.	.	.
Carex digitata	.	.	.	3.3	0.03	0– 1	.	.	.
Lilium martagon	.	.	.	3.3	0.03	0– 1	.	.	.
Mycelis muralis	.	.	.	3.3	0.03	0– 1	.	.	.
Rubus idaeus	.	.	.	3.3	0.03	0– 1	.	.	.
Rubus saxatilis	.	.	.	3.3	0.03	0– 1	.	.	.
Pulmonaria obscura	.	.	.	.	.	.	20.0	0.60	0– 5
Cardamine impatiens	.	.	.	.	.	.	16.7	0.17	0– 1
total number of species	43			38			32		
mean number of species/record	12.8			8.9			10.8		
sum of mean cover	64.03			55.35			64.51		

This secondarily acquired diversity of the herb layer may be explained by certain soil properties which changed during the development of the felling site, which has not yet attained the terminal composition and structure of the mature *Tilio-Carpinetum* forest.

The transitory status of the humus, especially its hydrophobic character, resulting from the delay in the process of aspen litter decomposition, explains the more xero- and heliophilous character of the herb layer in the secondary stand. The difference in the properties of the upper soil horizons of primary and secondary stands are periodically obliterated by heavy rooting of wild boar.

Regression

Regression is the process of gradual decline of a forest community and its components. This phenomenon has been locally observed since around 1959 with increased intensity in the period 1965–1975. It is manifested by massive self-thinning of spruce and pine. This process occurs especially in meso-oligotrophic *Pino-Quercetum* forests, but is spreading over wider forest areas. The fall of trees is preceded by their slow dying back. This is accompanied by noticeable changes in the herb layer, notably the appearance of heliophilous plants (*Calamagrostis arundinacea, Rubus idaeus*) which limits the occurrence of *Pino-Quercetum* components proper.

The causes of regression of this community are not sufficiently known, but are probably related to stand origin: these stands arose spontaneously as a response to previous forest exploitation (game, cattle, litter raking) and they consist of even-aged trees.

The intensity of the process may also be linked to the lowering of the ground water level in relation to the reclamation of the Polesie swamps adjacent on the east side to the Białowieża Forest.

Soon after disappearance of the old stand a rapid secondary succession began, with mass appearance of *Betula pendula* and *B. pubescens*, and in some parts *Picea* or *Carpinus*, especially on the oldest treeless surfaces (Faliński 1986c). This succession has been retarded again, notably since 1978 by snowbreak of birches.

Regression and secondary succession occur also in specific ecotones (Faliński 1986a). This implies:
1. Formation of oak bog forest (*Carici elongatae-Quercetum*) after spruce self-thinning in the ecotone between the fertile deciduous *Tilio-Carpinetum* and the meso-oligotrophic mixed *Pino-Quercetum*; under the surviving oak wetland species established in holes left by uprooted spruces and gave rise to a completely new community;
2. Expansion of the raised peat as the result of enlargement of the peat basin on the side of the *Tilio-Carpinetum* as a result of self-thinning of spruce.

Regression combined with secondary succession is described also for the thermophilous oak forest (*Potentillo albae-Quercetum*) which transformed gradually within the last 90 yr into a mesophilous *Tilio-Carpinetum*. This transformation has been interpreted as caused by the sudden cessation of game impact (Faliński 1986a). This transformation is associated with replacement of east-European 'continental' heliophilous species by central-European shade-tolerating ones, as well as with a decrease in the number of species (from 120–150 per 100 m^2 to 40–80 per 100 m^2).

Primary succession

In our area this process applies only to raised bogs (*Sphagnetum medio-rubelli*) the development of which started after the glaciation of free water surfaces in the Alleröd period (10000–9000 years B.C.). Its course can be followed only by palinological methods (Dąbrowski 1959). Development of raised peat bog towards pine bog forest (*Vaccinio uliginosi-Pinetum*) has been followed over 35 yr. Beside gradual stabilisation of the role of pine and penetration of spruce, arrival of terrestrial shrubs, herbs and bryophytes was noted, especially around the tree bases (Faliński 1986a). Establishment on ageing bogs of some undershrub species (*Vaccinium vitis-idaea*) went through an increase of plant numbers and cover, while pioneering individuals became fertile and accomplished complete development cycles and produced further generations (Falińska 1973a, 1973b, 1973c, 1986).

Conclusions

Interdependence of processes

In the initiation and course of nearly all processes constituting non-periodical vegetation dynamics a particular role is played by two groups of biotic-biocenotic factors:

1. The behaviour of the main forest community edificators: oak, hornbeam, spruce and pine, their interaction and the influence of some external factors such as climatic severity, human activity and fires (Fig. 8);
2. Primary environmental preferences of the big herbivorous animals (red-deer, roe-deer, elk, wild boar, European bison) and secondary changes in these preferences are caused by anthropogenic transformation of the forest environment (Faliński 1986a).

In the dynamics of natural forest ecosystems the basic process is fluctuation, within a state of permanence of the given ecological system. The forests acquire this permanence not by an unlimited viability and longlivedness of its biotic components (individuals), but by an unceasing exchange – in the first place unceasing rejuvenation of the tree stand.

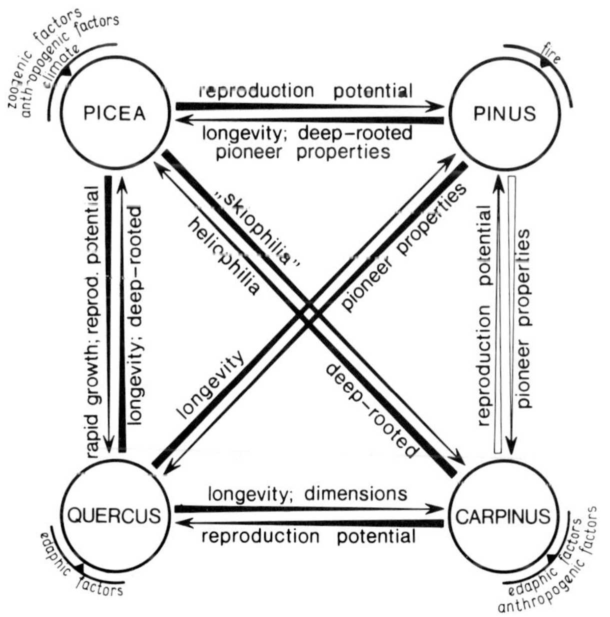

Fig. 8. Basis for interaction between the main forest-building trees in the Białowieża Forest: biological properties and favourable external factors. After J. B. Faliński (1986a).

Floristic-dynamic diagnosis of communities undergoing major ecological processes

The change from year to year of the total number of species is a good indication of the initiation and nature of some processes. This is visible for instance in the oscillation of species numbers in the herb layer of the *Tilio-Carpinetum* (cf. Fig. 4).

Similar behaviour of species over a longer time interval allows a formal description of dynamic groups of species. Six such groups have been distinguished, corresponding to six types of possible behaviour of these species in time.

G_1 – species present permanently from the beginning to the end of the observation period (*perm.*);

G_2 – regressive species that are present at the beginning of the observation period and disappearing gradually (*regress.*);

G_3 – progressive species appearing later or towards the end of the observation period (*progress.*);

G_4 – reoccurring species, appearing from time to time (*reoccur.*);

G_5 – transitory species, appearing for a certain time but persisting longer (*trans.*);

G_6 – ephemeral species (*ephem.*).

These groups may be distinguished in any observed community over longer time intervals. Establishment of the sequence of these groups, from the most numerous to the least numerous one, allows a formal diagnosis of the community from the point of view of the prevailing dynamical process (Fig. 9).

For a dynamic diagnosis of the community the first three groups are essential. The absence of one of the groups in any community requires a special interpretation.

Range and duration of major processes

In order to determine the duration of particular cycles of secondary succession leading to the forma-

FLUCTUATION
community in fluctuation
$G_1 > G_4 > G_6 > G_2 > G_3 > G_5$
perm > reocurr > ephem > regress > progress > trans

SECONDARY SUCCESSION
community in initial process stage
perm > ephem > regress > progress > reocurr
community in optimal process stage
trans > progress > regress > ephem > perm > reocurr
community in terminal process stage
perm > ephem > reocurr > regress > progress > trans

PRIMARY SUCCESSION
community in initial process stage
ephem > progress > reocurr
community in optimal process stage
trans > progress > ephem > perm > reocurr > regress
community in terminal process stage
perm > ephem > reocurr > regress > progress > trans

DEGENERATION
community in degeneration
perm > regress > ephem ...

REGENERATION
community in regeneration
perm > progress > ephem > trans > regress > reocurr

REGRESSION
community in initial process stage
perm > regress > reocurr > trans > ephem > progress
community in optimal process stage
regress > ephem > perm ...
community in terminal process stage
ephem > regress ...

Fig. 9. Floristic-dynamical diagnosis of plant communities undergoing different ecological processes.

tion of a permanent forest community, we need to know:

- sequence of arrival and disappearance of the particular woody species;
- time needed for the formation of the respective age- and spatial structure of the woody species populations;
- maximal age of individuals of the important species and their mean survival time;
- factors responsible for differentiation and integration of the structure of the phytocenosis.

In order to simplify our model, all factors may be omitted which may retard or accelerate the course of the given process.

The particular forest types, as seen from Fig. 10, differ significantly in the sequence of occurrence of the particular components, and above all, in the duration of one secondary succession cycle. The shortest cycle (100 yr) is characteristic for a secondary succession leading from a rush community to the *Carici elongatae-Alnetum*, and the longest cycle of 350 yr leads from the *Arrhenatheretum elatioris*

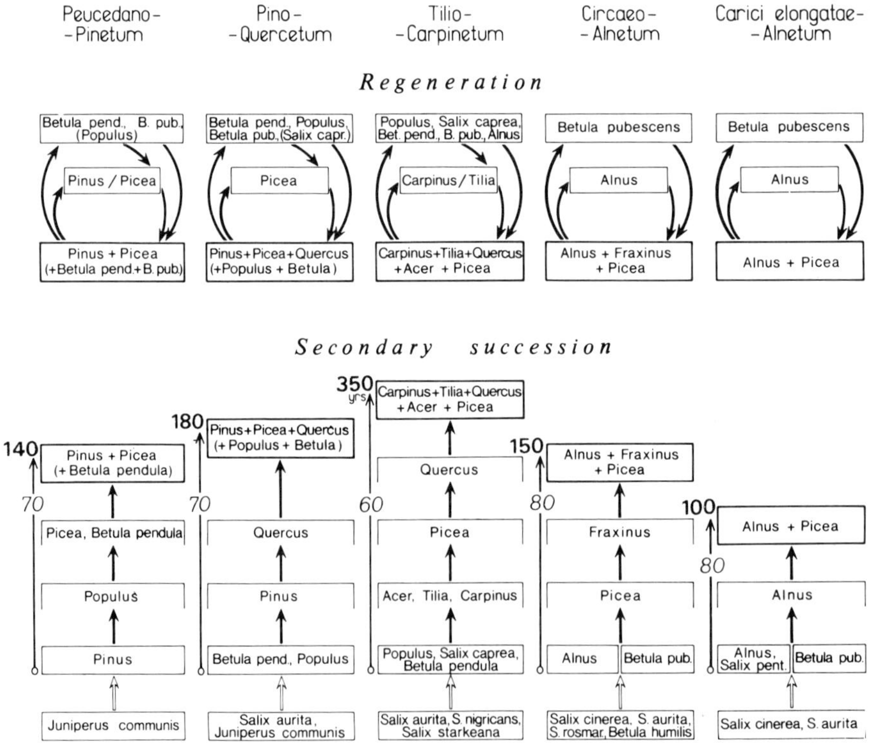

Fig. 10. Sequence of an appearance of woody species in the course of secondary recreative succession and regeneration in the main forest communities of the Białowieża Forest. Indicated are the time needed for recreation of forest formation with typical forest species (italic figures) and for recreation of forest with appropriate structure, dynamics and function (roman figures). (After Faliński 1986a).

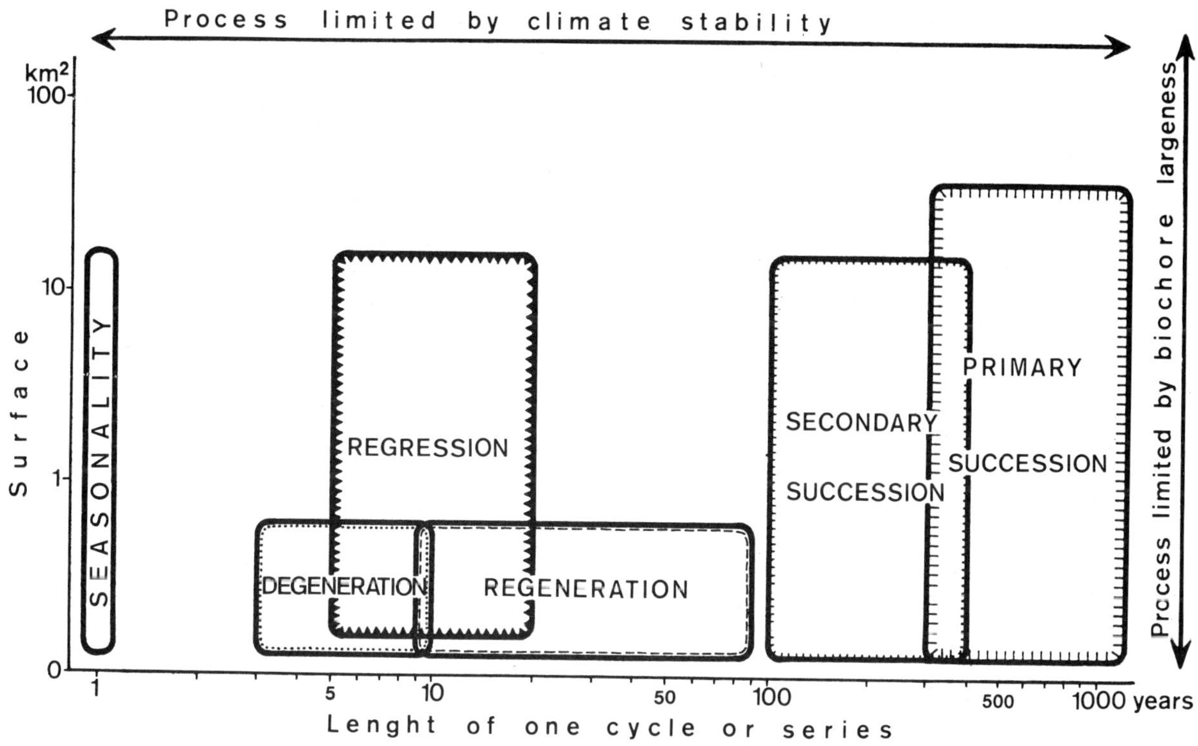

Fig. 11. Spatial range and time indispensable for realisation of individual cycle of some ecological processes. Generalisation of data from the Białowieża Forest.

meadow to the *Tilio-Carpinetum*.

In each dynamic community cycle two aspects should be distinguished: acquirement of a characteristic species and forest composition, and attainment of the specific structure, dynamics and function (Fig. 10).

The ranges of the particular processes are dependent on the total size of the area of the terminal phytocenosis. Determination of this size is generally not difficult, when the complexes of natural vegetation types are well preserved. Frequently, however, we need a reconstruction by using a phytosociological map, and a set of air photographs by direct field investigations. One large-surface deciduous forest site may exceed 20 km^2. The relation between surface and cycle length for the major processes is summarized in Fig. 11. Together with information on the duration of the particular cycles of major processes, it allows a certain generalisation (Fig. 11; see also Faliński 1984).

The diagram also includes information on the duration of the primary succession cycle on the basis of palynological data (Dąbrowski 1959). The duration of a single cycle depends on the properties of the habitat and the stability of the climate.

References

Dąbrowski, M. J. 1959. Późnoglacjalna i holoceńska historia lasów Puszczy Białowieskiej. Część I. Białowieski Park Narodowy. Late-glacial and Holocene history of Białowieża Primeval Forest. Part I. Białowieża National Park. Acta Soc. Bot. Pol. 28, 2: 197–248.

Falińska, K. 1973a. Dynamika sezonowa runa zbiorowisk leśnych Białowieskiego Parku Narodowego. Seasonal dynamics of herb layer in forest communities of Białowieża National Park. Phytocoenosis 2, 1: 1–120.

Falińska, K. 1973b. Flowering rhythms in forest communities in the Białowieża National Park. Ekol. pol. 21: 828–867.

Falińska, K. 1973c. Phenology of the species of the genus *Vaccinium* L. in the forest communities of the Białowieża National Park. Fragm. Florist. et Geobot. 19, 3: 327–338.

Falińska, K. 1986. Seasonal dynamics of forest communities. In: Faliński, J. B. Vegetation dynamics in temperate lowland primeval forests. Geobotany 8: 265–306.

Faliński, J. B. 1966. Antropogeniczna roślinność Puszczy Białowieskiej jako wynik synantropizacji naturalnego kompleksu leśnego. Végétation anthropogène de la Grande Forêt de Białowieża comme un résultat de la synanthropisation du territoire silvestre naturel. Dissert. Univ. Varsov. 13: 1–256.

Faliński, J. B. 1968. Stan i prognoza neofityzmu w szacie roślinnej Puszczy Białowoeskiej. The state and prognosis of neophytism in the plant cover of the Białowieża Primeval Forest. In: Faliński, J. B. (ed.), Synantropizacja szaty roślinnej. I. Neofityzm i apofityzm w szacie roślinnej Polski. Mat. Zakł. Fitosoc. Stos. UW 25: 175–216.

Faliński, J. B. 1976. Windwürfe als Factor der Differenzierung und der Veränderung des Urwald-biotopes im Licht der Forschungen auf Dauerflächen. Phytocoenosis 5, 2: 85–108.

Faliński, J. B. 1977. Research on vegetation and plant populations dynamics conducted by Białowieża Geobotanical Station of the Warsaw University in the Białowieża Primeval Forest 1952–1977. Phytocoenosis 6, 1/2: 1–148.

Faliński, J. B. 1978. Uprooted trees, their distribution and influence in the primeval forest biotope. Vegetatio 38: 175–183.

Faliński, J. B. 1980a. Vegetation dynamics and sex structure of the populations of pioneer dioecious woody plants. Vegetatio 43: 23–38.

Faliński, J. B. 1980b. Changes in the sex- and age-ratio in populations of pioneer dioecious woody species (*Juniperus, Populus, Salix*) in connection with the course of vegetation succession in abandoned farmlands. Ekol. pol. 28, 3: 327–365.

Faliński, J. B. 1984. Le temps et l'espace dans les recherches écologiques. Giorn. Bot. Ital. 118, 2: 67–68.

Faliński, J. B. 1986a. Vegetation dynamics in temperate lowland primeval forest. Ecological studies in Białowieża forest. Geobotany 8: 1–537. Junk. Dordrecht/Boston/Lancaster.

Faliński, J. B. 1986b. Sukcesja roślinności na nieużytkach porolnych jako przejaw dynamiki ekosystemu wyzwolonego spod długotrwałej presji antropogenicznej. Vegetation succession on abandoned farmland as a dynamical manifestation of ecosystem release from human impact. Cz. 1, 2, Wiad. Bot. 30: 25–50; 115–126.

Faliński, J. B. 1986c. La rélation: Végétation – formes et processus géomorphologiques. Les plus importantes procedures de recherche et les modes d'expression cartographique. In: Géhu, J.-M (ed.), Végétation et géomorphologie; Colloque Phytosociol., Lille 13: 541–552. J. Cramer, Vaduz.

Faliński, J. B. & Matuszkiewicz, W. 1963. La Grande Forêt de Białowieża. Der Urwald von Białowieża (2). Excursion internationale phytosociologique en Pologne N-E. Mat. Zakł. Fitosoc. Stos. UW 2: 31–60.

Knapp, R. 1974. Some principles of classification and of terminology in successions. In: Knapp, R. (ed.), Vegetation dynamics. pp. 167–177. Junk, The Hague.

Kowalski, M. 1972. Dynamika odnowień naturalnych w oddziale 319 Białowieskiego Parku Narodowego. Dynamics of natural regeneration in the compartment 319 of the Białowieża National Park. Folia Fores. Pol. A. 20: 93–109.

Kowalski, M. 1975. Kształtowanie się struktury odnowień w kilku zespołach lasu naturalnego. Formation of the structure of regeneration in several communities of a natural forest. Zesz. Nauk. AR, Leśnictwo 21: 43–55.

Kowalski, M. 1982. Rozwój drzewostanów naturalnych na powierzchni badawczej w Białowieskim Parku Narodowym. Rozprawy Naukowe i Monografie, Wyd. SGGW-AR.

van der Maarel, E. & Werger, M. J. A. 1978. On the treatment of succession data. In: Faliński, J. B. (ed.), Vegetation dynamics. Proc. of the Third. Symp. of the Working Group on Succ. Res. on Perm. Plots. Białowieża 1977. Phytocoenosis 7: 257–278.

Paczoski. J. 1930. Lasy Białowieży. Die Waldtypen von Białowieża. PROP, Monogr. nauk. 1: 1–575.

Włoczewski, T. 1954. Materiały do poznania zależności między drzewostanem a glebą w czasie i przestrzeni. Rocz. Nauk Leśn. 5: 161–241.

Włoczewski, T. 1972. Dynamika rozwoju drzewostanów w oddziale 319 Białowieskiego Parku Narodowego. Dynamics of the development of stands in the compartment 319 of the Białowieża National Park. Folia Fores. Pol. A. 20.

Vegetatio 77: 129–137, 1988
© Kluwer Academic Publishers, Dordrecht – Printed in the Netherlands

Plant population dynamics and species turnover on small islands near Karpathos (South Aegean, Greece)

D. Höner & W. Greuter
Botanischer Garten & Botanisches Museum Berlin-Dahlem, D-1000 Berlin 33, Federal Republic of Germany

Accepted 24.5.1988

Keywords: Islet chasmophyte, Islet specialist, Mass-population, Micro-population, *Silene holzmannii*

Abstract

The vascular flora of small offshore islets around Karpathos (Dodecanesos, Greece) has been surveyed and monitored during six consecutive years. The present study focuses on four islets that are inhabited by the 'islet specialist' *Silene holzmannii* (Caryophyllaceae), a species that has a widely scattered distribution in the Aegean where it exclusively occurs on very small islands. Repeated census of its populations, partly following controlled extirpation, reveals extreme variation in number of individuals. Sowing experiments involving various species demonstrate the low probability of successful new colonizations. Differences in the amount of apparent turnover of species as well as the occurrence of both stable micro-populations and mass-populations are discussed.

Nomenclature: Barclay (1986), Greuter *et al.* (1983, 1984, 1986) and Tutin *et al.* (1964–1980): Flora Europaea.

Introduction

Rechinger (1951: 201) was the first to recognize the floristic peculiarity that is a general feature of small islands in the Aegean Sea. He found that some plant species exclusively occur on such islets and are never found in similar habitats on larger islands. Besides these 'islet specialists' proper, Rechinger (1951: 167) also noticed that a number of so-called chasmophytes, i.e., taxa that are strictly limited to crevices of limestone cliffs on larger islands in the Aegean area, are able to colonize other habitats when occurring on small islands. In the following the term 'islet chasmophytes' will be used for this kind of species.

Rechinger's observations were since confirmed by others who had explored some of the countless islets of the Aegean Sea (Runemark 1969; Greuter 1972; Gustafsson & Snogerup 1974; Greuter & Pieper 1975; Snogerup *et al.* 1980; and Snogerup & Snogerup 1987).

In the course of the preparation of a checklist of the Karpathos flora (Greuter *et al.* 1983) the offshore islets surrounding the main island of Karpathos were thoroughly investigated. Eleven of them were selected for a detailed study of the quantitative and qualitative composition of their flora and of their changes in time and space (Fig. 1). Most of them are small enough to enable a complete inventory of their flora. Only three have been used for grazing purposes during the span of observation, and most are inhabited by at least some islet specialists and/or islet chasmophytes (Tables 1 and 2).

In this paper floristic changes observed on four of these islets (Nos 2, 4, 6 and 8) are discussed in detail, with special emphasis on the population dynamics of the islet specialist *Silene holzmannii* (caryophyl-

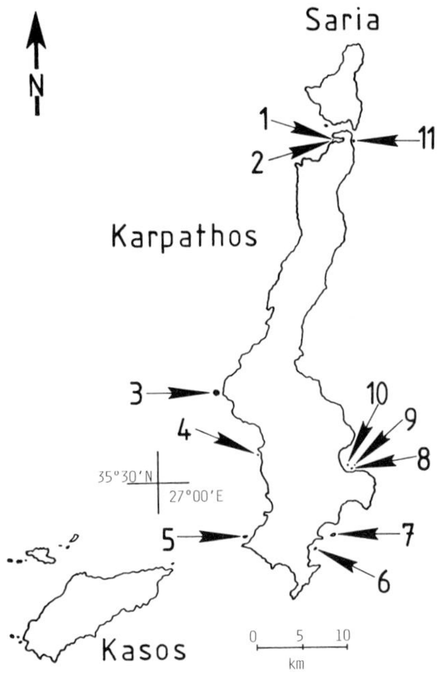

Fig. 1. Investigated islets near Karpathos (Dodecanesos, Greece). Numbers are used instead of names to designate the different islets. Generally known names exist for six islets only: No 3 = Sokastro, No 5 = Diakoptis, No 6 = Prasonisi, No 7 = Mira, No 9 = Despotiko, and No 11 = Ammoui.

Table 1. Plants confined to offshore islets in the Karpathos area ('islet specialists'). The numbers refer to the different islets as shown in Fig. 1.

	Plant name	Islets with positive records (disregarding sowing experiments)
	Salsola aegaea	6, 7, 9
Shrubs	*Suaeda vera*	3, 6
	Convolvulus oleifolius ssp. *scopulorum*	7, 8, 9
	Allium commutatum	6, 8, 9
Geophytes	*Asparagus stipularis*	3, 5, 6, 7, 8, 9
	Ornithogalum creticum	9
Biennials	*Lavatera arborea*	4, 7
	Silene holzmannii	2, 4, 6, 8
	Anthemis ammanthus ssp. *ammanthus*	3
	Anthemis ammanthus ssp. *paleacea*	7
Annuals	*Anthemis scopulorum*	2
	Hymenolobus procumbens ssp. *procumbens*	6, 7
	Trigonella rechingeri	2, 8, 10
	Orobanche sanguinea	8, 9

Table 2. Chasmophytic plant species confined to cliffs on Karpathos but growing outside of cliffs on small islands ('islet chasmophytes'). The numbers refer to the different islets as shown in Fig. 1.

	Plant name	Islets with positive records (disregarding sowing experiments)
	Achillea cretica	11
Shrubs	*Helichrysum orientale*	1
	Medicago arborea	4
Geophytes	*Allium bourgeaui* ssp. *bourgeaui*	4
Biennials	*Matthiola sinuata* ssp. *glandulosa*	4

laceae), a species that these four islets have in common and that has not so far been found elsewhere on or around Karpathos.

The area

Karpathos is the second largest island of the Dodecanesos. It is situated in the South Aegean Sea about halfway between Crete and Rhodes (coordinates see Fig. 1). The climate may be characterized as typically Mediterranean with a hot, dry summer and a cool, rainy winter. Normally rainfall is heaviest in January and February. The total amount of annual precipitation may vary considerably: Within a seven-year period (1971 to 1977) a maximum annual rainfall of 560 mm and a minimum of 230 mm were measured (Fischer 1980). There are only few days of calm weather in the course of the year. Strong winds are blowing almost permanently, mostly from the north-west but sometimes turning into warm southerly storms. Wind force frequently reaches values of 5–6 on the Beaufort scale (Fischer 1980).

The islets surrounding Karpathos mainly consist of calcareous rocks, sometimes overlain by sandstones (Christodoulou 1963). Rocky depressions are filled with clayey or sandy soil or with pebbles of varying size. Taking into consideration the present sea depths and the postulated values of eustatic lowering of the sea level during the past glaciations (Pfannenstiel 1951), as has been done by Greuter

(1970), one must conclude that all offshore islets of Karpathos have been fused to the main island up to the late Pleistocene. There is no evidence for post-glacial catastrophic events that might have impaired the islet floras. In particular, the influence of the eruption of the Santorin volcano (situated ca 180 km NW of Karpathos, eruption ca 1500 B.C.) on the eastern Mediterranean area should not be overrated (Olausson 1971).

Some important features of the islets I 2, I 4, I 6 and I 8 are listed in Table 3. The total number of vascular plant species for each of the islets (according to the floristic surveys in 1987) may be seen from Fig. 2.

Table 3. Some important features of four islets off Karpathos that are inhabited by *Silene holzmannii*. The islet numbers refer to Fig. 1.

	I 2	I 4	I 6	I 8
Maximum elevation	6 m	8 m	18 m	12 m
Size	ca 9000 m^2	ca 2500 m^2	ca 9500 m^2	ca 2000 m^2
Smallest distance from the mainland	ca 35 m	ca 50 m	ca 150 m	ca 90 m
Distribution of vegetation	Mainly restricted to the easterly slopes of the top area (ca 1000 m^2)	Mainly restricted to the easterly slopes of the top area (ca 800 m^2)	About evenly distributed all over the islet	About evenly distributed all over the islet
Exposure to winds and breakers	Fully exposed to north-westerly winds and breakers	Partly sheltered from north-westerly winds and breakers	Fully exposed to most winds but sheltered from north-westerly breakers	Partly sheltered from most winds and breakers
Presence of man and disturbing animals	Not observed	Not observed	Inhabited by a large colony of silver gulls; occasional visits by bird-shooters	Inhabited by a single rat (1986, 1987); occasional visits by fishermen

Material and methods

Contributors to the field work for this study have been W. Greuter, D. Höner, R. Pleger, S. Potthoff and Th. Raus (all Berlin-West). The islets have been visited at regular intervals since 1982 (additional visits to I 1, I 2 and I 11 in 1981) generally in April or May (rarely in June). This is the best time to find most annuals alive and fully developed. With the exception of the year 1984, when the islets were visited only in September/October (no visits to I 1, I 2 and I 11), autumn-flowering geophytes may often have been overlooked, but in some cases bulbs have been dug out or withered remains have been identified.

In this paper, observed changes in the composition of the flora are based exclusively on the presence or absence of aerial organs and disregard underground parts, in particular the seed-bank in the soil. Therefore the term 'apparent turn-over' sensu Nilsson & Nilsson (1983) is appropriate, to indicate the total amount of observed introductions and extinctions. All visitors to the islets have been careful to

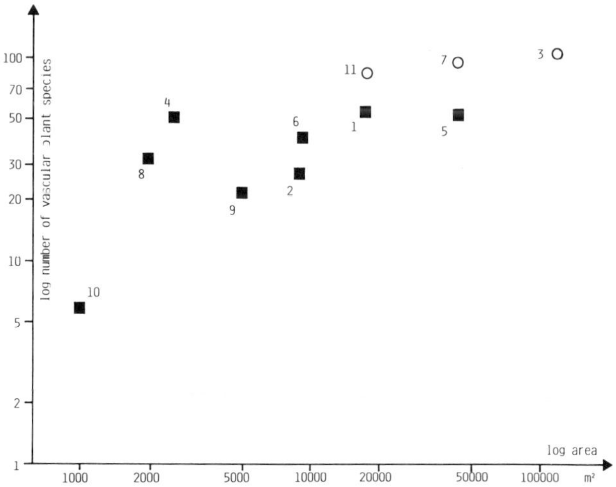

Fig. 2. Species-area relation of the investigated islets (according to the counts of 1987. The numbers refer to the different islets as shown in Fig. 1. Circles have been used for grazed islets, islets where no grazing has been observed are designated by black squares.

interfere as little as possible with the vegetation. Vouchers have been collected only to a very limited extent (all are deposited in the Herbarium of the Botanisches Museum Berlin-Dahlem-B). However, certain deliberate, controlled changes have been effected in order to test the probability of successful introductions of species. A total of 46 sowing experiments have been carried out on these four islets: Seeds or propagules of species that had been collected in the Karpathos area were sown and covered by soil in habitats suitable for germination on islets where these species did not occur previously. The list of these species comprises halophytes of the littoral zone, elements of the main vegetation type dominated by dwarf shrubs ('phrygana'), chasmophytic species and elements of the supralittoral zone (including islet specialists). Twenty-four different taxa have been tested using various amounts of seeds or propagules, from a single one up to unnaturally high quantities.

Whenever there was enough time available (which was usually the case) counts were made of the populations of the islet specialist *Silene holzmannii* (Caryophyllaceae). *Silene holzmannii* is a tall annual with somewhat weedy tendencies. It may occasionally reach a size of up to 50 cm and is particularly well suited for quantitative population studies since it may not easily be overlooked. Most of the capsules open very late and usually the dead remains of fruiting plants will remain upright for another year. More detailed information concerning the ecology of *Silene holzmannii* was presented by Greuter (1972). On I 4 an extirpation experiment was carried out in 1982, when all but three of the ca 250 individuals of the *Silene holzmannii* population were removed.

Results and discussion

Apparent turnover of species and sowing experiments

It has been emphasized before that the measure of apparent turnover only refers to observed floristic changes. Although we made every effort to achieve complete coverage by our floristic surveys, we cannot completely exclude that we might have overlooked one or the other of the present species in some cases so that the total number of recorded introductions is too low and/or the total number of recorded extinctions too high. Some of the islet specialists are minute indeed, such as *Hymenolobus procumbens* ssp. *procumbens*, a species that is so far only known from I 6, I 7 and a further islet near Kasos for the whole Karpathos island group (Karpathos, Kasos and Saria) but which, according to Greuter & Raus (1986: 105), 'may easily have been overlooked elsewhere due to its early flowering habit'. It was indeed found on sandy sea-shores on two places of the Nomos Argolis (Peloponnisos) and on the island of Gavdhos (S. of Crete) by Runemark (pers. comm.) but has not been observed on the numerous other islets in the Aegean investigated by Lund botanists, so that it is not an exclusive islet specialist throughout its Aegean range.

Figure 3 reveals that several species have been found to be temporarily missing on a given islet. Nearly all such cases concern micro-populations of annuals, so that three explanations are possible:

- extinction followed by introduction of the same species;
- successful survival of the species in the seed-bank of the soil, but without overground appearance in one or more of the vegetation periods;
- early or delayed germination (sprouting) of the species, so that it was overlooked or was indeed invisible at the time of the floristic survey (sampling error).

Considering the low success of sowing experiments that have been carried out on the islets (see below) the first explanation appears to be the least probable, whereas the assumption of 'pseudoturnover' sensu Lynch & Johnson (1974), as under the second and third explanation, is quite reasonable.

The apparent turnover rate so far observed on I 2 is very low as compared to those on the other islets within the same period of time. Two reasons may be responsible for this fact:

- No trace of permanent or frequent disturbances by man or animals (as has been the case on I 6 and I 8) has been found on I 2 and I 4. The pronounced effect of a silver gull colony on an islet flora is evidenced by the case of I 6 where a great number of introductions (mostly ruderals and

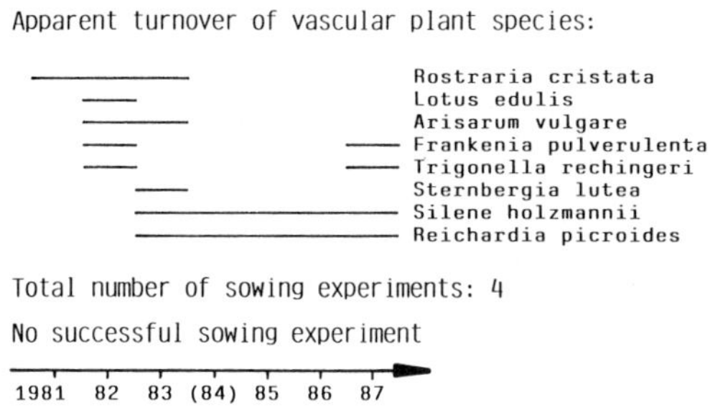

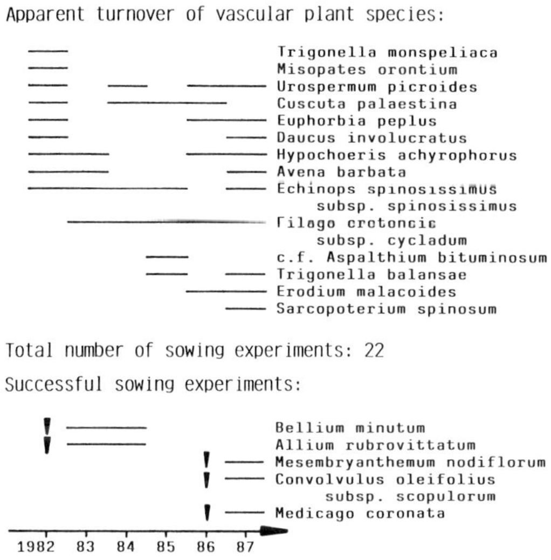

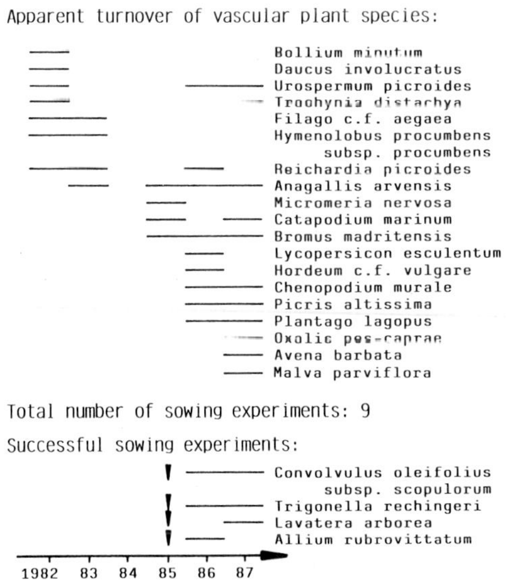

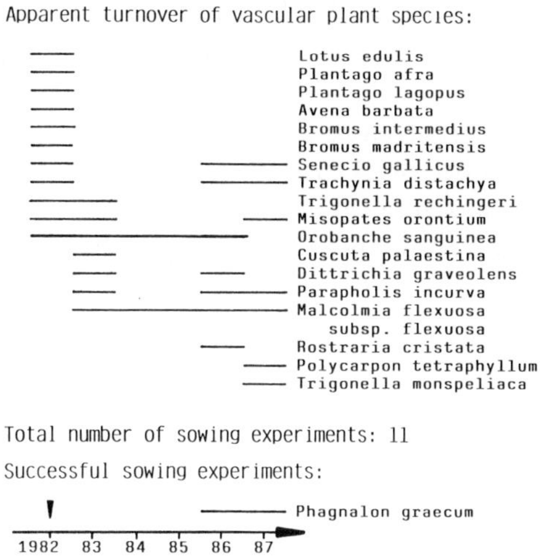

Fig. 3. Apparent turnover of plant species and introductions by sowing experiments on a) islet No 2, b) islet No 4, c) islet No 6 and d) islet No 8. Only species that have not been observed consistently have been listed. Presence in a given year is represented by a solid line, that is interrupted in years in which the species was not observed. (Note that the flora of islet No 2 has not been checked in 1984.) Successful sowing experiments are denoted by black arrowheads.

cultivated plants) has been recorded. Gillham (1961) emphasized the role of gulls in introducing seeds of alien annuals, and Abbott (1977) found significantly higher immigration rates of alien species on islets with gull rookeries. Raus (1986) exploring the flora of the volcanic island Nea Kaimeni in the Aegean also observed increased introductions of ruderals close to a gull colony.

As has been noted above, I 2 is fully exposed to the prevailing north-westerly winds and to salt spray. A comparatively small number of species inhabits this islet (see Fig. 2), suggesting that only a few species well adapted to the extreme maritime conditions are actual or potential colonizers of such an islet. This explanation would account for the fact that most of the changes that have been observed on the other three islets concern ruderals or phrygana elements, i.e., species that are not or only imperfectly adapted to the conditions of coastal habitats. I 4, which is similarly undisturbed by man and animals, is partly sheltered from wind and waves, and indeed it has a more diverse flora and a higher rate of apparent turnover.

The results of the sowing experiments can be summarized as follows: When using a small amount of seeds or propagules (up to 9) no germination or sprouting could be observed in any case. But even using unnaturally high amounts of diaspores (up to a few hundred at one time and in a single place) success was far from overwhelming. The best success has been noted on I 6, which has a rather undersaturated flora in terms of species/area ratio, particularly considering its elevation and sheltered position. The two initially successful introductions of *Allium rubrovittatum* followed by extinction after one (I 6) or two years (I 4) reveal that germination or sprouting of an introduced species by no means guarantees a successful colonization in the long run (establishment). Actually, only those of our experimental introductions are at all likely to result in successful establishment that concern perennials, i.e., *Phagnalon graecum* (a dwarf shrub of rocky places and cliffs) on I 8 and the islet specialist *Convolvulus oleifolius* ssp. *scopulorum* on I 4 and I 6.

Dynamics of the populations of Silene holzmannii

The results are summarized in Fig. 4. On I 2 *Silene holzmannii* was unknown prior to 1983 when five individuals were counted. Experimental studies have shown that the capsules of *Silene holzmannii* are buoyant for up to several weeks and that the germination rate decreases only slowly under salt-water treatment, which is suggestive of possible long-range dispersal by sea currents (Greuter 1972; Potthoff in prep.). However, sea spray is unlikely to reach the site where this species has first been noticed on I 2, so that we now believe that birds were most likely the vectors for this introduction. Abbott & Black (1980) have pointed out that islands with sand and gravel beaches (lacking on the islets off Karpathos) are the ones that are likely to be reached by propagules over the sea. The *Silene holzmannii* population on I 2 grew steadily until 1986 when an explosive increase took place. Extraordinary propagation rates could also be observed on I 4 and I 6 – but not on I 8 – where a rat bit off nearly all the capsules of this species thereby reducing the population from 42 individuals (1985) to a single one (1987). The rapid decline on I 8 shows an extreme vulnerability of *Silene holzmannii* with respect to grazing. The parallel increases of the populations observed on the other three islets, situated at considerable distances from each other and covered by quite different vegetation types, have obviously been the consequence of unusually favourable weather conditions. Native inhabitants of Karpathos have told that autumn and early winter (1986/87) had been abnormally calm and dry. Another islet specialist that similarly profited from the unusual weather conditions is the shrubby *Convolvulus oleifolius* ssp. *scopulorum* of which more than 10000 seedlings per islet have been found in 1987 (I 8 and I 9).

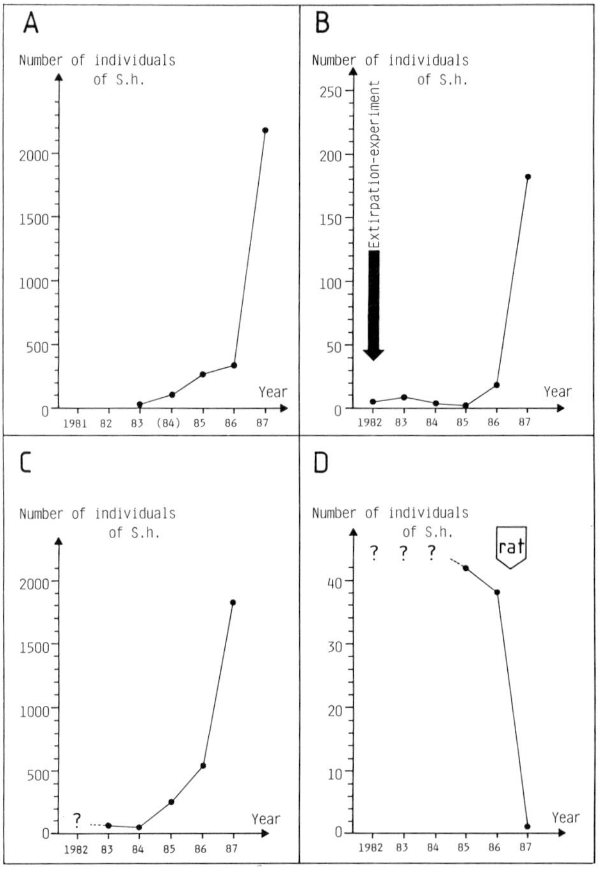

Fig. 4. Population dynamics of *Silene holzmannii* (S. h.) on a) islet No 2, b) islet No 4, c) islet No 6 and d) islet No 8. Note: On islet No 2 the population size for 1984 was determined by counting the withered fruiting inflorescenses in 1985.

The experimental extirpation of *Silene holzman-*

nii on I 4 first resulted in an astonishingly stable micro-population that did not exceed seven individuals over a period of four subsequent years. Due to high germination rates and lack of dormancy (Greuter 1972; Potthoff in prep.) the seed reserve in the soil must be virtually nil, a fact that might easily result in the extinction of the species in the event of its being decimated by, e.g., grazing.

Micro-populations and mass-populations

As in the case of *Silene holzmannii* on I 4, micro-populations of several species have been observed to maintain virtual stability in number and distributional pattern through several years. Often such micro-populations are found in isolated depressions of the rock filled with clayey soil or in the wind-screen provided by big stones. Such clay pans are of considerable interest in that they constitute in a way islets on the islet, their surroundings of bare rock obviously functioning as a kind of isolation barrier. In these sheltered places propagules of species already present on the spot accumulate, whereby an invasion by other species becomes more difficult. These micro-habitats present obvious analogies with cliff habitats found on the main island of Karpathos, and it is not too surprising to find that a number of islet chasmophytes form sizeable populations on some of the small islands (Table 2).

The principle of the first colonizing species being favoured regardless of their competitive properties has been called 'contingent exclusion' by Shmida & Ellner (1984: 42). This principle probably accounts for the stability of isolated micro-populations as well as for the build-up and maintenance of mass-populations that often dominate the vegetation of small islands. Examples of mass-populations on islets off Karpathos are provided by the two genera *Allium* and *Anthemis*. Three species of *Allium* tend to form mass-populations, viz., *Allium rubrovittatum* (an element of rocky, open places) on I 2, *Allium commutatum* (an islet specialist) on I 6, I 8 and I 9, and *Allium bourgeaui* ssp. *bourgeaui* (an islet chasmophyte) on I 4. Four taxa of *Anthemis* show the same strategy, viz., *Anthemis ammanthus* ssp. *ammanthus* (an islet specialist) on I 3, *Anthemis ammanthus ssp. paleacea* (an islet specialist) on I 7, *Anthemis rigida* (a common supralittoral element) on I 4 and *Anthemis scopulorum* (an islet specialist) on I 2. In none of these cases we have ever noticed a natural co-occurrence of two or more species of the same genus, nor has the establishment of additional taxa of the same genus by experimental sowing been successful so far. Even distances of a few metres separating an islet from the main island obviously constitute effective barriers to colonization. A big proportion of the islet specialists and islet chasmophytes has been found to form stable mass-populations on the offshore islets around Karpathos. Similar observations were made by Lund botanists on islets of the Central Aegean over a period of more than 14 years (Snogerup & Snogerup 1987).

Concluding remarks

The distributional limitation of certain species to small islands is a phenomenon that has been first reported from the Aegean area, where it is widespread and characteristic. However, islet specialists have meanwhile been shown to occur elsewhere. Abbott & Black (1980) recorded *Hymenolobus procumbens* and *Lavatera arborea* on small islands near Perth (Western Australia), Stearn (1984: 178) observed mass-populations of *Allium commutatum* that covered more than 50% of the ground on three Pityusic islets (Spain), and Lanza & Poggesi (1986) exploring 96 different islets off Corsica in the Mediterranean Sea found *Hymenolobus procumbens* on three islets, *Allium commutatum* on 28, and *Lavatera arborea* on 37 islets.

Some of such species are endemic to small areas, such as *Euphorbia margalidiana* (Kuhbier 1978) and *Daphne rodriguezii* (Lucas & Synge 1978: 525) in the Balearic islands (Spain), or *Silene velutina* (Lanza & Poggesi 1986), *Buphthalmum inuloides* (Valsecchi 1977) and *Asperula deficiens* (Diana-Corrias 1980) around Corsica and Sardinia (France/Italy). Similarly, a great number of Aegean islet specialists as well as elements of the characteristic cliff flora are endemic to this area, and they are supposed to be remnants of a competitively weak flora that occurred around the Sea of Crete in the Pliocene (Runemark 1969). The ecological factors that characterize the cliff habitat, viz., absence of a well-developed soil layer, severe drought and extreme daily temperature ranges (Snogerup 1971) also influence the open vegetation of islets. Reduction or absence of grazing is another important factor that is common to vertical cliffs and the smallest islands. However, the obviously most important common feature of islets and cliffs is the low level of interspecific competition. According to Carlström (1986: 87) the same is true for open serpentine communities on Rhodes (Greece) and on the Marmaris peninsula (Turkey), where she found a number of usually strictly chasmophytic plants to occur outside of cliffs. One of the reasons why competition on small islands plays a minor role might be the low flux of foreign propagules that is a consequence of the surface compartmentation on the islets (already mentioned before); another reason is the under-saturation of the flora. Whitehead & Jones (1969) explain the notoriously low species numbers on small islands by the restricted pool of species that may find suitable habitats on them, and, in addition, small plant populations (such as are expected to occur frequently on islets) are exposed to a great risk of random elimination (Runemark 1969). Recurrent environmental changes will select for species with a broad ecological amplitude that are able to conquer new space quickly if the conditions are favourable. This has been evidenced for the islet specialist *Silene holzmannii*, and might quite likely also be the case for other small-island species.

Acknowledgements

We gratefully acknowledge the support by the Deutsche Forschungsgemeinschaft

References

Abbott, I., 1977. Species richness, turnover and equilibrium in insular floras near Perth, Western Australia. Austral. J. Bot. 25: 193–208.

Abbott, I. & Black, R., 1980. Changes in species composition of floras on islets near Perth, Western Australia. J. Biogeogr. 7: 399–410.

Barclay, C., 1986. Crete – checklist of the vascular plants. Englera 6.

Carlström, A., 1986. The phytogeographical position of Rodhos. Proc. Roy. Soc. Edinburgh 89B: 79–88.

Christodoulou, G., 1963. Geological map of Greece – sheet Karpathos 1:50000. Institute for Geology and Subsurface Research, Athens.

Diana-Corrias, S., 1980. Le piante endemiche della Sardegna: 74–77. Boll. Soc. Sarda Sci. Nat. 19: 289–309.

Fischer, U., 1980. Karpathos als Besonderheit unter den Ägäischen Inseln. Exkursionsber. Inst. Geogr. Univ. Salzburg 6: 175–200.

Gillham, M. E., 1961. Alteration of the breeding habitat by seabirds and seals in Western Australia. J. Ecol. 49: 289–300.

Greuter, W., 1970. Zur Paläogeographie und Florengeschichte der südlichen Ägäis. Feddes Repert. 81: 233–242.

Greuter, W., 1972. L'écueil à *Silene holzmannii,* en Crète, et son peuplement végétal. Saussurea 3: 157–166.

Greuter, W., Burdet, H. & Long, G. (eds), 1984, 1986. Med-Checklist. A critical inventory of vascular plants of the circum-mediterranean countries, Vol. 1 and 3. Genève.

Greuter, W. & Pieper, H., 1975. Notiz zur Biogeographie der landfernen südägäischen Klippe Avgo. Candollea 30: 7–11.

Greuter, W., Pleger, R. & Raus, Th., 1983. The vascular flora of the Karpathos island group (Dodecanesos, Greece). A preliminary checklist. Willdenowia 13: 43–78.

Greuter, W. & Raus, Th. (eds), 1986. Med-Checklist Notulae, 13. Willdenowia 16: 103–116.

Gustafsson, M. & Snogerup, S., 1974. Studies in the Aegean Flora XXII. The flora of the island of Skantzoura. Bot. Not. 127: 364–372.

Kuhbier, H., 1978. *Euphorbia margalidiana* – eine neue Wolfsmilch-Art von den Pityusen (Balearen/Spanien). Veröff. Überseemus. Bremen Reihe A, 5: 25–37.

Lanza, B. & Poggesi, M., 1986. Storia naturale delle isole satelliti della Corsica. L'universo 66. Firenze.

Lucas, G. & Synge, H. (eds), 1978. The IUCN plant red data book. IUCN, Morges, Switzerland.

Lynch, J. F. & Johnson, N. K., 1974. Turnover and equilibria in insular avifaunas, with special reference to the California Channel Islands. Condor 76: 370–384.

Nilsson, S. G. & Nilsson, I. N., 1983. Are estimated species turnover rates on islands largely sampling errors? Amer. Nat. 121: 595–597.

Olausson, E., 1971. Tephrochronology and the late Pleistocene of the Aegean Sea. Opera Bot. 30: 29–39.

Pfannenstiel, M., 1951. Quartäre Spiegelschwankungen des Mittelmeeres und des Schwarzen Meeres. Vierteljahresschr. Naturf. Ges. Zürich 96: 81–102.

Raus, Th., 1986. Floren- und Vegetationsdynamik auf der Vulkaninsel Nea Kaimeni (Santorin-Archipel, Kykladen, Griechenland). Abh. Landesmus. Naturk. Münster Westfalen 48: 373–394.

Rechinger, K. H., 1951. Phytogeographia aegaea. Akad. Wiss.

Wien Math.-Naturwiss. Kl. Denkschr. 105 (2, 2).
Runemark, H., 1969. Reproductive drift, a neglected principle in reproductive biology. Bot. Not. 122: 90–129.
Shmida, A. & Ellner, S., 1984. Coexistence of plant species with similar niches. Vegetatio 58: 29–55.
Snogerup, S., 1971. Evolutionary and plant geographical aspects of chasmophytic communities. In: Davis, P. H., Harper, P. C. & Hedge, L. C. (eds), Plant life of South-West Asia: 157–170. Botanical Society of Edinburgh, Edinburgh.
Snogerup, S., von Bothmer, R. & Gustafsson, M., 1980. Floristic report from the island of Psathura (Greece). Bot. Not. 133: 145–148.
Snogerup, S. & Snogerup, B., 1987. Repeated floristical observations on islets in the Aegean. Pl. Syst. Evol. 155: 143–164.
Stearn, W. T., 1984. The genus Allium (Alliaceae) on the Pityusic Islands. In: Kuhbier, H., Alcover, J. A. & Guerau d'Arellano Tur, C. (eds), Biogeography and ecology of the Pityusic Islands: pp. 171–184. The Hague, Boston, Lancaster.
Valsecchi, F., 1977. Le piante endemiche della Sardegna: 8–11. Boll. Soc. Sarda Sci. Nat. 16: 295–313.
Whitehead, D. R. & Jones, C. E., 1969. Small islands and the equilibrium theory of insular biogeography. Evolution 23: 171–179.

Vegetatio 77: 139–147, 1988
© Kluwer Academic Publishers, Dordrecht – Printed in the Netherlands

Vascular plant colonization and vegetation development on sea-born volcanic islands in the Aegean (Greece)

Th. Raus
Botanischer Garten und Botanisches Museum Berlin-Dahlem, Königin-Luise-Straße 6–8, D-1000 Berlin 33, Federal Republic of Germany

Accepted 22.8.1988

Keywords: Extinction, Flora, Invasion, Island biogeography, Mediterranean, Pseudoturnover, Santorini

Abstract

Inside the seawater-filled Santorini caldera (S Cyclades) the uninhabited islands Palea Kaimeni (PK) and Nea Kaimeni (NK) have been formed by submarine/subaerial activity. About 197 B.C. PK (0.54 km^2) emerged, harbouring (in 1987) 178 vascular plant species, and covered by a mosaic of evergreen sclerophyllous and halophytic shrub communities and short-lived therophytic plant communities. NK, actually colonized by 156 vascular taxa, emerged ca 1570 A.D., its surface having enlarged to 3.44 km^2 owing to four subsequent eruptive phases of the island's volcano until 1950. At present, the most evolved vegetation type on NK is a steppe-like pioneer community rich in grasses and legumes apart from some *Ficus carica* trees which survived this century's volcanic eruptions. The beginnings of shrub vegetation on NK, originating from PK's consolidated woody canopy, are documented and evidence is shown of several ways of zoogenic and anthropogenic import of new vascular plants to the island. Comparative plant censuses demonstrate that permanent immigration of plants into NK overbalances by far the loss of species caused by the repeated destructions of the island's plant cover. Cases of species turnover (verified extinctions and invasions), of pseudoturnover (due to sampling errors), of long distance dispersal and of successful stabilization of founder-populations are exemplified.

Nomenclature: Raus, Th. 1986.

Introduction

Sea-born volcanic islands offer the opportunity to record the colonization by organisms of a primary substrate (primary succession) and to observe the subsequent population dynamics and species turnover of plants and animals. The Krakatau explosion of August 26, 1883 and the later formation of Anak Krakatau stimulated this kind of investigation (Backer 1929; Ernst 1934; Docters van Leeuwen 1936; Flenley & Richards 1982). The most recent and best monitored case is the island of Surtsey, S of Iceland, emerged on November 15, 1963 (Guenther & Klug 1970; Fridriksson 1975). In contrast, similar events in the Santorini island group (Kikladhes, Greece) have been neglected by biologists. No systematic programmes of botanical and zoological field studies have ever been set up there and many basic data on the flora and fauna are lacking (see Schmalfuss *et al.* 1981). This paper brings together what is known about the developing new flora and vegetation under isolation of these islands in a semi-arid mediterranean climate.

Study area

Topography and environmental factors

The Santorini archipelago is the southernmost island group of the Cyclades, situated in the Aegean Sea 260 km SE of Athens and 120 km N of the island of Crete. Origin and present configuration of its five islands (Fig. 1) are the result of repeated volcanic activities (see Pichler & Friedrichs 1980 for details). Around 1500 B.C. a former single island (ca 30 km diam.) exploded and a caldera was formed. The caldera wall was perforated and the caldera became filled with seawater. Additional collapsing of the caldera wall left three islands forming a ring, viz. Thira (75 km^2, 568 m alt.), Thirasia (9.4 km^2, 295 m alt.) and Aspronisi (0.13 km^2, 71 m alt.). About 1300 years after the big eruption, a magmatic refilling of the caldera began and in the central part of the caldera two sea-born islands grew up. These are called the Kaimeni Islands (orthography following Pichler & Kussmaul 1980; elsewhere given as 'Kammeni' Islands, i.e. the burnt islands), namely Palea Kaimeni (i.e. the old one) and Nea Kaimeni (i.e. the young one). Palea Kaimeni (PK) emerged in 197 B.C. It covers an area of 0.54 km^2 today after having been enlarged, in 1866 A.D., by ascending submarine lava, and is 103 m high. Nea Kaimeni (NK) achieved its present shape through eight eruptive phases between ca 1570 A.D. and 1950 A.D. (Fig. 2). It is thus more than 1800 yr younger than PK and it came into existence as a single island (ca 2 km diam.) only after August 17, 1925, when the two former islands Mikra Kaimeni (emerged ca 1570) and Old Nea Kaimeni (emerged from 1707 onwards) were united by a large lava-flow (Washington 1926; Reck 1936). The most recent eruption took place in 1950; it covered a part of the older surfaces with fresh ashes. At present, NK covers an area of 3.44 km^2 and reaches an altitude of 124 m.

Only lithosols, at most exhibiting immature A-C-profiles, are developed on the Kaimenis. In general, raw lava-blocks with crevices, and coarse ashes serve as the substrate for higher plants. On both islands there is a complete lack of permanent surface water and freshwater springs. A considerable part of the annual rainfall gets lost by runoff (Schmalfuss *et al.* 1981: p. 8). The macroclimate is of a dry-mediterranean type with a mean temperature of ca 25 °C in the hottest- and of ca 11 °C in the coldest

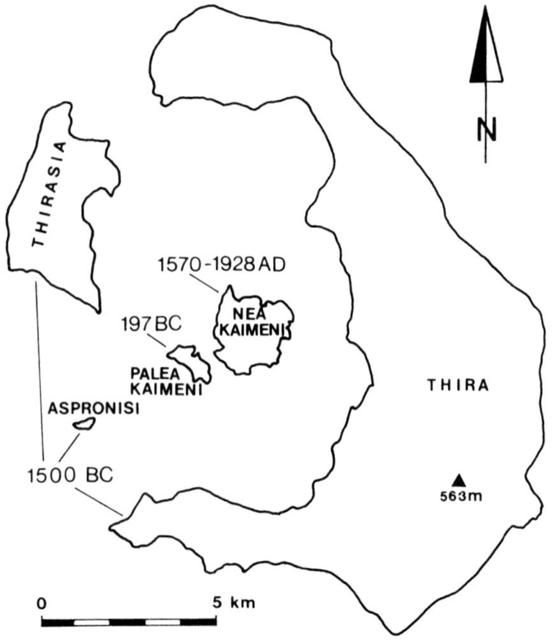

Fig. 1. Topography of the Santorini archipelago (after Schmalfuss *et al.* 1981). Years indicate time of origin of actual land-surfaces (after Pichler & Kussmaul 1980).

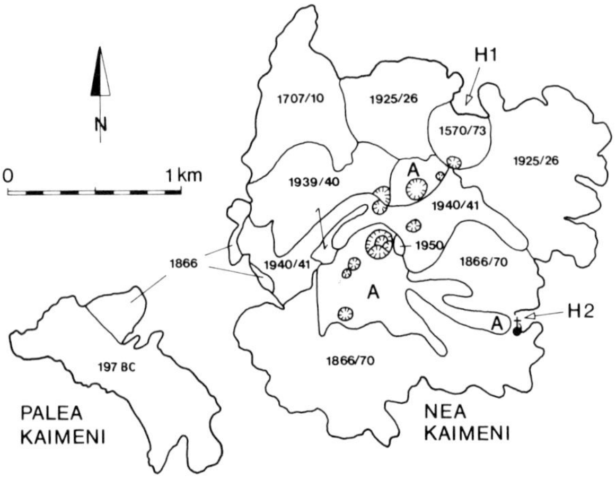

Fig. 2. Geological arrangement of the Kaimeni Islands (after Pichler & Kussmaul 1980). Years indicate time of origin of actual land-surfaces. A = young ashes of 1939/41 and 1950, covering older lava-flows. H1 = northern harbour. H2 = eastern harbour with chapel.

month, in general without frost in winter. Mean precipitation is 357 mm/yr only and the dry season lasts from April to October (see the Walter-Lieth climate diagram in Schmalfuss *et al.* 1981: p. 7). Frequent overheating of the mostly black-coloured surfaces of the volcanic substrate is probably the most outstanding feature of the microclimate. It is a stress factor for sessile organisms like plants, especially seedlings (Richter 1984). In addition, steady winds in summer increase evaporation.

Both islands are uninhabited and actually ungrazed, but grazing had happened regularly before Easter on PK until 1979 (local comm.). On NK, some still active fumaroles attract thousands of tourists every year; they enter the island in a small harbour on its north coast for a walk to the island's top and back to the harbour again. A second harbour on the east coast is frequented by local fishermen only who erected a chapel there, which is the only building on NK. A similar chapel was built at the north coast of PK.

Vascular flora

From the ring-islands 548 species of vascular plants are known, 50 of them still unpublished (leg. Raus, vouchers at Berlin). PK harbours 178 vascular plant species (110 unpublished). 156 species are reported from NK (Raus 1986, and unpubl.). These figures are cumulative and include a number of old records which have not been reconfirmed by later collectors (Hansen 1971). The number of taxa of higher plants, now known to have been recorded as occurring on the Santorini archipelago as a whole amounts to about 562 (checked misidentifications excluded). This confirms the statements of Rackham (1978: p. 759) and Schmalfuss & Schawaller (1984) that the flora and fauna of Santorini is not at all impoverished as might be supposed in view of the local history of strong volcanic devastations (Diapoulis 1971: p. 240). Taking the Kaimenis as a whole, 94.5% of the actual vascular flora occur on the ring-islands, too, and are supposed to have invaded from there. 5.5% do not occur on the ring-islands. These are supposed cases of long distance dispersal, among them four pteridophytes, *Asplenium obovatum, A. onopteris, Ophioglossum lusitanicum* and *Polypodium cambricum* (their spores being transported by winds), and two littoral seed-plants, viz. *Cymodocea nodosa* and *Ruppia cirrhosa* (their seeds being transported by seawater-currents).

The percentage of annuals increases with decreasing age of land-surface available for plant colonization (see Table 1). After the big explosion of 1500

Table 1. Statistics on the vascular flora of the Santorini archipelago.

	Ring islands	Palea Kaimeni	Nea Kaimeni
Maximal age of land-surface (prior 1985)	c. 3500 yr	c. 2180 yr	c. 415 yr
Volcanic impact (after 1900)	no	no	very strong
Human impact (after 1900)	strong	negligible	locally moderate
Size of area	84.53 km^2	0.54 km^2	3.44 km^2
Number of vascular plant species (cumulated)	548	178	156
Annual herbs	65.6%	72.2%	80.3%
Biennial and perennial herbs and dwarf shrubs	31.1%	25.0%	16.9%
Woody species	3.3%	2.8%	2.8%

B.C. the ring-islands were covered with a thick layer of pumice and ashes, locally many m deep, burying and wiping out the pre-existing plant cover. Thus, modern vegetation and most of the present flora inhabit the ring-islands for at most 3500 yr and are the result of a nearly complete resettlement of vascular plants in historical times. A few bulbous species and chasmophytes of rock-crevices may have been able to escape the catastrophe (Rackham 1978: p. 760; see also Schmalfuss & Schawaller 1984: p. 14, for certain cases of invertebrates).

Vascular plant colonization on Nea Kaimeni

Plant censuses and floristical changes

The first vascular plant census on NK was done in April 1911 by F. Vierhapper *et al.* (see Vierhapper 1914, 1919 for details), about 40 yr after the vegetation-destroying eruptions in 1866–1870. In 1911 (Table 2, col. V), Vierhapper counted 83 vascular plant species on Mikra Kaimeni and Old Nea Kaimeni which then still consisted of two separate islands besides PK (see above). Cammerloher (1935; Table 2, col. C) repeated the census in 1933, on the now united NK, and encountered only 24 species of higher plants, i.e. less than a third of the stock of 1911. The plant cover had been destroyed nearly completely by volcanic events in 1925–1928. Hansen (Table 2, col. H) collected 69 species on NK in April 1967, among them 18 new records, while 31 species mentioned by Vierhapper and Cammerloher were missing now (Hansen 1971). This loss may be due to the destruction of the island's plant cover during the outbursts of 1939/41 and in 1950. Alas, detailed plant censuses concerning the floristic effects of these phases of volcanic activity on NK are lacking. After Hansen's count of 1967, only scattered and incomplete observations are obtainable (Anagnostidis & Economou-Amilli 1978; Diapoulis 1971; Rackham 1978; Steidel unpubl., vouchers at STU!).

In 1984 (Table 2, col. R 1) I undertook a complete census again, and found 118 species of vascular plants on NK (Raus 1986, with *Lathyrus cicera* and *Sonchus oleraceus* added, and *Scrophularia heterophylla* replaced by *S. lucida* ssp. *filicifolia*). There had been no volcanic impact on the island's plant cover since Hansen's census and 35 species, not enumerated by previous authors, immigrated on to NK obviously after 1967.

While all previous inventories including my own took place in spring, I now visited NK on September 23 and 24, 1986, in order to judge possible sampling errors caused by seasonal omissions. At the end of the same growing-period, on April 7, 1987 (Table 2, col. R2), I obtained the most recent figure of 130 vascular plant species through a nearly complete census. Actually, 11 species of the Vierhapper census of 1911 have still not been found again, whereas 74 species have newly arrived on NK since then, 52 among them with repeatedly confirmed success of establishment. In other words, permanent immigration of vascular plants into NK overbalances by far the loss of species caused by repeated volcanic destructions of the local plant cover.

Extinctions

Atriplex prostrata, Cakile maritima, Frankenia hirsuta, Medicago litoralis, Spergularia salina and *Suaeda maritima,* enumerated by Vierhapper only (Table 2, col. V), form an ecologically defined group of coastal species indicating that around 1911 a coastal sandy habitat occurred on NK which was probably destroyed during the volcanic eruptions of 1925/28. (See further Cammerloher 1935.) Some further comments may be presented on evident extinctions which occur in the process of vascular plant colonization of NK and which are not caused by direct thermal or mechanical volcanic impact.

In 1987 I revisited most of the small populations of species which had been recorded on the island for the first time three years before. Seven species, brought in by birds or man (Raus 1986: p. 386), were not found again in their known localities (Table 2, col. R2, -signs). For example, two rich-flowering individuals of *Fumaria judaica* had been imported 1984 by means of weed-contaminated waste earth of ornamental flower-pots which fishermen annually arrange around the chapel of the eastern harbour (Fig. 2: H1). These plants evidently failed in propagating and expanding form this spot by seed

Table 2. Checklists of vascular plant censuses on Nea Kaimeni since 1911 (nomenclature following Raus 1986: 377 – 381; four previous misidentifications, given in inverted commas, corrected). Single entries, given by a pair of letters, must not be added to the column in which they appear (see text).

	V	C	H	R1	R2
Pteridophyta					
Polypodiaceae					
Adiantum capillus-veneris	+	.	.	.	+
Anogramma leptophylla	+	.	+	+	+
Asplenium ceterach	.	.	.	+	+
!A. onopteris	.	.	.	.	+
Cheilanthes maderensis	.	.	.	+	+
Ch. pteridioides	+	+	.	+	+
Pteridium aquilinum	+	.	+	.	+
Gymnospermae					
Ephedraceae					
Ephedra campylopoda	.	.	.	+	+
Dicotyledones					
Anacardiaceae					
!Pistacia lentiscus	.	.	.	.	+
Boraginaceae					
Echium angustifolium	+	.	.	.	.
E. arenarium	.	.	.	+	?
Myosotis incrassata	+	.	.	+	+
Caryophyllaceae					
Cerastium comatum	+	.	+	+	+
C. semidecandrum	+		+	+	+
Paronychia echinulata	.	.	Ra	.	.
Polycarpon tetraphyllum s.l.	+	+	+	+	+
Sagina apetala	+	.	+	+	+
S. maritima	.	.	.	+	+
Silene colorata	.	.	+	+	+
S. gallica	+	.	Dp	+	+
S. sedoides	+	.	.	+	+
Spergula arvensis	.	.	.	+	+
Spergularia salina	+	.	.	.	.
Stellaria pallida	.	.	.	+	+
Chenopodiaceae					
Atriplex halimus	+	.	.	+	+
A. prostrata	+	.	.	.	.
!Chenopodium murale	.	.	.	.	+
Suaeda maritima	+	.	.	.	.
Compositae					
!Aetheorhiza bulbosa * microcephala	.	.	.	.	+
Andryala integrifolia	+	+	+	+	+
Anthemis rigida * rigida	.	.	.	+	–
Bellium minutum	.	.	.	+	+
Calendula arvensis	+	.	.	.	.
Carduus pycnocephalus	.	.	St	+	+
Carlina corymbosa * graeca	+	.	+	+	+
Chondrilla juncea	.	.	+	+	+
Chrysanthemum coronarium	.	.	St	+	+
!Conyza bonariensis	.	.	.	.	+
Crepis foetida	.	.	.	+	?
C. multiflora	+	+	+	+	+
Dittrichia graveolens	.	.	.	+	+
D. viscosa	.	.	+	+	+
Echinops spinosissimus	+	+	Dp	+	+
Filago contracta (F."pygmaea")	.	.	+	.	.
F. cretensis * cretensis	.	.	+	+	+
F. eriocephala (F."vulgaris")	.	.	.	+	+
F. gallica	+	.	+	+	+
Hedypnois rhagadioloides	.	.	Dp	+	–
Helichrysum italicum	+	+	+	+	+
Hypochoeris achyrophorus	.	.	+	.	+
H. glabra (H."radicata")	+	+	+	+	+
Phagnalon graecum	+	.	+	+	+
Reichardia picroides	+	.	+	+	+
Senecio gallicus	+	+	+	+	+
S. vulgaris	+	.	+	+	+
Sonchus asper * glaucescens	.	.	.	+	–
S. oleraceus	.	.	.	+	?
S. tenerrimus	+	.	.	+	+
Tolpis barbata	+	.	Dp	+	+
Urospermum picroides	+	.	+	+	+
Convolvulaceae					
Cuscuta palaestina	+	.	.	+	+
Crassulaceae					
Crassula alata	+	.	+	+	+
Sedum litoreum	+	.	Dp	+	+
Umbilicus horizontalis	+	.	.	.	.
U. rupestris	+	+	+	+	+
Cruciferae					
Biscutella didyma	+	.	+	+	+
Brassica tournefortii	.	.	.	+	+
Cakile maritima	+	.	.	.	.
Clypeola ionthlaspi	+	.	.	+	+
Hirschfeldia incana	.	.	+	+	?
!Malcolmia chia	.	.	.	.	+
Matthiola sinuata * glandulosa	.	.	.	+	+
Raphanus raphanistrum	.	.	St	.	.
Sisymbrium orientale	+	.	.	.	.
Euphorbiaceae					
Euphorbia peplus	+	.	+	+	+
Mercurialis annua	.	.	.	+	+
Frankeniaceae					
Frankenia hirsuta	+	.	.	.	.
Geraniaceae					
Erodium cicutarium	+	+	+	+	+
E. neuradifolium	+	.	.	+	+
Geranium molle	+	.	+	+	+
Labiatae					
Ballota acetabulosa	.	.	.	+	+
Leguminosae					
Lathyrus aphaca	.	.	+	+	+
L. cicera	.	.	St	+	?
L. clymenum	+	.	.	+	+
Lupinus angustifolius	+	+	+	+	+
Medicago litoralis	+	.	.	.	.
Melilotus indica	+	.	.	.	.
Ononis diffusa	+	.	.	.	.
Ornithopus compressus	+	.	+	+	+
O. pinnatus	+	.	.	+	+
Trifolium arvense	+	(+)	+	+	+
T. campestre	+	.	+	+	+
T. glomeratum	.	.	.	+	+
T. infamia-ponertii	.	.	.	+	+
T. scabrum	+	.	+	+	+
T. stellatum	+	.	.	.	+
T. suffocatum	.	.	.	+	+
T. tomentosum	.	.	+	+	+
!Vicia cretica * cretica	.	.	.	.	+
V. lathyroides	+	.	+	+	+
!V. sativa * nigra	.	.	.	.	+
!V. villosa * eriocarpa	.	.	.	.	+
Malvaceae					
!Lavatera cretica	.	.	.	.	+
!Malva parviflora	.	.	.	.	+
Moraceae					
Ficus carica	+	+	+	+	+
Orobanchaceae					
Orobanche pubescens	.	.	.	+	+
Papaveraceae					
Fumaria judaica	.	.	.	+	–
Plantaginaceae					
Plantago afra	.	.	.	+	?
P. arenaria (=P. indica)	.	.	+	.	.
P. weldenii	.	.	+	+	+
Polygonaceae					
Rumex bucephalophorus * aegaeus	+	+	+	+	+
!R. tuberosus * creticus	.	.	.	.	+
Rubiaceae					
Galium aparine	.	.	.	+	+
G. recurvum	+	.	+	+	+
Valantia hispida	+	.	+	+	+
Scrophulariaceae					
Linaria pelisseriana	+	.	+	+	+
Misopates orontium	+	.	+	+	+
Scrophularia heterophylla	+	.	+	.	+
S. lucida * filicifolia	+	+	Dp	+	+
Solanaceae					
Solanum nigrum	.	+	+	+	+
Tamaricaceae					
Tamarix parviflora (cult.)	.	.	+	+	+
Umbelliferae					
Daucus involucratus	.	.	.	+	?
!Tordylium apulum	.	.	.	.	+
Torilis leptophylla	.	.	+	+	?
Urticaceae					
Parietaria cretica	+	+	+	+	+
P. judaica	.	.	.	+	–
Urtica urens	.	.	.	+	–
Monocotyledones					
Gramineae					
Aira cupaniana	.	.	.	+	+
A. elegantissima	.	.	.	+	+
Avellinia michelii	+	.	+	+	+
Avena barbata	+	+	+	+	+
A. sterilis	.	.	An	.	.
Briza maxima	+	.	.	+	+
Bromus haussknechtii	.	.	.	+	+
B. intermedius	.	.	+	+	+
B. madritensis	+	.	+	+	+
B. rigidus * ambigens	+	.	.	+	+
B. rubens	+	+	Dp	+	+
B. tectorum	+	.	+	+	+
Corynephorus articulatus	+	+	+	+	+
Holcus setiglumis	.	.	.	.	Dm
Hordeum leporinum	+	+	+	+	+
Hyparrhenia hirta	+	.	+	+	+
Lagurus ovatus	+	.	+	+	+
Lamarckia aurea	+	.	+	+	+
!Lolium rigidum * lepturoides	.	.	.	.	+
L. rigidum * rigidum	.	.	.	+	?
Psilurus incurvus	+	.	+	+	+
Rostraria cristata	+	+	+	+	+
Stipa capensis	.	.	+	+	+
Trachynia distachya	.	.	+	+	+
Vulpia ciliata	+	.	+	+	+
V. fasciculata	+	.	+	+	+
V. muralis (V."bromoides")	.	.	+	.	+
Liliaceae					
Aloe vera (cult.)	.	.	.	+	–
Asparagus aphyllus	.	+	Dp	.	.
A. stipularis	+	+	+	+	+

Signs and abbreviations:

* = subspecies
V = plant census 1911 (Vierhapper 1914, 1919)
C = plant census 1933 (Cammerloher 1935)
H = plant census 1967 (Hansen 1971)
R1 = plant census 1984 (Raus 1986)
R2 = plant census 1986/87 (Raus orig.)

An = leg. Anagnostidis 1977 (Anagnostidis &Economou-Amilli 1978)
Dm = leg. Diamantopoulos 1985 (unpubl., det. Raus)
Dp = leg. Diapoulis 1969 (Diapoulis 1971)
Ra = leg. Rackham 1977 (Rackham 1978)
St = leg. Steidel 1978 (unpubl., det. Hansen)
! = invasion after 1984 (new floristic record)
– = confirmed extinction 1986/87
? = known localities not controlled in 1986/87

dispersal. The 1984 immigration attempts of *Aloe vera* (planted), *Anthemis rigida, Hedypnois rhagadioloides, Parietaria judaica, Sonchus asper* and *Urtica urens* were evidently unsuccessful, too. The same may apply to *Avena sterilis, Filago contracta, Holcus setiglumis, Paronychia echinulata, Plantago arenaria* and *Raphanus raphanistrum* which have been seen only once on NK without being reconfirmed by later observers (Table 2, symbols An, Dm, Ra, St). Ephemeral invasion of vascular plant species with immediate failure of persistence is obviously a regular event on NK.

Invasions and stabilization of founder-populations

The floristic stock least damaged by volcanic impact and always met with on NK since Vierhapper's count in 1911, consists of 17 species (or 18 if one assumes *Trifolium* 'spec.' of Cammerloher (1935) to be *T. arvense,* the locally most unpretending species of the genus and one of the first invaders of bare lava stands on NK). Most of them are annuals and only three perennial species (viz. *Asparagus stipularis, Ficus carica* (woody) and *Helichrysum italicum*). Cammerloher (1935: pp. 86–88) documented a *Ficus* tree surviving the outburst of 1925, which recovered vegetatively from overground combustion. Later, a second even more mature *Ficus* tree was discovered elsewhere on the island (Raus 1986: map 3, 1). *Asparagus stipularis,* as a bulbous geophyte, may also have survived overground heat and burying with an ash-layer in situ (Cammerloher 1935). *Helichrysum italicum* may well have newly invaded NK after 1925/28 as is probable in all cases of continuously censused annual species.

After 1925/28, Cammerloher (1935) added 17, Hansen (1971) 42, Diapoulis (1971) 3, Steidel (unpubl.) 3, and Raus (1986) 40 continuously reconfirmed species to NK's flora (the question marks in Table 2, col. R2 refer to occurrences taken for granted). The 14 most recent additions (after 1984; see exclamation marks in front of the plant names in Table 2) are mapped in Fig. 3 in order to enable future monitoring. In 1987, the annual species *Chenopodium murale, Lavatera cretica, Lolium rigidum* ssp. *lepturoides, Malva parviflora, Tordylium apulum* and, in autumn 1986, *Conyza bonariensis* grew in places which had been well investigated by me in 1984. They therefore can be considered as real invaders during this 3 years' period. The same is true for the geophyte *Rumex tuberosus* ssp. *creticus* and the woody *Pistacia lentiscus,* the latter probably representing a crucial invasion in terms of future vegetation development on NK (see below). *Trifolium stellatum,* which had been missing since 1911, should also be mentioned in this context.

A number of 1984 newcomers were able to stabilize their founder-populations or even to expand on NK. Nicely verified cases in this respect are represented by *Galium aparine, Mercurialis annua* and *Stellaria pallida* which entered NK as single escaped individuals in 1984 by means of waste flowerpot earth like the above mentioned *Fumaria judaica.*

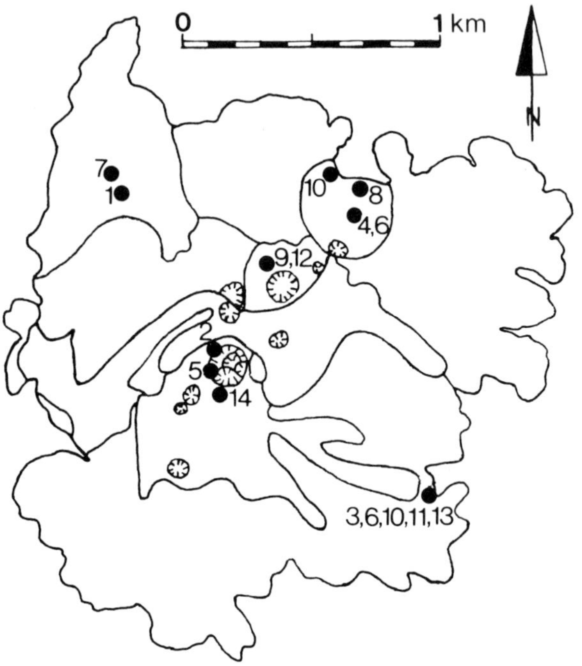

Fig. 3. Invasion of vascular plants to Nea Kaimeni after 1984 (Raus obs. 1986/87). Localities of founder-populations of 1 *Asplenium onopteris,* 2 *Pistacia lentiscus,* 3 *Chenopodium murale,* 4 *Aetheorhiza bulbosa,* 5 *Conyza bonariensis,* 6 *Malcolmia chia,* 7 *Vicia cretica,* 8 *Vicia sativa* ssp. *nigra,* 9 *Vicia villosa* ssp. *eriocarpa,* 10 *Lavatera cretica,* 11 *Malva parviflora,* 12 *Rumex tuberosus* ssp. *creticus,* 13 *Tordylium apulum,* 14 *Lolium rigidum* ssp. *lepturoides.*

Contrary to *Fumaria* (now extinct) the three species increased their number of individuals to 5–10 on the monitored spot within three years.

In the case of *Chrysanthemum coronarium,* ca 10 individuals were observed near the northern harbour (Fig. 2: H1) in 1984 (Raus 1986: p. 386) and ca 300 plants were in full flower there in 1987. *Myosotis incrassata* and *Valantia hispida* expanded agressively along tourists' paths between the northern harbour and the top of the island as was indicated and predicted in the case of the latter species already by Raus (1986: p. 386). Two shrubby invaders, monitored since 1984, viz. *Atriplex halimus* and *Ballota acetabulosa,* did not enlarge their number of individuals but their number of flowering and fruiting branches.

In 1986/87 I newly detected several seedlings and saplings of *Ficus carica,* in the shelter of deep crevices or at the base of big lava-blocks, on the lava field of 1570/73 (see Fig. 2) and on the very top of the island. They probably represent offspring of NK's two resident, old, well-fruiting *Ficus* trees (see above) but may well have been brought in, too, by gulls feeding on *Ficus* fruits on the ring-islands (see Raus 1986: pp. 385–386).

Sampling errors

No doubt all censuses are affected by sampling errors owing to the large dimensions and the difficult topography of NK's lava flows. To mention just one example: a small population of *Adiantum capillus-veneris,* refound in 1984 for the first time since 1911, grows in the center of the 1570/73 area (see Fig. 2) in such an inaccessible place that its absence on the checklists of 1967 and 1984 surely represents a sampling error. This is a case of pseudoturnover in the sense of Nilsson & Nilsson (1983).

This kind of sampling error is quantitatively more or less 'self-compensating' because it applies about equally to each of the counts. Mis-identifications (examples given in inverted commas after the accepted plant names in Table 2) do not necessarily affect quantitative aspects of comparison either. Double enumeration of species under different synonyms, however, as found with Diapoulis (1971), Anagnostidis & Economou-Amilli (1978: p. 708) and Steidel (in Schmalfuss *et al.* 1981: p. 9), does influence the results.

Vascular plant colonization on Palea Kaimeni

In contrast to NK, there is only one (and recent) nearly complete vascular plant census from 1984 (Raus, unpubl.). Existing species turnover on PK, however, is evidenced by *Glaucium flavum* (*G. 'leiocarpum'*), *Hyoscyamus albus, Sisymbrium polyceratium, S. orientale* and *Trifolium resupinatum* which were reported from PK by previous collectors (Diapoulis 1971; Hansen 1971) but were not reconfirmed later (Raus, unpubl.; Steidel, unpubl.). 63 species of PK's flora have not yet been reported from NK, among them some woody perennials like *Artemisia arborescens, Calicotome villosa, Crithmum maritimum, Erysimum senoneri, Limonium graecum, Lotus cytisoides, Lycium intricatum, Prasium majus, Sarcopoterium spinosum, Satureja thymbra* and *Thymelaea hirsuta.* This clearly indicates PK's longer history of vascular plant colonization as compared with NK.

Vegetation development on the Kaimeni Islands

Palea Kaimeni

PK's actual vegetation chiefly consists of an evergreen sclerophyllous shrub community with dominating *Pistacia lentiscus,* rarely accompanied by scattered individuals of *Ballota acetabulosa, Calicotome villosa* and *Prasium majus.* The shrubs remain low and are heavily affected by wind and salt spray; they may represent the climax vegetation on the island. Between the patches of the evergreen shrubs a mosaic of herb communities, rich in therophytes occurs. In the supralittoral zone a shrub community is growing which is dominated by *Atriplex halimus,* a woody chenopod, mixed with single spiny shrubs of *Lycium intricatum.* This halophytic plant community also represents the presumably final stage of the supralittoral vegetation succession according to the prevailing environmental and floristic conditions. A more open, low-growing *Limonium graecum*-community colonizes the SE tip of the island. Scattered, inconspicuous chasmophytic fern communities occupy shady crevices of cliffs and slopes of loose lava-blocks mainly along the N coast and in the center, with *Anogramma leptophylla, Asplenium ceterach, A. obovatum, Cheilanthes pteridioides, Ch. vellea,* and *Polypodium cambricum* in varying combinations. In 1984 and, again in 1986/87 the plant cover of the old part of PK appeared consolidated and dramatic changes are not expected for the next decades. The nearly bare NW part of PK, added to the island in 1866 (see Fig. 2) had a flora of 32 vascular plant species in 1984, among them only 6 long-lived perennials, viz. *Asparagus stipularis, Atriplex halimus, Helichrysum italicum, Hyparrhenia hirta, Phagnalon graecum,* and *Pistacia lentiscus* which are slowly entering the area from its southern periphery.

Nea Kaimeni

After several outbursts of the island volcano in our century, the re-establishment of NK's plant cover has reached different stages in different parts of the island, according to age and physical structure of the relevant substrata (for details see Raus 1986: pp. 387-391). From a distance one can distinguish these different stages of vegetation development by the pattern of yellow-green and black colours of the land-surface. The most evolved and closed vegetation type on NK at present is a steppe-like herb community with the perennial grass *Hyparrhenia hirta* and the annual legume *Lupinus angustifolius* co-dominating. For detailed description through relevés see Raus (1986: Table 1). In 1933 *Hyparrhenia hirta* was lacking on NK, after the volcanic eruption of 1925/28 (Cammerloher 1935: p. 86). Thus the island's closed *Hyparrhenia* grasslands actually have an age of at most ca 50 yr, and probably rather less than that.

The beginning of a shrub vegetation on NK probably dates from 1984 when I found two successfully established individuals of *Atriplex halimus* on the SW coast of NK, just opposite PK. The seeds obviously came by seawater transport, for the plants are growing at the drift line. In autumn 1986 I detected two single, established seedlings of the evergreen *Pistacia lentiscus,* PK's dominant shrub species, just on the summit of NK. The berries from which these plants have sprouted had obviously been brought in by birds which picked up the fleshy fruits on PK and transported them to the top of NK. Of course, nothing can be said about earlier invasions of *Pistacia lentiscus* on NK which have failed. But at present, these plants represent a kind of starting point of a possible, at least partial, colonization of NK by *Pistacia lentiscus.* With respect to 1) the small distance, of 300 m only, between both islands, 2) the high potential of PK to serve as a donor of diaspores of woody species, and 3) the fact that there is no grazing pressure on NK, it is astonishing that the leading local shrub species has difficulties to immigrate successfully on NK. The main ecological minimum factors are supposed the hazardous freshwater support for the young seedlings and the extremely hot microclimate of the black-coloured lithosols (with surface temperatures of more than 80 °C in summer which may kill the seedlings; see Richter 1984).

Outlook

There is an urgent need of systematic investigations on the Kaimeni Islands in terms of population dynamics, calculation of turnover rates, colonizing mechanisms and development and consolidation of plant communities (including cryptogams, see Anagnostidis & Economou-Amilli 1978). Ferro & Furnari (1968, 1970), Nilsson & Nilsson (1978, 1982, 1983) and Snogerup & Snogerup (1987) provide stimuli for this work. As a first step J. Diamantopoulos (Thessaloniki) initiated permanent square monitoring on NK in 1985.

Acknowledgements

I am much indebted to J. Diamantopoulos, Thessaloniki, A. Hansen, Copenhagen, and H. Schmalfuss and C. Steidel, Stuttgart, for providing me with unpublished details of their valuable field work on the Kaimeni Islands, and to H. Pieper, Kiel, for facilitating access to important literature. Ch. Hillmann-Huber, Berlin, kindly assisted in preparing the figures.

References

Anagnostidis, K. & Economou-Amilli, A. 1978. Microorganisms from the volcano of Nea Kammeni Island. In: Doumas, C. (ed.), Thera and the Aegean World, I: 707-723. London.

Backer, C. A. 1929. The problem of Krakatoa as seen by a botanist. Surabaya (Batavia).

Cammerloher, H. 1935. Ein Beitrag zur Flora von Kaimeni (Santorin). Österr. Bot. Z. 84: 81-90.

Diapoulis, C. 1971. The development of the flora of the volcanic islands Palaia Kammeni and Nea Kammeni. In: Kaloyeropoulou, A. (ed.), Acta 1st Internat. Congress Volcano Thera 1969: 238-247. Athens.

Docters van Leeuwen, W. M. 1936. Krakatau 1883-1933. Ann. Jard. Bot. Buitenzorg 46/47: 1-506.

Ernst, A. 1934. Das biologische Krakatauproblem. Vierteljschr. Naturf. Ges. Zürich 79 (Beibl.): 1-187.

Ferro, G. & Furnari, F. 1968. Flora e vegetazione di Stromboli (Isole Eolie). Arch. Bot. Biogeogr. Ital. 44: 21-45, 59-87.

Ferro, G. & Furnari, F. 1970. Flora e vegetazione di Vulcano (Isole Eolie). Pubbl. Ist. Bot. Univ. Catania 1970: 1-64.

Flenley, J. R. & Richards, K. 1982. The Krakatoa centenary expedition final report. Dep. of Geography, University of Hull, Misc. Ser. 25: 1-196.

Fridriksson, S. 1975. Surtsey. Evolution of life on a volcanic island. Butterworths, London.

Guenther, E. W. & Klug, H. 1970. Surtsey, Island. Natürliche Erstbesiedlung (Ökogenese) der Vulkaninsel. Schr. Naturwiss. Vereins Schleswig-Holstein, Sonderband 1970. Kiel.

Hansen, A. 1971. Flora der Inselgruppe Santorin. Candollea 26: 109-163.

Nilsson, I. N. & Nilsson, S. G. 1982. Turnover of vascular plant species on small islands in Lake Möckeln, South Sweden 1976-1980. Oecologia (Berlin) 53: 128-133.

Nilsson, S. G. & Nilsson, I. N., 1978. Species richness and dispersal of vascular plants to islands in Lake Möckeln, southern Sweden. Ecology 59: 473-480.

Nilsson, S. G. & Nilsson, I. N., 1983. Are estimated species turnover rates on islands largely sampling errors? The American Naturalist 121: 595-597.

Pichler, H. & Friedrich, W. L. 1980. Mechanisms of the Minoan eruption of Santorini. In: Doumas, C. (ed.), Thera and the Aegean World, II: 413-427. London.

Pichler, H. & Kussmaul, S. 1980. Comments on the geological map of the Santorini Islands (with a coloured Geological Map 1:20000 and two profiles). In: Doumas, C. (ed.), Thera and the Aegean World, II: 413-427. London.

Rackham, O. 1978. The flora and vegetation of Thera and Crete before and after the great eruption. In: Doumas, C. (ed.), Thera and the Aegean World, I: 755-764. London.

Raus, Th. 1986. Floren- und Vegetationsdynamik auf der Vulkaninsel Nea Kaimeni (Santorin-Archipel, Kykladen, Griechenland). Abh. Landesmus. Naturk. Münster 48: 373-394.

Reck, H. 1936. Santorin - Der Werdegang eines Inselvulkans und sein Ausbruch 1925-1928. 3 vol. Reimer, Berlin.

Richter, M. 1984. Vegetationsdynamik auf Stromboli. Zur Geoökologie trocken-mediterraner Standorte. Aachener Geogr. Arb. 16: 41-110.

Schmalfuss, H. & Schawaller, W. 1984. Die Fauna der Ägäis-Insel Santorin. Teil 5 Arachnida und Crustacea. Stuttgarter Beitr. Naturk. ser. A, no. 371: 1-16.

Schmalfuss, H., Steidel, C. & Schlegel, M. 1981. Die Fauna der Ägäis-Insel Santorin. Teil 1. Stuttgarter Beitr. Naturk. Ser. A, no. 347: 1-14.

Snogerup, S. & Snogerup, B. 1987. Repeated floristical observations on islets in the Aegean. Pl. Syst. Evol. 155: 143-164.

Vierhapper, F. 1914, 1919. Beiträge zur Kenntnis der Flora Griechenlands. Bearbeitung der anläßlich der zweiten Wiener Universitätsreise im April 1911 in Griechenland gesammelten Pflanzen. A. Anthophyta und Pteridophyta. Verh. Zool.-Bot. Ges. Wien 64: 239-270 and 69: 102-312.

Washington, H. S. 1926. Santorini eruption 1925. Geol. Soc. Amer. Bull. 37: 349-384. New York.

Vegetatio 77: 149–158, 1988
© Kluwer Academic Publishers, Dordrecht – Printed in the Netherlands

Vegetation development on the outer islands of the Bothnian Bay

Terttu Vartiainen
Munkkiniemen puistotie 1 B 39, SF-00330 Helsinki, Finland

Accepted 3.9.1988

Keywords: Baltic Sea, Land uplift, Ordination, Sea shore vegetation, Sea water level, Succession

Abstract

This paper summarizes succession studies on the outer islands of the Bothnian Bay (Finland). The study area is characterized by consistent and relatively rapid uplift (vertical rise on average 75 cm per century). The spatial and temporal succession has been deduced from the sequence of the littoral and epilittoral vegetation as well as from the historical records of the island development.

Four types of successional series are distinguished. The relationships between substrate type (boulder, gravel, sand and clay – silt shores) and 25 typical plant communities are outlined and treated as successional sequences.

Nomenclature: Tutin, T. G. *et al.*, 1964–1980. Flora Europaea. Cambridge University Press. Cambridge; Ehrendorfer, F. 1973. Liste der Gefässpflanzen Mitteleuropas. 2. Aufl. I–XII; Koponen, T. Isoviita, P. & Lammes, T. 1977. The bryophytes of Finland: an annotated checklist. Flora Fennica 6: 1–77.

Introduction

The glacio-isostatic phenomenon of land uplift is characteristic of the coasts of Finland, Sweden, Norway and Canada. Direct influence of land uplift is reflected in changes of the ratio between land and water. Botanically its influence is evident in the alternation of plant cover in response to continuous changes in sea- and ground water levels (cf. Ericson 1980, 1981; Cramer 1986; Cramer & Hytteborn 1987).

Botanical research in the archipelago of Finland along the gulfs of the Baltic Sea has both floristic and phytosociological traditions, which go back to the beginning of this century (Luther 1979 and references therein). Ecological factors affecting island vegetation were examined especially in the archipelago of Åland in SW-Finland by Palmgren (1925) who discussed the role of land uplift as a phytogeographical factor. Later on several other botanists have investigated the vegetation and flora of the islands from different points of view.

From the 1930s onwards the Gulf of Bothnia, where new islands are continuously emerging from the sea, came under intensive botanical study. Attention has mainly focused upon the correlation between the number of species and elevation and area of the islands (Valovirta 1937; Brunberg-Schwanck & Bärlund 1948; Palomäki 1963; Vartiainen 1967; Schwanck 1974).

The primary aim of this paper was to study the response of littoral and epilittoral plant communities to the elevation and site age gradients on different shore types on the outer islands of the northernmost Gulf of Bothnia, the Bothnian Bay. These islands offer a unique possibility to examine vegetation dynamics in an area, where the influence of man has been minimized since the 1940s and where the rate of land uplift is high, about 75 cm per century on average (Kääriäinen 1975).

Another aim was to examine the relationships between plant communities and substrate using ordination techniques, and a third one was to give an example of the historical development of island vegetation in relation to the enlargement of islands.

Study area

The study area extends 140 km in N-S direction and 20–30 km seaward from the mainland between 64°12′–65°46′N and 23°33′–25°12′E (Fig. 1) in the middle boreal vegetation zone (Ahti *et al.* 1968). The area comprises 150 mostly outlying low islands situated either as a broken zone around a large nuclear island or as a chain in NW-SE direction or scattered.

The islands vary in elevation (0.05–10 m), area (0.01–180 ha), topography, soil and vegetation and are either treeless or wooded. Four island types can be distinguished from boulder skerries with scarce vegetation through grassy skerries and grassy islands with more or less open plant cover and bush-like trees to wooded islands which are amalgamations of smaller ones. The fifth type is represented by the smallest islets in shallow waters lying close to larger islands. This geomorphological-botanical classification was made prior to the phytosociological survey on the basis of visual estimates.

Bedrock comprises Svecofennian sedimentary rock, Karelian fyllite, quartzite and dolomite. The relief is low due to the flatness of the peneplain, and it is covered by glacial and post-glacial deposits. Wave action has sorted these into boulder, gravel, sand and clay-silt substrates (Alestalo 1982). Accordingly, four shore types are distinguished: boulder, gravel, sand and clay-silt shores. Shore types alternate with each other and may change in temporal sequence by means of land uplift and of influence of marine forces, erosion and accumulation. This involves also changes in successional sequences of vegetation. Solar radiation, wind, thermal and hygric conditions determine exposure factors. From June to August the mean cloudiness is less than on the mainland. Southwestern winds prevail in the growing season, which lasts from the end of May to mid-October, i.e. about 140 days. Precipitation is

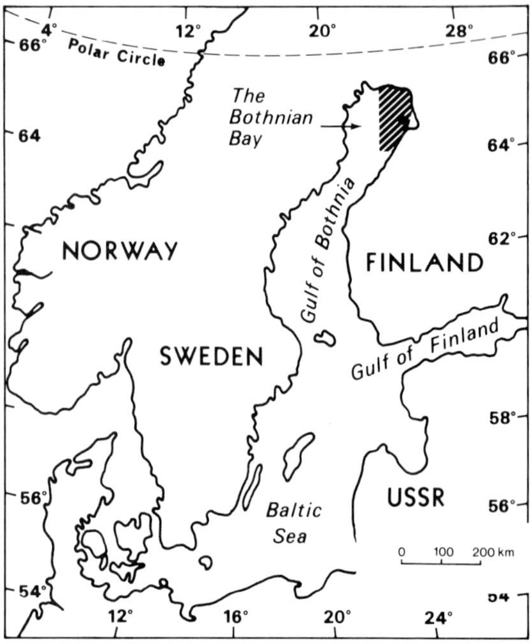

Fig. 1. The study area.

500–550 mm/yr (Atlas of Finland 1986). In the Bothnian Bay the salinity is 3–4‰ at the bottom and 1.5–3.5‰ for the surface water. Ice-cover remains from November–December to May (Atlas of Finland 1987).

The island vegetation includes 28 common plant community types selected from 69 stand types and representing about 90% of sample plots studied (Vartiainen 1980). Twenty-five of the major types are taken into consideration in this paper. The vegetation of the interior of wooded islands consists of deciduous woods, swampy meadows, juniper heaths and stone ridges with lichen and bryophyte species. The island flora is composed of 281 vascular plant-, 96 bryophyte- and 93 lichen species (see Vartiainen 1980).

Methods and terminology

The field work was performed on 150 islands in 11 summers during 1947–1969. Belt transects 2 m wide, 547 altogether, with 2300 sample plots, were set up from the water-line across each treeless island and to the shore bush of wooded ones. The lowest and highest elevations of each sample plot were measured using an altimeter. The common plant communities were described for different shore types (Vartiainen 1980).

The species with ≥30% presence in at least one stand type were included in the numerical analysis, using the frequency scale 1–10 (with 10% intervals). This vegetation data matrix (25 stand types with 62 vascular plant and 24 bryophyte species) was studied with reciprocal averaging (RA) ordination using the program DECORANA (Hill & Gauch 1980).

The terminology of the littoral used is that of Du Rietz (1930). The successional terms follow Braun-Blanquet (1964), Daubenmire (1968) and Ricklefs (1973). The successional sequences (series) are derived from the belt transects on different shore types. Vertical extension and abundance (means and standard deviations) of species of each adjoining plant community within the respective transects were tabulated. The successive plant communities are called seral stages (Vartiainen 1980).

Results

Vegetation dynamics on different shore types

The general zone pattern of the littoral and epilittoral vegetation on different shore types in relation to changing site elevation is illustrated in Fig. 2 (cf. Luther 1961; Ericson 1980, 1981; Ericson & Wallentinus 1979). The mean vertical extension of stands is varying. The populations of vascular plant species are partly overlapping, partly there are gaps between the adjoining stands.

The belts, which include one or several zonal plant communities, form different successional series. These can be separated into meadow- and heath-series depending on the substrate (Table 1).

In the hydrolittoral of alluvial shores *Phragmites australis* forms the densest stands, but it appears also on other shore types as a belt around wooded islands. In the lower geolittoral the most typical establishments in primary succession are found: *Phrag-*

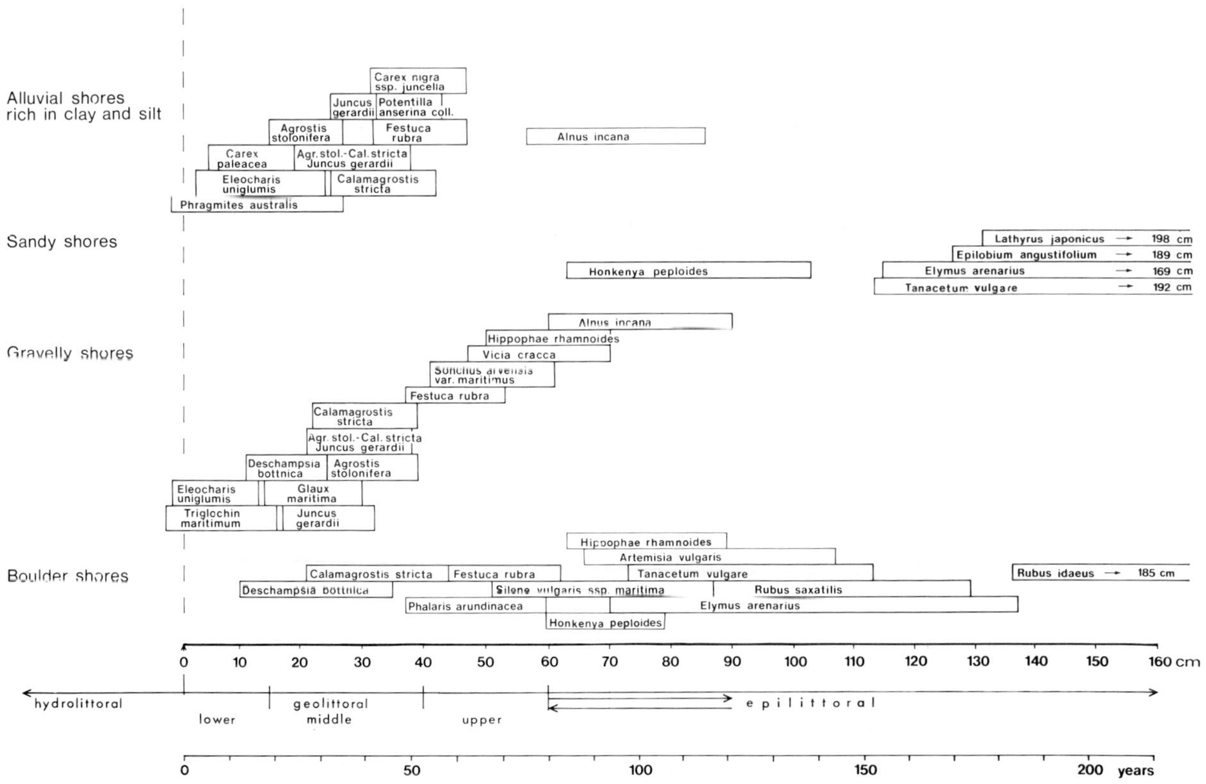

Fig. 2. Hydrolittoral-epilittoral sequence of plant communities on different shoretypes. (Modified after Vartiainen 1980).

Table 1. Scheme of successional series on different substrates.

Forests		*Milium-Cornus* grass-herb forest	*Empetrum-Vaccinium* forest	*Cornus-Lycopodium* forest	*Rubus-Deschampsia* forest
Series of belts		Meadow series	Heath series Subseries of		
Shore types Zones		Shores rich in clay and silt	Sandy shores	Gravelly shores	Boulder shores
Epilittoral		Belt of shore scrub	Dune vegetation *Elymus arenarius* belt	Vegetation on gravelly and stony substrates	Belt of shore thickets
Geolittoral	upper	*Juncus gerardii* belt (*Festuca rubra* *Vicia cracca*)	*Honkenya peploides* belt	Belt of shore scrub and thickets *Sonchus arvensis* var. *maritimus* belt	*Silene vulgaris* ssp. *maritima* belt *Angelica archangelica* ssp. *litoralis* belt
	middle	*Juncus gerardii* belt		*Juncus gerardii* belt	
	lower	*Carices distigmaticae* belt *Eleocharis uniglumis* belt		*Deschampsia bottnica* belt	
Hydrolittoral		Emergent vegetation: *Phragmites australis* belt			

mites, Eleocharis uniglumis, or *Carex paleacea* and on gravelly shores the same *Eleocharis*-species or *Triglochin maritimum.* In the middle geolittoral these are followed by plant communities common to both shore types with smoothly rising slopes. Establishments of *Deschampsia bottnica,* endemic in the Gulf of Bothnia area, and *Glaux maritima* are characteristic of gravelly shores. In the upper geolittoral of both shore types *Festuca rubra* stands appear. *Carex nigra* spp. *juncella* and *Potentilla anserina* coll. stands characterize the alluvial shores and *Sonchus arvensis* spp. *maritimus,* as well as shore thickets, the gravelly shores.

Mean elevation amplitudes of all the stands mentioned change on both shore types between 10–20 cm corresponding with a 15–30 yr cycle.

In the epilittoral there are differences in the occurrence of *Alnus incana* shrubland although the amplitude of the elevation is the same (30 cm) on alluvial and gravelly shores. On the former ones there may appear *Myrica gale* or *Salix phylicifolia* as individual bushes between meadows and shore bush, as well as wrack populations. *Hippophae rhamnoides* forms thickets on gravelly shores at a lower elevation than *Alnus incana* (cf. Cramer 1986), but disappears before the latter one. Both have a cycle of 40 yr.

On sandy shores the first establishments of *Honkenya peploides* appear at the border between geo- and epilittoral. In epilittoral dunes many alternative plant communities may form the next successional stage in the *Elymus arenarius* belt.

On boulder shores, mostly open plant communities appearing in the geolittoral are characteristic particularly of other shore types. *Silene vulgaris* ssp. *maritima* establishments are typical of the upper geolittoral of treeless islands forming its own belt, as well *Rubus idaeus* stands in the epilittoral. Correspondingly, an *Angelica archangelica* ssp. *litoralis* belt characterizes the boulder shores of wooded islands.

The elevation amplitudes of plant communities are bigger on exposed shores than on sheltered ones, varying from 35 – 75 cm in the upper geolittoral and from 50 – 140 cm in the epilittoral, corresponding to 45 – 100 and 65 – 185 yr cycles respectively.

The woodland of the interior of the islands can be considered as a successional stage of the meadow or heath series (Table 1). The classification of the insular forests partly follows that presented by Havas (1967) in the coastal area of the Bothnian Bay. The forests have a relatively homogeneous tree layer, the dominant tree being *Betula pubescens.* There are differences in the field layer and dominance (for further details, see Vartiainen 1980).

Interpretation of the ordination

The RA ordination (Fig. 3) shows the successional relationships between different plant community types (site types). The first axis can be interpreted in terms of a gradient from wet to dry substrate, while the second axis relates to soil type (size of soil particles).

The site types can be grouped in relation to the four shore types (see also Table 2). On clay-silt shores (I) the *Eleocharis uniglumis* and *Carex paleacea* communities are found closest to the shore line and the *Festuca rubra-Vicia cracca* community on the opposite site close to the shore bush.
On gravel shores (III) *Glaux maritima* and *Deschampsia bottnica* communities indicate a moisture tendency similar to those of alluvial clay-silt shores.

In the middle of the first axis we find plant com-

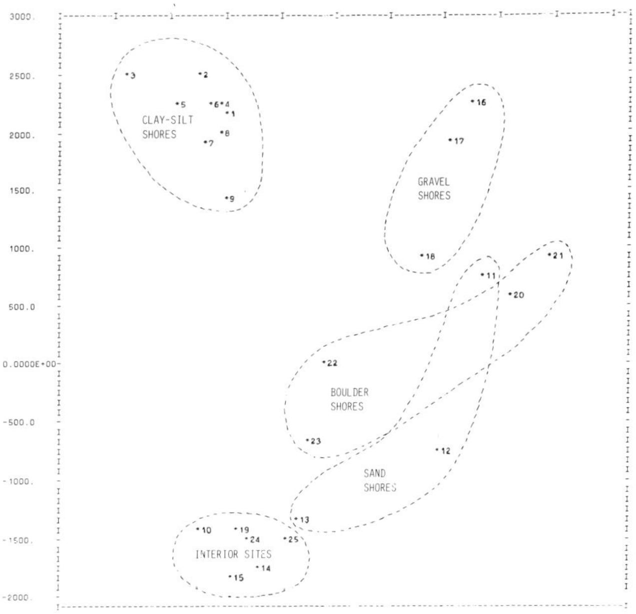

Fig. 3. Ordination of the 25 major plant community types with Reciprocal Averaging. Axis 1 vertical, axis 2 horizontal (for numbers see Table 2).

munities (types 11, 18, 20, 21) which belong to three shore groups (II, III, IV). All appear at the upper limit of flooding. *Sonchus arvensis* var. *maritimus* communities appear in the driest site in their own group (III), while *Honkenya peploides* and *Silene vulgaris* var. *maritima* as well as *Angelica archangelica* ssp. *litoralis* communities appear on relatively wet sites (II, IV). Epilittoral heath (23) and dune communities (12) have the same moisture requirements. In the interior group we find four forest types, epilittoral heaths and boulder crests. *Milium-Cornus* forests are found on the most fine-grained substrate behind alluvial meadows and boulder crests on the highest (oldest) and driest sites.

Species richness

The successional trend from wet to dry is evident also in the species richness of the major plant community (site) types (Table 2).

The number of vascular species increases in all site types of the heath series from wet to dry conditions. In the meadow series the direction is not as clear. Es-

Table 2. Major plant community types (1 – 25) and their species richness.

Shore type (I – IV) Plant community type (1 – 25)	Nr of sample plots	Nr of vascular plant species/ sample plot (median)	Nr of bryophyte species/ sample plot (median)
I. Clay-silt shores			
1. Phragmites australis	11	7	0
2. Eleocharis uniglumis	23	3	0
3. Carex paleacea	23	3	<1
4. Agrostis stolonifera	18	6	<1
5. Calamagrostis stricta	25	5	<1
6. Juncus gerardii	15	6	<1
7. Agr. stol. – calam. str. – Juncus ger.	31	16	2
8. Festuca rubra	29	10	1
9. Festuca rubra – Vicia cracca	17	14	1
10. Milium-Cornus forest	40	10	1
II. Sand shores			
11. Honkenya peploides	17	4	0
12. Elymus arenarius/Tanacetum vulgare	30	8	0
13. Deschampsia flexuosa/Rubus saxatilis	32	17	3
14. Empetrum nigrum	27	15	5
15. Empetrum – Vaccinium forest	17	20	6
III. Gravel shores			
16. Glaux maritima	7	3	0
17. Deschampsia bottnica	36	4	0
18. Sonchus arv. var. maritimus	13	5	<1
19. Cornus-Lycopodium forest	14	28	7
IV. Boulder shores			
20. Silene vulgaris ssp. maritima	11	8	0
21. Angelica arch. ssp. litoralis	10	5	0
22. Hippophae rhamnoides	35	13	<1
23. Epilittoral heath	53	14	1
24. Rubus – Deschampsia forest	21	22	6
25. Interior boulder crest	25	13	11

tablishments of hydrolittoral on the alluvial shores have the smallest number except *Phragmites australis* stands with its greater vertical extension. This can be explained by the mosaic pattern of the vegetation with many alternations at the same level.

The forest types of the heath series are characterized by a greater number of vascular plant species than that of the *Milium-Cornus* forest representing the meadow series. In the former case there is more small-scale variation in the topography and instability in hygric conditions resulting in a greater number of vascular plant species than in the *Milium-Cornus* forest with a thick humus layer and more stable hygric conditions.

Historical development: a case study

Historical maps of different development stages of the islands can be used to reconstruct the succession of the plant cover (Fig. 4, see also Vartiainen 1980).

There were only small boulder skerries about 900 yr ago (according to an uplift of 75 cm/100 yr) where now the wooded island of Maakrunni is situated with an area of 150 ha and an elevation of about 7 m above MWL. By 1770 there were heaths on former boulder and sandy habitat sites and meadows between them due to the accumulation of organic and mineral material on sheltered shores. Around 1880 land uplift had resulted in a doubled

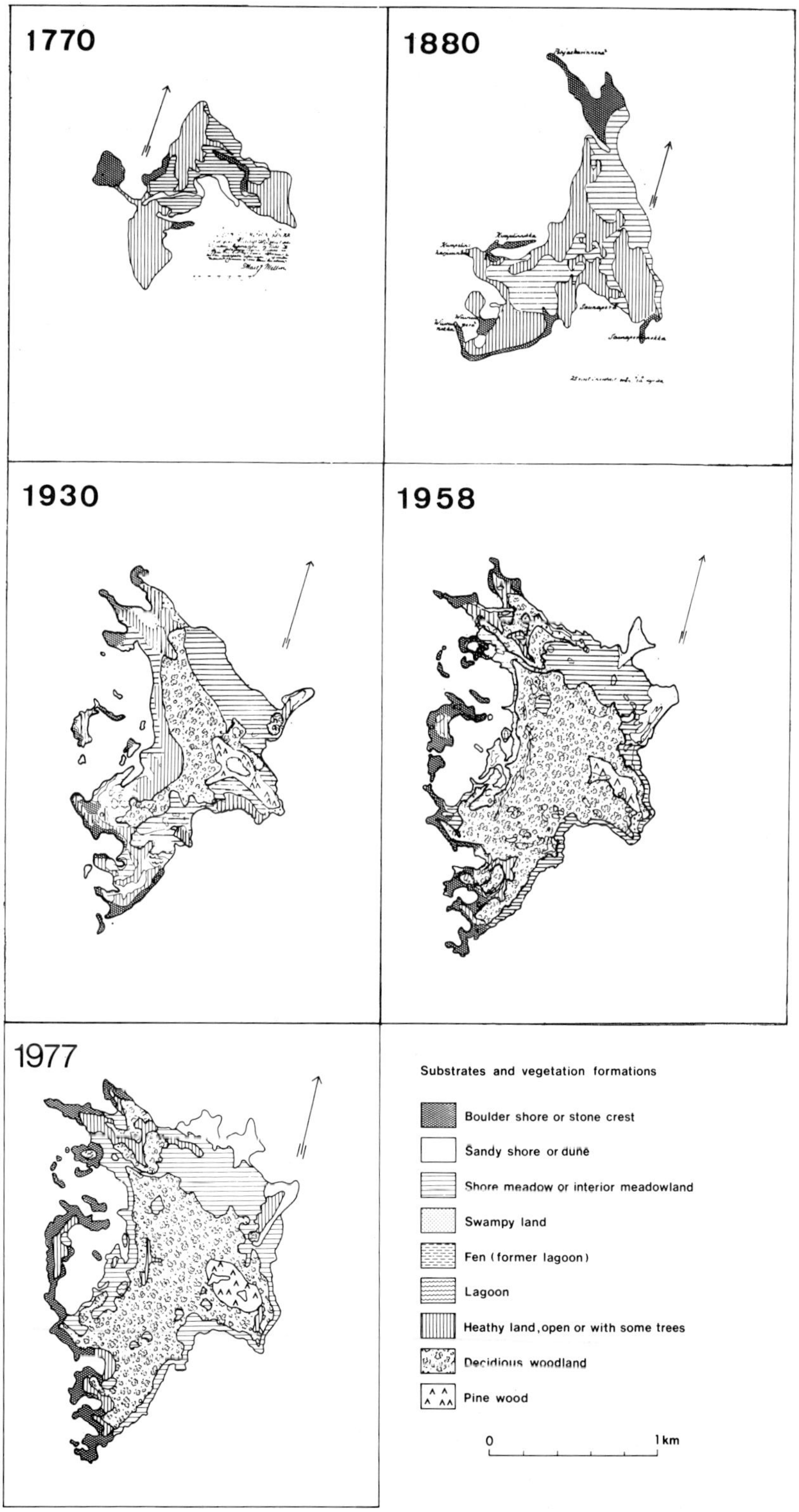

Fig. 4. Development of plant cover on Maakrunni from 1770 to 1977. Four maps of situations prior to 1958 have been drawn by the author on the basis of the old land estate maps in the archives of the surveyor's office of Oulu county in Oulu; the 1977 map was made on the basis of an aerial photo taken by the National Board of Survey, Helsinki (cf. Vartiainen 1980).

area. With this increase in width and height, boulder cape tips appeared, heathy land developed on former stone crests and all low-lying shore area has become meadow. It may be assumed that shore bush had developed as well. Steep boulder shores subjected to strong wave action seem to have remained the same for long periods.

The most distinctive vegetation changes in this century are due to increased elevation, larger distance to the sea water level and other environmental changes (cf. Miles 1979): the former meadows have turned into deciduous woods and the highest and sheltered heathlands into interior pine woods. Some open sandy fields (former dunes) are still to be seen. The eastern alluvial meadow has enlarged and sand banks have extended. Stony and boulder covered western cape points have partly become connected with each other through a small island in between them. In the future there will obviously be a large western bay connected with the sea only through some narrow channels. The next developmental stage will probably be a lagoon, where *Phragmites* will advance from the shore towards the centre. Gradually swampy land with e.g. *Carex nigra* ssp. *juncella* may develop into a part of a shore meadow, the successional sequence of which differs from that of a lagoon. Later on changes towards a vegetation dominated by scrubs and trees can be expected resulting in the enlargement of former woodland (cf. Luther 1961).

Discussion and conclusions

Investigations concerning the development of sea shore vegetation are numerous, while recently studies also use dynamic and numerical approaches (e.g. van der Maarel *et al.* 1985; Cramer 1986).

In the land uplift area the adjacent stands representing the same successional series move seawards as the land continues to emerge (Ericson 1981).

In the present study the long-term development of vegetation was analyzed by means of spatial and temporal age gradients (Figs 2 and 4). Elevation and distance from the sea water line are the most important factors affecting zonal structure. Elevation together with the substrate determines the hygric conditions involving both sea- and ground-water level. On the same basis the successional sequences can be separated.

The sequences may be disturbed by environmental changes connected first of all with the seasonal changes of sea-water level (cf. Cramer & Hytteborn 1987). From spring to autumn the physical processes caused by the sea, affect the shore vegetation in different ways on different shore types. On low-lying gravel and sheltered alluvial shores rich in clay and silt, flooding accumulates organic and mineral soil resulting in different seral stages within the sequence in question (Ericson 1981).

Wave action and surf-spray affect exposed sandy and boulder shores by erosion of sandy habitats and thus creation of new habitats (Willers 1987).

Beside the seasonal fluctuations of sea-water level there are year-to-year variations in weather factors (see also Cramer & Hytteborn 1987). The indirect influence of wind is connected with the sea; changes in sea-water level particularly affect the physiognomy of the vegetation and may also determine the composition of the flora (cf. Ericson 1981). Water may rise 1.5 m above the MWL during periods with prevailing W and SW winds; during N and E winds water level may equally fall.

Growing conditions are hard in open places such as heaths and stony ridges, because of the intensity of sunshine and the drying effect of SW winds in the growing season.

Ice damage affects shore bushes, e.g. *Alnus incana* and *Hippophae rhamnoides* mostly on gravel and boulder shores. Ice movement may displace shore material and produce new habitats (Willers 1987). Fluctuations in water level may accentuate the effect of ice (Luther 1951).

In the study area elevation amplitudes of plant communities are small on clay-silt and gravel shores: changes in most communities are gradual with decreasing abundance of the dominant species in the following stage higher up (Fig. 2). On boulder shores the large elevation amplitudes are due to the uneven distribution of stones and of finer material accumulated by chance.

In the series on shores rich in clay and silt the number of common vascular species is greatest and on gravelly shores smallest, which is due to the homogeneity of the substrate of the former shore and on the heterogeneity of the latter ones (Crocker & Major 1955). The same kind of hygric conditions on sandy and boulder shores result in high frequencies of common species (Table 2).

The habitats of the interior of the large wooded islands (cf. Fig. 3) differ in dominant species and frequencies of the most common species, and moreover have a large number of occasional species, which are mostly relics from former stages.

Flooding, surf-spray and pack-ice have no direct influence on the vegetation of the interior woods. However, shore line displacement continuously changes the ground-water level and hygric conditions of wooded sites.

The differences in successional sequences are mainly related to the sea-shore displacement caused by land uplift. The edaphical and topographical changes can interfere with the successional trend (Miles 1979; Cramer 1986). Due to the regularity in land uplift, succession of plant communities can generally be predicted by comparing the vegetation of islands of different developmental stages, especially on alluvial and clay-silt shores. On sand, gravel and boulder shores, the prediction is more difficult because of the heterogeneity of the sea bottom in front of them.

Acknowledgements

This work was finalized at the Department of Botany, University of Helsinki. I thank Prof. Y. Vasari and Dr. P. Pakarinen as well as two anonymous reviewers for comments on the manuscript. I am grateful to Drs D. and C. Norris for checking parts of the English text. Finally I thank Ms Pirkko Numminen and Ms Tuija Jantunen for drawing the figures and Ms Pirkko Dookie for typing the manuscript.

References

Ahti, T., Hämet-Ahti, L. & Jalas, J. 1968. Vegetation zones and their sections in northwestern Europe. Ann. Bot. Fenn. 5: 169–211.

Alestalo, J. 1982. Morphologisk succession vid österbottnisk landhöjningskuster. Landhöjning och kustförändring. Symposiepubl. 1: 85–94.

Atlas of Finland 1986. Water. Folio 132. Appendix 210. National Board of Survey. Helsinki.

Atlas of Finland 1987. Climate. Folio 131. Appendix 210. National Board of Survey. Helsinki.

Austin, M. P. 1977. Use of ordination and other multivariate descriptive methods to study succession. Vegetatio 35: 165–175.

Braun-Blanquet, J. 1964. Pflanzensoziologie. 3rd ed.

Brunberg-Schwanck, B. & Bärlund, U. 1948. Vegetation och landhöjning. En analys av växtligheten på några nybildade små öar i mellersta Österbotten och på Åland. Skärgårdsboken. Nordenskiöld-samfundet i Finland: 257–291.

Cramer, W. 1986. Vegetation dynamics on rising sea shores in estern central Sweden. Thesis Uppsala.

Cramer, W. & Hytteborn, H. 1987. The separation of fluctuation and long-term change in vegetation dynamics of a rising seashore. Vegetatio 69: 157–167.

Crocker, R. & Major, J. 1955. Soil development in relation to vegetation and surface age at Glacier Bay, Alaska. J. Ecol. 43: 427–488.

Daubenmire, R. 1968. Plant communities. New York.

Du Rietz, G. E. 1930. Vegetationsforschung auf soziationsanalytischer Grundlage. Handb. Biol. Arbeitsmeth. 11: 293–480.

Ericson, L. 1980. The Downward migration of plants on a rising Bothnian sea-shore. Acta Phytogeogr. Suec. 68: 61–72.

Ericson, L. 1981. Aspects of the shore vegetation of the Gulf of Bothnia. Wahlenbergia 7: 45–60.

Ericson, L. & Wallentinus, H.-G. 1979. Sea-shore vegetation around the Gulf of Bothnia. Wahlenbergia 5: 1–142.

Havas, P. 1967. Zur Ökologie der Laubwälder, insbesondere der Grauerlen, an der Küste der Bottenwick. Aquilo Ser. Bot. 4: 1–36.

Hill, M. O. & Gauch, H. 1980. Detrended correspondence analysis, an improved ordination technique. Vegetatio 42: 47–58.

Kääriäinen, E. 1975. Land uplift in Finland on the basis of the sea level records. Rapp. of the Finnish Geodetic Institute 75: 5.

Luther, H. 1951. Verbreitung und Ökologie der höheren Wasserpflanzen im Brackwasser der Ekenäs-Gegend in Südfinnland. I. Allgemeiner Teil. Acta Bot. Fenn. 49: 1–232.

Luther, H. 1961. Veränderungen in der Gefässpflanzenflora der Meeresfelsen von Tvärminne. Acta Bot. Fenn. 62: 1–100.

Luther, H. 1979. Åttio år botanisk skärgårdsforskning i Finland. Skärgård 2: 3–7.

Miles, J. 1979. Vegetation dynamics. Chapman and Hall. London.

Palmgren, A. 1925. Die Artenzahl als pflanzengeographischer Character sowie der Zufall und die säkulare Landhebung als pflanzengeographische Faktoren. Acta Bot. Fenn. 1: 1–142.

Palomäki, M. 1963. Über den Einfluss der Landhebung als ökologischer Faktor in der Flora flacher Inseln. Fennia 88(2): 1–75.

Ricklefs, R. E. 1973. Ecology. Nelson. London.

Schwanck, B. 1974. Vegetationens förändring på de framväxande skären. Skärgård i omvandling: 79–86. Nordenskiöld-samfundet.

Valovirta, E. J. 1937. Untersuchungen über die säkulare Landhebung als pflanzengeographischer Faktor. Acta Bot. Fenn. 20: 1–173.

van der Maarel, E., Boot, R., van Dorp, D. & Rijntjes, J. 1985. Vegetation succession on the dunes near Oostvoorne. The Netherlands; a comparison of the vegetation in 1959 and 1980. Vegetatio 58: 137–187.

Vartiainen, T. 1967. Observations on the plant succession of the Islands of Krunnit in the Gulf of Bothnia, Finland. Aquilo Ser. Bot. 6: 158–171.

Vartiainen, T., 1980. Succession of island vegetation in the land uplift area of the northernmost Gulf of Bothnia, Finland. Acta Bot. Fenn. 115: 1–105.

Verwijst, T. & Cramer, W. 1986. Age structure of woody species in primary succession on a raising Bothnian sea-shore. In: Fanta, J. (ed.), Forest dynamics research in western and central Europe, pp. 145–163. Pudoc, Wageningen.

Willers, T. 1987. Die finnische Küste mit besonderer Berücksichtigung der Vegetation der Geröll-, Block- und Felsufer. Hamburger Vegetationsgeogr. Mitt. 2.

Vegetatio 77: 159–168, 1988
© Kluwer Academic Publishers, Dordrecht – Printed in the Netherlands

Primary succession over sixty years on hundred-year old islets in Lake Hjälmaren, Sweden

Håkan Rydin & Sven-Olov Borgegård
Institute of Ecological Botany, Uppsala University, Box 559, S-751 22 Uppsala, Sweden

Accepted 24.5.1988

Keywords: Beta-diversity, Convergence, Divergence, Long-term change, Shading

Abstract

New islands were created as a result of an artificial lowering of the water table of Lake Hjälmaren in 1882–1886 and the floras of 40 such islands have been monitored ever since plant colonization started in 1886. Data used in this paper are from belt-transects across ten of these islands which were analysed in 1927–29 and reanalysed in 1986. During this 60-year period shading by trees has increased and shading and litter accumulation have affected the ground flora. Succession includes aspects of both convergence and divergence. Species of dry, open sites and shore species have been excluded from the island interiors, making these parts of the islands more similar, but there has simultaneously been an increased variability of vegetation types of shaded, mesic habitats. There have been no consistent trends in beta-diversity between the two surveys. Differences in successional patterns between island interiors and shore habitats are attributed to an outward migration of plant species and to differences in disturbance regimes.

Nomenclature: Krok, Th. O. B. N. & Almquist, S., 1984. Svensk Flora, fanerogamer och ormbunksväxter. 26th ed., rev. L. Jonsell & B. Jonsell. Esselte Studium, Uppsala, for vascular plants; Nyholm, E., 1954–1969. Illustrated moss flora of Fennoscandia. II. Musci. Swedish Natural Science Research Council, Stockholm, for mosses.

Introduction

In this paper we report on vegetation changes during late stages of a primary succession on islands in Lake Hjälmaren (cf. Rydin & Borgegård in press). The islands were formed in 1882–1886 when the water level of the lake was lowered ca 130 cm. The external environment affecting the islands has remained largely unaltered, so this 'experiment' allows for the study of the intrinsic properties of primary succession. The insular environment also means that the surroundings are similar for different islands, in contrast to mainland environments where successional pathways of each patch are strongly influenced by the abiotic conditions and species composition of surrounding habitats. There might be, for example, higher species richness on the mainland due to influx of individuals into habitats where the population is not self-maintaining (mass effect, see Shmida & Ellner 1984; Shmida & Wilson 1985).

We will make use of transect analyses made on the islands 1927–1929 (ca 40 years after the emergence of the islands) and corresponding analyses made in 1986 (after 100 years) to discuss the following issues: Are there any trends relating to convergence or divergence, i.e. is the present vegetation composition as described by means of the transects more or rather less similar between islands as compared with 1927?

The same question is raised for changes within islands as reflected by beta-diversity measures (Wilson & Shmida 1984). We will also discuss the processes and environmental changes that has led to the observed changes in species composition.

Study area

Hjälmaren is a large (478 km^2), eutrophic lake (Willén 1984) at 22 m a.s.l. situated in the boreonemoral zone (Sjörs 1963) in south central Sweden. The annual mean temperature is 5–6 °C (mean July temperature 17 °C, mean February temperature −3 °C), mean annual precipitation 550–600 mm (Atlas över Sverige 1953) and the length of the growing season (threshold +5 °C) is 180–190 days (Tuhkanen 1980). Before 1880 the water level fluctuation (between normal high water and normal low water) was 64 cm and from 1886 to 1932 53 cm (Statens Meteorologisk-Hydrografiska Anstalt 1935).

Immediately after the second stage of the lowering of the water table, in the autumn of 1886, a number of new or enlarged islands were visited by a botanist (Callmé 1887), and subsequent botanical surveys have been made in 1892 (Grevillius 1893) and 1903–04 (Birger 1905). Detailed, but yet mainly unpublished surveys were made in 1927–33 by Erik Julin and Ragnar Morander. We have made field studies in 1984–86. In total 40 islands have been included in the studies and in this paper we use data from vegetation analyses made on ten of them (Fig. 1).

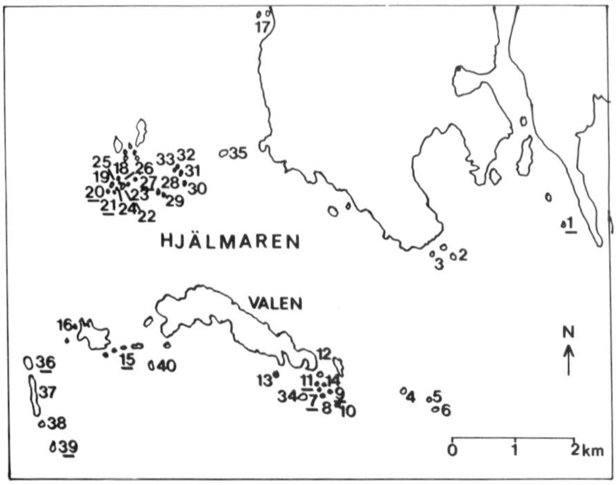

Fig. 1. The central basin of Lake Hjälmaren (Storfjärden) with the studied islands. Numbers 1–40 refer to islands in Rydin & Borgegård (in press) and underlined numbers (1, 7, 9, 11, 15, 20, 21, 27, 36 and 39) indicate islands included in this study. Numbers 36–39 are esker islands.

Methods

Julin and Morander analysed the vegetation in transects of contiguous quadrats. This was done in 1927–29, and the information used here was obtained from their unpublished field notes. Plots of 1 × 1 m were used except for islands 7 and 20 where plot size was 2 × 2 m. In general the transects were placed across the widest part of the island, perpendicular to the longitudinal direction of the island. This means that the direction is most often east-west. The transects were not marked so relocation was based on field notes of the direction of the transect, length from shore to shore and descriptions including fixed points (such as boulders) and vegetation maps. In some cases the morphological changes were great (accumulation of organic debris had increased the size of the island) and we had to give up a few transects which could not be relocated. On the basis of field measures and Julin and Morander's maps we estimate that our transects are placed within 2 m distance from those of Julin and Morander. In sheltered positions we had to make the transect 1–3 m longer to cover newly vegetated parts of the islands (probably due to an outward migration of shore plants). In any case, it was ascertained that the full length of the Julin and Morander's transects was included in the 1986 transects.

The plots were analysed with the Hult-Sernander cover scale (1 – <1/16; 2 – 1/16–1/8; 3 – 1/8–1/4; 4 – 1/4–1/2; 5 – >1/2). Julin and Morander distinguished five strata: dominant and subordinate tree layers, shrub layer, field layer and bottom layer (epigeic mosses and lichens). This stratification was hardly reproducible, so we have lumped the tree layers (>3 m) and included shrubs in the field layer. Whenever Julin and Morander had noted a species in two strata that were to be lumped, we use the highest of the two cover values.

Vegetation data were analysed with detrended correspondence analysis (DCA, Hill & Gauch 1980; computer program DECORANA, Hill 1979). All plots (212 plots from 1927–29 and 217 from 1986) were included in one ordination, and the option to down-weight rare species was applied. Changes in tree cover were analysed separately and tree data were not included in the ordination. On the islands there is a marked difference between shore and island interior vegetation, and we assigned the plots to two groups: 'shore plots' and 'terrestrial plots'. The borderline was drawn at the stems of the outermost trees (in general *Alnus glutinosa*). This border coincides well with the upper limit of drift material (Rydin & Borgegård in press). The terrestrial habitats are thus sheltered from disturbance by wave actions and ice pressure. In total there were 67 shore- and 145 terrestrial plots from 1927–29 and 72 shore- and 145 terrestrial plots from 1986.

The distinction between *Poa nemoralis* and *P. palustris* made in the 1927 survey seems to be unreliable in some cases so we had to lump these species. We have also lumped *Betula pendula* and *B. pubescens* as these were not separated by the earlier investigators.

We apply two measures of beta diversity. The first was proposed by Whittaker (1960) and recommended in the comparison made by Wilson & Shmida (1984); $B_W = s/a - 1$, where s = total species number in the transect and a = average species number in the plots (alpha diversity). The second one is the range of DCA scores of the plots of each island (cf. Christensen & Peet 1984).

Økland (1986) has argued that separation of samples along DCA axes is a superior measure of floristic (or ecological) distance, and we use it to discuss patterns of convergence or divergence during succession. For such comparisons we calculate the mean DCA coordinates of the terrestrial and shore plots of each island at the two surveys – to compare individual plots would be meaningless as they were not permanently marked.

For each island we calculate the mean plot frequency of all species, and for the species in the tree layer also their mean cover. This was made separately for the terrestrial and shore parts of the islands. When calculating changes in cover, Hult-Sernander cover values were transformed into per cent, using the midpoint of each cover class (cover 1 – 3%; 2 – 9%; 3 – 19%; 4 – 38%; 5 – 75%).

Results

During the time between the surveys shading by trees has increased on the islands (Table 1). Converted to per cent values, the mean tree cover for the ten transects was 19.6% in 1927 and 92.3% in 1986. Even if island no 27 (which was clearcut in 1926–27) is discarded the increase is highly significant ($t = 19.3$, paired comparison, $n = 9$, $P < 0.001$). *Alnus, Betula, Populus* and *Sorbus* have increased in cover, but *Betula* has decreased in mean plot frequency on the islands.

Table 2 presents changes in plot frequency of species (in shrub-, field- and bottom layers) which have been observed in at least six of the ten transects in 1927 or 1986. There is a notable decrease in both terrestrial and shore parts of the wetland species *Alnus glutinosa, Phalaris arundinacea, Phragmites australis* and *Salix cinerea*. Increasing are *Calamagrostis canescens, Epilobium angustifolium, Geranium robertianum, Rubus idaeus* and *Sorbus aucuparia*. A number of typical shore species have decreased in frequency in the island interiors and increased on the shores (*Carex acuta, Lycopus europaeus* and *Scutellaria galericulata*).

In the DCA ordination the first axis reflects a continuum from species typical of dry, open habitats (like *Galium verum* and *Pimpinella saxifraga*) over mesophytes of more shaded sites to shore species (such as *Lysimachia vulgaris, Lycopus europaeus* and *Carex acuta*) and helophytes (*Phragmites australis*) (Fig. 2). This can also be seen in the separation of terrestrial and shore plots (Fig. 3a, b), but the plots assigned to the shore subgroup show great variability in both surveys. The divergence or convergence of plots of different types can be seen in the change in range of DCA-scores (Table 3). These comparisons make use of extreme plots only, and therefore dispersion of plots along the DCA axes was analysed by comparing the 1927 and 1986 data with an F-test of equality of variance (Zar 1984). Along the first axis the terrestrial plots (for island interiors) occupy a narrower range in 1986 than in 1927 (cf.

Table 1. Change in mean cover and frequency of the four most abundant species in the tree layer of terrestrial and shore habitats of the ten transects. $n = 10$ (number of transects) for terrestrial habitats, $n = 9$ for shore habitats; no shore plots were analysed on island no 7.

Species	Habitat	Cover (%) 1927	Cover (%) 1986	Change	Frequency (%) 1927	Frequency (%) 1986	Change
Alnus glutinosa	terrestrial	9	38	29	37	61	24
	shore	5	52	47	24	87	63
Betula spp.	terrestrial	8	20	12	52	40	−12
	shore	1	6	5	20	13	−7
Populus tremula	terrestrial	7	23	16	45	52	7
	shore	4	25	21	18	38	20
Sorbus aucuparia	terrestrial	<1	6	6	1	18	17
	shore	0	6	6	0	13	13
Total tree	terrestrial	24	92	68	–	–	–
	shore	10	91	81	–	–	–

Table 2. Change in mean plot frequency in terrestrial and shore parts of the transects for species occurring in at least six of the ten transects in 1927 or 1986. Data for the tree layer are not included. Islands where the individual species did not occur neither in 1927 nor in 1986 were excluded from calculations of mean values and changes. '*n*' refers to the number of islands on which the calculations were based.

Species	Terrestrial parts Frequency (%) 1927	1986	Change	*n*	Shore parts Frequency (%) 1927	1986	Change	*n*
Alnus glutinosa	15	7	−8	5	21	2	−19	5
Betula spp.	55	<1	−54	6	6	8	1	4
Calamagrostis canescens	36	54	18	9	11	43	32	27
Carex acuta	25	10	−15	5	28	50	22	8
Epilobium angustifolium	22	47	25	9	16	26	9	6
Galium palustre	66	19	−47	10	38	38	0	8
Geranium robertianum	16	47	31	9	16	23	8	5
Lycopus europaeus	15	1	−14	4	19	20	1	6
Lysimachia vulgaris	12	22	10	6	4	35	31	5
Lythrum salicaria	10	10	0	5	24	29	5	7
Moehringia trinervia	29	45	16	9	13	3	−10	6
Phalaris arundinacea	41	31	−10	8	63	34	−29	9
Phragmites australis	25	11	−14	5	42	34	−8	5
Populus tremula	55	56	1	8	17	41	24	5
Rubus idaeus	19	57	38	10	23	61	38	6
Salix cinerea	29	7	−21	9	43	13	−30	5
Scrophularia nodosa	37	12	−24	7	13	20	6	5
Scutellaria galericulata	37	18	−19	7	11	42	31	7
Solanum dulcamara	27	11	−15	8	26	21	−5	9
Sorbus aucuparia	14	35	20	9	2	22	20	3
Stachys palustris	28	5	−23	3	10	14	3	6
Taraxacum spp.	20	10	−10	6	10	25	15	6
Tussilago farfara	9	8	0	3	10	41	30	7
Urtica dioica	13	24	10	7	20	10	−10	4
Bryum spp.	8	11	2	4	8	19	11	7
Calliergonella cuspidata	66	0	−66	6	27	31	3	4
Climacium dendroides	18	3	−16	6	40	0	−40	1

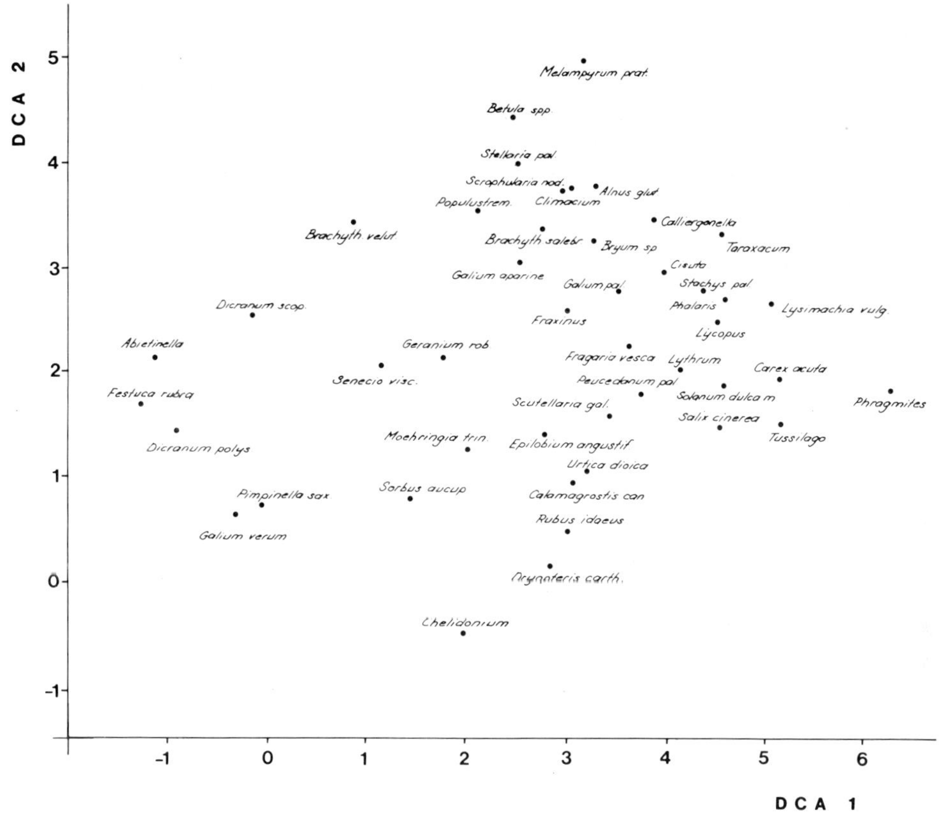

Fig. 2. DCA scores for species occurring on at least 5 islands or in at least 20 of the 429 plots. In addition, a few species at the extreme left are indicated. Names of woody species refer to field or shrub layer specimen – tree data were not included in the ordination.

Fig. 3 a and b), so there has been a decrease in overlap between the two types, even though the mean scores of the shore plots of each island have moved in direction towards the terrestrial cluster (Fig. 4). The convergence of terrestrial plots along this first DCA axis is at least partly due to contrasting changes in esker islands (36 and 39) and other, morainic, islands (Fig. 4). Several esker plots were of the dry open type in 1927 (Fig. 3a). These parts of the esker islands are now more shaded and have a flora more similar to other parts of these and other islands (Fig. 3b). Along the second DCA axis, which mainly reflects a separation among woodland species, there has been a divergence of plots of both types (Fig. 3, Table 3).

Even though tree data were not included in the ordination, the mean score of the terrestrial plots of each island is correlated with the tree layer composition of the transect (Table 4). These relations hold for the first axis in 1927 (positively correlated with proportion of *Alnus* and negatively with *Populus*) and the second axis in 1986 (negatively correlated with proportion of *Alnus* and positively with *Populus*).

When testing for within-island divergence or convergence, the patterns are not so obvious; there are no clear trends of changes in the range of DCA scores covered by shore or terrestrial parts of individual islands (Wilcoxon paired sample tests (Zar 1984) were non-significant in all cases), and the same holds for changes in beta-diversity as reflected by the total range of DCA scores of each island and by B_W.

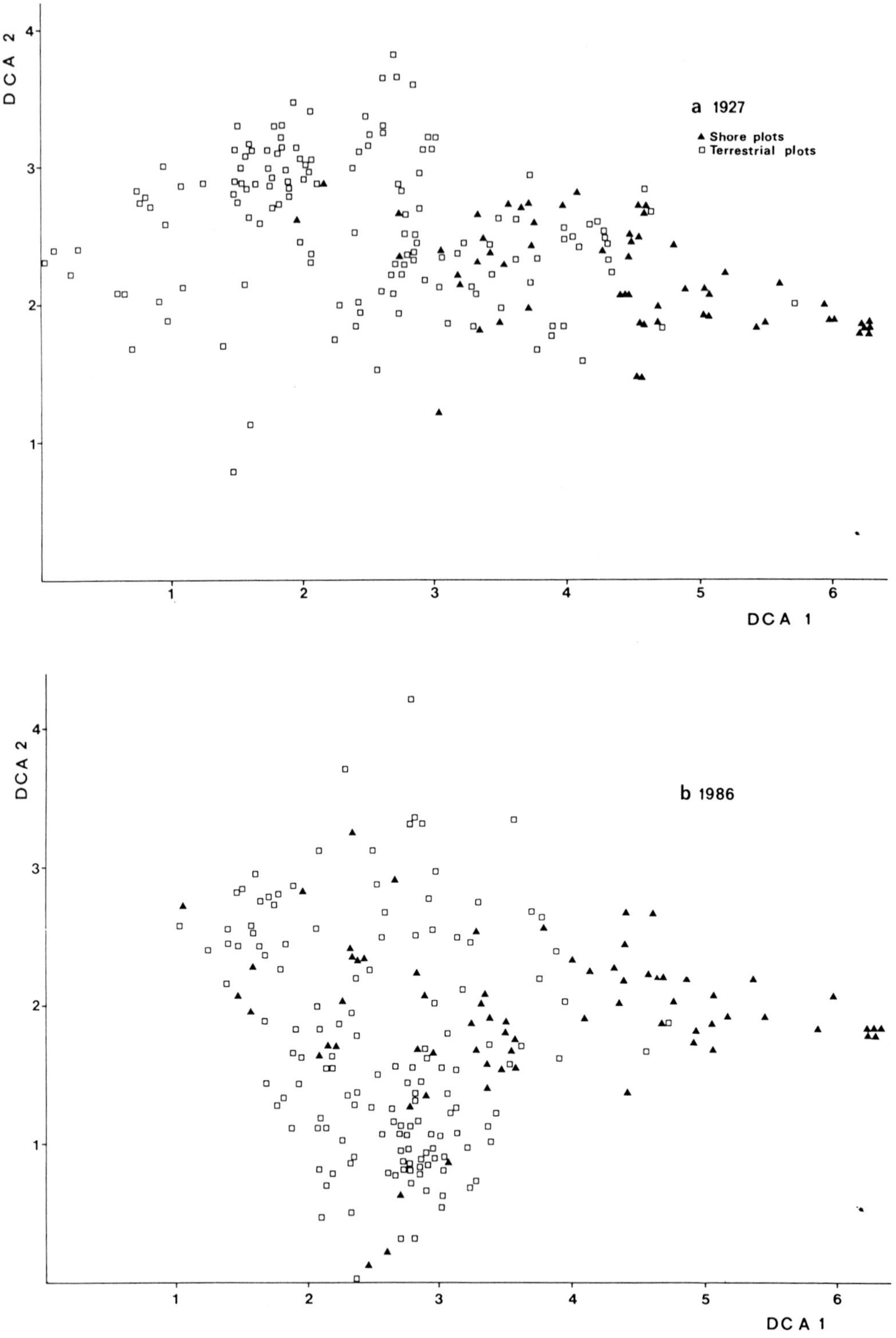

Fig. 3. Sample scores in the DCA ordination of the 429 plots. All plots were included in one ordination, but the two surveys are presented separately (a, b). Triangles: shore plots; squares: terrestrial plots.

Table 3. Variability in floristic composition in shore-, terrestrial- and all plots from the two surveys, as reflected by range of DCA scores (standard deviation units). '*F*-test' is test of equality of variance – significant *F*-test indicates a change in variance of scores between the two surveys. All plots were included in one ordination (the same as in Figs 2–4).

	DCA axis 1			DCA axis 2		
	Range 1927	Range 1986	*F*-test	Range 1927	Range 1986	*F*-test
Shore plots	4.32	5.21	n.s.	1.64	2.59	**
Terrestrial plots	5.78	2.60	***	3.01	4.22	***
All plots	6.26	5.25	***	3.01	4.22	***

n.s. non-significant; ** $0.01 \geq P > 0.001$; *** $P \leq 0.001$

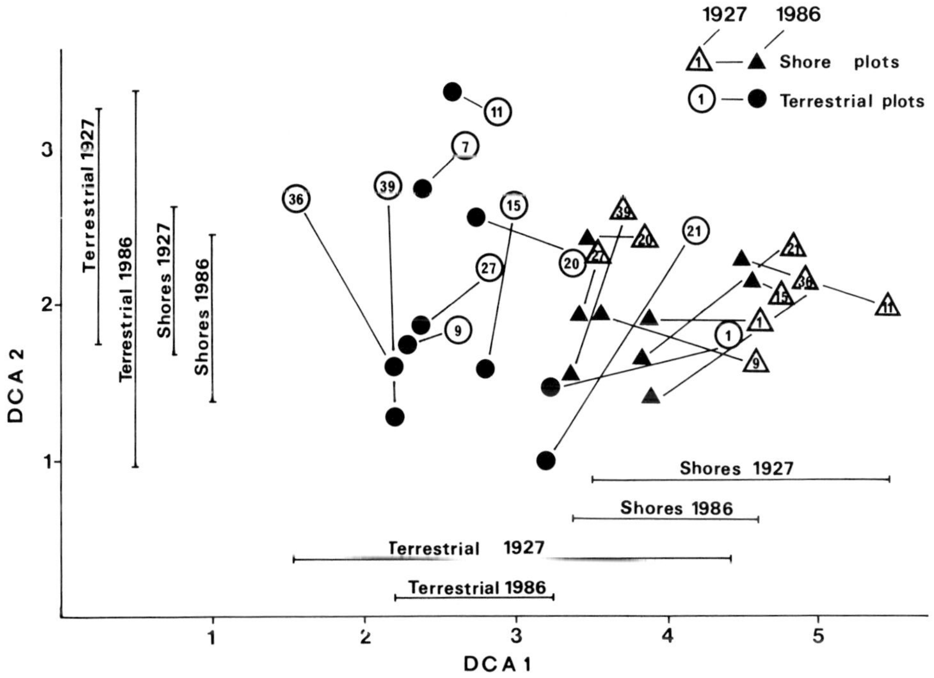

Fig. 4. Mean DCA scores of shore and terrestrial plots from the ten islands. Data are from the same ordination as in Figs. 2–3. No shore plots were analysed on island no. 7.

Table 4. Correlations between mean DCA scores of plots from the terrestrial parts of the islands and relative abundance of *Alnus glutinosa, Betula* spp. and *Populus tremula*. Relative abundance was calculated as proportion of total tree cover (after conversion from Hult-Sernander values to per cent cover). DCA scores from the same ordination as in Figs. 2–4. Island 27 was clearcut in 1926–27 and is excluded from the correlation analyses of the first survey. DCA 1 and DCA 2 refer to correlation with the scores of the first and second DCA axes, respectively.

	1927 (*n* = 9)				1986 (*n* = 10)			
	DCA 1		DCA 2		DCA 1		DCA 2	
Alnus glutinosa	0.82	**	−0.46	n.s.	−0.50	n.s.	−0.88	***
Betula spp.	−0.52	n.s.	0.52	n.s.	−0.37	n.s.	0.60	n.s.
Populus tremula	−0.71	*	0.13	n.s.	−0.20	n.s.	0.74	*

n.s. non-significant; * $0.05 \geq P \geq 0.01$; ** $0.01 > P > 0.001$; *** $P \leq 0.001$

Discussion

In 1927 the interiors of the islands were separated along a floristic gradient with dry meadow plants and shore plants as endpoints. There are two obvious reasons for this. First it reflects the size of the islands; the most 'terrestrial' islands (36 and 39) are the largest ones, both in area and in distance from shore to shore across the island. The most 'shore-like' islands (1 and 21) are the smallest ones by the same measures. Secondly, the largest ones are esker islands and have a substrate which is more permeable and has a lower water holding capacity than that of the morainic islands. During the sixty-year period between the surveys there has been a steady accumulation of organic debris making the wet depressions of the morainic islands more open for colonization of terrestrial plants. This has led to an increase in species richness of the small islands (Rydin & Borgegård in press), and is also reflected by, for instance, a decrease in *Carex acuta* in the island interiors at the same time as *C. acuta* is increasing along the shores. At the same time, this accumulation of organic matter will increase the water holding capacity of the esker islands and reduce the chances for survival of dry site species like *Festuca rubra, Galium verum, Pimpinella saxifraga* and various bryophytes. This process, in combination with the increased shading, will thus have an opposite effect on esker and morainic islands and promote convergence.

Simultaneous with this convergence of island interiors mediated by the exclusion of species of dry, open sites and of shore species there has been a divergence separating sites within the broad category of mesic, shaded conditions. This divergence is reflected along the second DCA axis, where the positions are correlated with the proportions of *Alnus* and *Populus* in the tree layer. The lack of correlation with *Betula* is explainable by the fact that the two species included (*B. pendula* and *B. pubescens*) have slightly contrasting relations to moisture conditions. *Melampyrum pratense*, which did not occur in the transects in 1927, is a characteristic species on islands with low cover of *Alnus* (7, 11 and 20) and is absent from the other islands, which are generally moister sites with a luxuriant vegetation of *Calamagrostis canescens, Rubus idaeus* and *Urtica dioica*. Small differences in average height of the islands lie behind this pattern, so succession seems to be connected with a shift from substrate quality to topography as the factor controlling moisture conditions, and thereby floristic composition, of the site.

If this interpretation is correct it also gives some clues to the relative importance of allogenic and autogenic processes. That tree layer succession is a driving force for changes in field and bottom layer was stressed by Malmer *et al.* (1978) and Brewer (1980). They dealt with a secondary forest succession where increased shading and litter fall was important. In our case the convergence of the terrestrial plots along the first DCA axis is interpreted as a result of accumulation of organic matter and increased shading, thus mainly caused by the vegetation itself. The divergence along the second axis is interpreted as an adjustment to local environment, thus at least partly caused by allogenic forces.

The search for an answer to the question of community convergence or divergence during succession is a classical exercise in plant ecology, but different answers have been obtained (e.g. Olson 1958; Matthews 1979; Christensen & Peet 1984; van der Maarel *et al.* 1985). Beside the methodological difficulties to define and measure convergence, our study gives a hint to why unanimity has not been reached. Not many are today in favour of strict monoclimax arguments. However, from our study we can not argue that divergence would take place in absence of environmental heterogeneity. The islands in Hjälmaren differ in texture and structure of the mineral soil (esker vs. morainic islands) but the importance of this difference has decreased during succession. On the other hand, the subtle differences in elevation (and possibly other factors accounting for differentiation in water availability) have become more important. Succession is thus not a process of monotonic change in one single driving variable. As long as the driving forces and environmental factors controlling vegetation differ over succession, it seems unlikely that the question of convergence or divergence can be given an unambiguous answer.

According to the ordination the vegetation of the shores has become more 'terrestrial-like' during succession. This indication of an outward migration of plant species is most obvious on the smallest (and wettest) islands (1 and 21) where the position of shore plots in the ordination today is similar to (or even at a more 'terrestrial' position than) the position of the terrestrial plots in 1927 (Fig. 4). It is also reflected by the decrease in plot frequency of several shore species (as *Carex acuta* and *Scutellaria galericulata*) on the island interiors and simultaneous increase on the shores. Also, these changes are likely to be an effect of increased shading and/or above mentioned changes in soil properties, possibly in combination with interspecific competition.

Some of the vegetational changes on the shores are consequences of the exposure to disturbance by ice push and wave actions in these habitats. Several pioneer (or early successional) species were widely distributed over the whole islands in the first years following the emergence of the islands (Callmé 1887; Grevillius 1893) but are now mainly found on the shores. The transect data do not allow for quantitative comparisons in many species, but the increase in plot frequency of *Taraxacum* and *Tussilago farfara* on the shores is striking.

Christensen & Peet (1984) suggest that changes in beta-diversity could be brought about by dominance of a few species in late stages (decreased beta-diversity) or by decrease in the realized niches (increased beta-diversity). Trends of increased beta-diversity over succession was reported by Matthews (1979) and Christensen & Peet (1984). Our analysis of beta-diversity (which refers to the differentiation within islands, i.e. within a rather small area) does not show any such trends, neither with Whittaker's B_W nor with changed range of DCA scores. The latter measure is dependent upon extreme plots in each case and this indicates that the gradient length of each island has not changed in any consistent manner.

Acknowledgements

We are grateful to Ragnar Morander and Erik Julin for letting us use their unpublished data and to Bo Eriksson for support during field work. We thank Colin Prentice and Eddy van der Maarel for comments on the manuscript, Erik Sjögren for identification of bryophytes and Agneta Nordgren and Folke Hellström for technical assistance. Financial support was obtained from the Swedish Natural Science Research Council.

References

Atlas över Sverige. 1953. (National Atlas of Sweden) Sheet 32. Svenska Sällskapet för Antropologi och Geografi, Stockholm.

Birger, S. 1905. De 1882–1886 nybildade Hjälmaröarnas vegetation. Arkiv för Botanik 5(1): 1–152.

Brewer, R. 1980. A half-century of changes in the herb layer of a climax deciduous forest in Michigan. J. Ecol. 68: 823–832.

Callmé, A. 1887. Om de nybildade Hjelmar-öarnes vegetation. Bihang Kungl. Svensk. Vetenskaps–Akad. Handl. Band 12, afd. 3, No 7, pp 1–44.

Christensen, N. L. & Peet, R. K. 1984. Convergence during secondary forest succession. J. Ecol. 72: 25–36.

Grevillius, A. Y. 1893. Om vegetationens utveckling på de nybildade Hjelmar-öarne. – Bihang Kungl. Svensk Vetenskaps–Akad. Handl. Band 18, Afd 3, No. 6, pp 1–110.

Hill, M. O. 1979. Decorana – A FORTRAN program for detrended correspondence analysis and reciprocal averaging. Cornell University, Ithaca, N.Y.

Hill, M. O. & Gauch, H. G. 1980. Detrended correspondence analysis, an improved ordination technique. Vegetatio 42: 47–58.

Malmer, N., Lindgren, L. & Persson, S. 1978. Vegetational succession in a South Swedish deciduous wood. Vegetatio 36: 17–29.

Matthews, J. A. 1979. A study of the variability of some successional and climax plant assemblage-types using multiple discriminant analysis. J. Ecol. 67: 255–271.

Økland, R. H. 1986. Rescaling of ecological gradients. I. Calculation of ecological distance between vegetation stands by means of their floristic composition. Nord. J. Bot. 6: 651–660.

Olson, J. S. 1958. Rates of succession and soil changes on southern Lake Michigan sand dunes. Bot. Gaz. 119: 125–170.

Rydin, H. & Borgegård, S-O. Plant species richness on islands over a century of primary succession: Lake Hjälmaren, Sweden. Ecology, in press.

Shmida, A. & Ellner, S. 1984. Coexistence of plant species with similar niches. Vegetatio 58: 29–55.

Shmida, A. & Wilson, M. V. 1985. Biological determinants of species diversity. J. Biogeogr. 12: 1–20.

Sjörs, H. 1963. Amphi-Atlantic zonation, nemoral to Arctic. In: Löwe A & Löwe, D. (eds), North Atlantic biota and their history, pp 109–125. Pergamon Press, Oxford.

Statens Meteorologisk-Hydrografiska Anstalt. 1935. Förteckning över Sveriges vattenfall 61: Svartån-Eskilstunaån. Statens Meteorologisk-Hydrografiska Anstalt, Stockholm.

Tuhkanen, S. 1980. Climatic parameters and indices in plant geography. Acta Phytogeogr. Suec. 67.

van der Maarel, E., Boot, R., van Dorp, D. & Rijntjes, J. 1985. Vegetation succession on the dunes near Oostvoorne, The Netherlands; a comparison of the vegetation in 1959 and 1980. Vegetatio 58: 137–187.

Whittaker, R. H. 1960. Vegetation of the Siskiyou Mountains, Oregon and California. Ecol. Monogr. 30: 279–338.

Willén, E. 1984. The large lakes of Sweden: Vänern, Vättern, Mälaren and Hjälmaren. In: Taub, F. B. (ed.), Lakes and reservoirs. Ecosystems of the world 23, pp 107–134. Elsevier, Amsterdam.

Wilson, M. V. & Shmida, A. 1984. Measuring beta diversity with presence-absence data. J. Ecol. 72: 1055–1064.

Zar, J. H. 1984. Biostatistical analysis. 2nd ed. Prentice Hall, Englewood Cliffs, N.J.

Vegetatio 77: 169–175, 1988
© Kluwer Academic Publishers, Dordrecht – Printed in the Netherlands

The dynamics of the dune vegetation on the Polish Baltic coast

H. Piotrowska
Department of Plant Ecology and Nature Protection, University of Gdańsk, Czołgistów 46, 81-378 Gdynia, Poland

Accepted 1.6.1988

Keywords: Abrasion, Aeolian process, Anthropogenic change, Deflation, Historical change, Regression, Succession

Abstract

This paper deals with the Baltic coastline in Poland. Features of the climate, geomorphology, substratum, soils and vegetation are outlined. The historical forest succession is based on the results of palynological studies of fossil soils. The present-day dynamics of the plant communities are examined from the viewpoint of primary succession, regression, regeneration and secondary succession. The primary succession on moving dunes is described with particular reference to deflation fields. Attention is drawn to the spatial changes in the communities from the 16th to the 20th century as a result of natural and human factors. This study shows that during the 19th and 20th centuries and especially in the last 25 years, the dynamics of the dunes and their vegetation have been much reduced, owing to the widespread stabilization of the substratum by forests and the ongoing abrasion of the coastal dunes.

Introduction

The vegetation dynamics are nowhere so apparent as on dune and saline sea-shores (Ranwell 1972; Ellenberg 1986). This has been amply demonstrated in papers on the coasts of western Europe (e.g. Doing 1975; van der Maarel 1978; van Dorp *et al.* 1985; van der Maarel *et al.* 1985; van der Meulen & Wanders 1985). However, on the Polish Baltic Sea coast, the problem of vegetational changes on the dunes has only occasionally been a principal research object (Wojterski 1964a; Tobolski 1980; Piotrowska & Stasiak 1982 a,b; Piotrowska, in press).

Habitat conditions and natural communities

The climate of the southern Baltic coast has an oceanic element. The oceanic character diminishes eastwards, with distinct changes at the mouths of the Oder, Vistula and Niemen rivers. On the Polish coast the mean annual temperature is 7.6 °C, the mean range is 18.7 °C/year; the annual precipitation is from 550 to 700 mm with a maximum in summer and a minimum in spring. The prevailing winds are westerly and south-westerly, often strong (8–14 m/s) and, during winter storms, very strong (>14 m/s). In comparison with inland areas, winter on the coast is relatively short and mild, the spring cool and the autumn prolonged.

The shore-line consists of alternating stretches of dunes and cliffs. A characteristic feature are the long sand-bars, formed from dunes, which enclose bays becoming lagoons. There are about 13 of these bars, from under 100 to ca 3000 m wide (mostly about 1000 m), and from a few to 70 km long. Their lengths increase eastward. Away from the bars, there are dunes here and there along the coastal margins of the lowlands or Pleistocene uplands. The dunes lying parallel to the beach are usually 3–8 m in height, occasionally over 10 m. The moving dunes, both stabilized and unstabilized, are much higher: 30 m a.s.l. on the Vistula Bar and 40–56 m on the Leba Bar. The groundwater is only exceptionally accessible to plants, e.g. in the deflation hollows amongst the moving dunes. The dune sand consists for 90% of fine or medium grained quartz. It is extremely poor in nutrients, with scarcely a trace of calcium, and very susceptible to acidification. The pH of the sand on the beach and yellow dune is 7.0–6.0, on the grey dune it is 5.5–5.0 and in the forest soils 4.5–3.5. The dominant soil-forming process is podzolization, but mature forest soils are rare.

The most important natural communities on the Polish dunes are: *Elymo-Ammophiletum honckenyetosum* on the fore dune, *festucetosum arenariae* or *typicum* on the yellow dune; *Helichryso-Jasionetum typicum* on the seaward side of the grey dunes, and *cladonietosum* on the landward side (Piotrowska & Celiński 1965; Piotrowska 1984); *Empetro nigri-Pinetum cladonietosum, piroletosum, typicum* and *ericetosum* – depending on the moisture content of the oligotrophic habitats (Wojterski 1964a); *Betulo-Quercetum* – in mesotrophic habitats within range of low dunes.

The study was carried out between 1978 and 1986 along the whole Polish coast, where dunes extend over a total of 400 km, i.e. 4/5 of the entire shoreline. Along with literature data (only the most important sources are quoted here) the author's own results have been used in this paper.

Names of plant communities are used following Matuszkiewicz (1981). The expressions 'dune' and 'dune coast' are used in their widest sense: they refer to a substratum of dune origin, regardless of the nature of the present-day plant cover or of the extent of stabilization.

Historical changes in the vegetation

The history of the vegetation has been studied on the Leba Bar using modern palynological methods; fossil soils were the object of study here (Tobolski 1980). The Leba Bar got about its present-day shape in the Subboreal period. It was covered by low, gently sloping rampart dunes lying roughly parallel to the shore. The first forest community to become dominant on the bar was mesotrophic oak forest. Around 1940 B.P. this oak woodland was in large part destroyed by a fire, probably started by man. Beech expanded rapidly on the mature forest soils of the former oakwoods, and later, acidophilous beech forest covered much of the bar. A side effect of the fire was the local intensification of aeolian processes and the gradual rise in importance of pine; this tree began to colonise the bar sands around 1540 B.P. In later times, frequent fires and forest clearing led to the almost total destruction of the beechwoods. As a result, pine expanded further, and aeolian processes intensified: from the 16th century onwards, huge moving dunes began to grow and wander about. They destroyed the last vestiges of the deciduous forests and the younger pinewoods as well.

Present-day changes in the communities

Primary succession

On the coastal dunes with a zonal distribution of habitats and vegetation, the succession passes from initial forms of the *Elymo-Ammophiletum* through the *Helichryso-Jasionetum* to forest communities. Its initial stages are similar along the whole coast, but subsequent development depends on local ecological conditions (cf. Westhoff & Schouten 1979). The series bifurcates at the *Helichryso-Jasionetum* stage: lichen-dominated phytocenoses prepare the habitat for the *Empetro nigri-Pinetum* (Wojterski 1964a), whereas phytocenoses dominated by bryophytes will become *Betulo-Quercetum*. The coastal succession has been presented in a number of papers (e.g. Wojterski 1964a; Piotrowska & Celiński 1965), so a few comments will suffice here. The distribution patterns presented in these papers are mainly based on the spatial sequence of ecological zones. This generalisation of the situation leaves much unsaid and is difficult to trace in the field. The vegetational development depicted can take place only when sand accumulates continuously, which causes the dunes to grow out into the sea and the zones to shift their positions. Nowadays this process takes place only exceptionally on a few, short stretches and it is usually ephemeral. A complete sequence of zones reflecting exactly the coastal vegetation succession is very rare on the Polish coast (cf. Wojterski 1964b).

The primary succession with the full range of possible stages is found on the moving dunes on the Leba Bar. Three succession series are developing here. One of them is associated with the lee slopes of the moving dunes. The initial stage comprises mainly *Ammophila arenaria* (cf. Ranwell 1960). On sheltered spots, the second – and last – stage of this series forms a sward with *Corynephorus canescens*.

Two other series have been found on the deflation fields whose floors have a very specific microrelief (Fig. 1): low rib-like ridges of sand have built up transversally to the long axis of the field, while between these ridges, troughs are visible (Miszalski 1973) – these are in fact the original field floor. Deflation fields grow in an easterly direction in the wake of the receding dunes (Fig. 2). The succession begins on the bare, moist sand at the field front, and ends at the western, i.e. the oldest, edge. The ridges and troughs are composed of the same substratum, but differ in their moisture content. The 0.5–2.5 m high ridges are relatively dry, whereas the troughs are moist (the ground water level here varies by 0–60 cm/year). On these two alternating habitats the vegetation has developed in quite different ways (Fig. 3) as regards floristic composition, number of

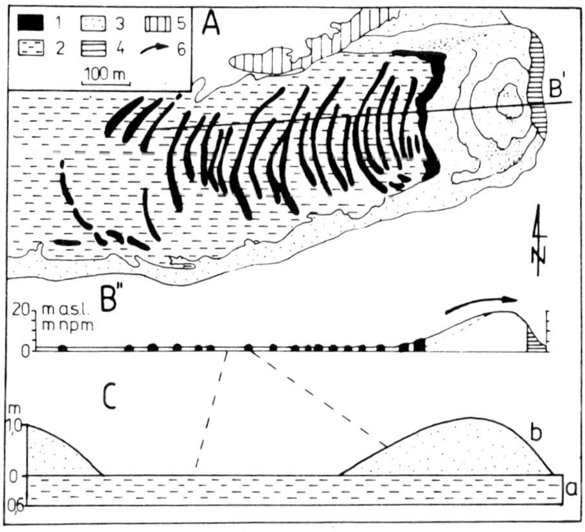

Fig. 1. Example of a deflation field (A, B) and microrelief of its floor (C). 1) ridges, 2) troughs, 3) moving dune and its arms, 4) dune front, 5) other dunes, 6) direction of dune movement (adapted from Miszalski 1973); a) oscillation of the ground water level on the original moist floor of the field, b) dry ridges (original).

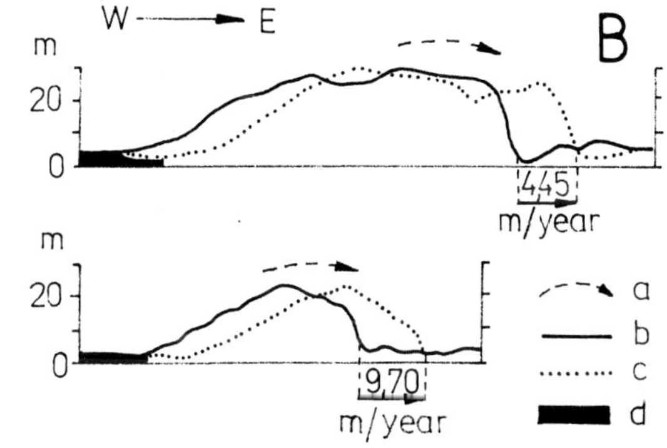

Fig. 2. The movement of moving dunes on the Łeba Bar. a) direction of dune movement; b) state in 1958, c) state in 1968; d) deflation hollows (adapted from Miszalski 1973).

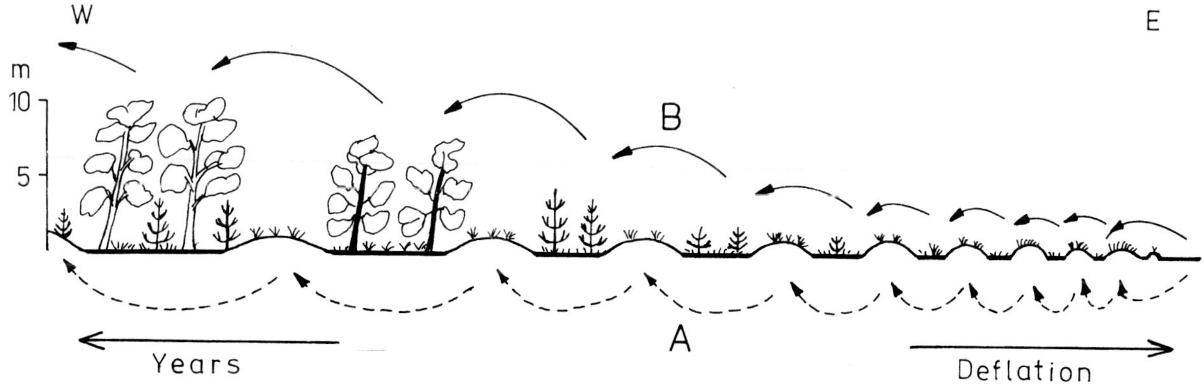

Fig. 3. Two simultaneous succession series on the deflation field. A) succession on the ridges, B) succession in the troughs (original).

species, cover of the vegetation layers, number of stages, and the rate and direction of succession. On the ridges, the succession is slow and the forest stage is never reached. A few psammophytes and lichens are involved. *Ammophila arenaria* and *Corynephorus canescens* make up the initial stage and this is followed by three more stages composed of a psammophilic sward and a lichen layer of gradually changing composition. In the troughs succession proceeds faster and involves three times as many vascular plant species and one and a half times as many cryptogamic species as on the ridges; the terminal community is pine forest. Bryophytes, particularly *Polytrichum commune* var. *perigoniale*, are prevalent among the cryptogamic plants. The pioneers of this succession are *Juncus articulatus* subsp. *litoralis* and *Agrostis stolonifera* (cf. Ranwell 1960); the phytocenoses are then taken over by Droseraceae- and Ericaceae species, and finally by *Pinus sylvestris*, which transforms the heathland into a young stage of *Empetro nigri-Pinetum ericetosum* or -*typicum*. Eight stages of development have been distinguished (Piotrowska in press).

In the deflation hollows, the development of the vegetation is interrupted at various stages by the encroachment of the dunes following behind. That is why there are no old forests or mature soils here.

Regression, regeneration and secondary succession

Apart from the well-known floristic, structural and habitat differences, the herbaceous and forest communities on the dunes display variations with respect to the proximity of the sea, the intensity of aeolian processes, the extent of substratum stabilization and of soil development, not to mention the nature and intensity of human activities (cf. Westhoff 1985). That is the reason why the reaction of these communities to destructive agencies varies strikingly.

The components of the *Elymo-Ammophiletum* are naturally adapted to the aeolian processes occurring on the dunes (Ranwell 1972; Huiskes 1979; Ellenberg 1986), as long as these processes are not too rapid. They require a gradual accumulation of sand and they themselves contribute to this. They tolerate moderate deflation, but a sudden accumulation of a large amount of sand or excessive deflation will destroy these phytocenoses. Abrasion of the dunes will eliminate them together with the substratum. Extreme habitat conditions and the lack of competing species from other habitats will usually allow the phytocenoses to regenerate themselves (cf. Wojterski 1964b). A fairly level surface, newly stripped or covered by wind-blown sand will always be colonized; for abrasion slopes to be colonized, however, is rare, as seeds roll down on to the beach.

Regression in *Helichryso-Jasionetum* phytocenoses is caused by erosion of heavy accumulation of sand. Both processes are frequent on the coast as the sea has abraded the yellow dunes and is now about to abrade the remaining belt of psammophilic sward. Regression means the destruction of initial soil stages. The bare sand is colonized by *Ammophila arenaria* and *Corynephorus canescens*; regeneration has thus started, but will continue only if these phytocenoses remain at an appropriate distance from the sea. On the grey dunes, secondary succession is also possible if fossil soils are present in the shallow substratum left after erosion.

In the forest communities the causes and effects of regression are highly diversified, mainly as a result of various human activities. Natural causes of regression are exceptional and they exert only local effects. They may include erosion of the substratum in *Empetro-Pinetum* phytocenoses with open stands of trees, or blown-down trees in *Betulo-Quercetum* phytocenoses which, after abrasion of the shore, suddenly found themselves in the vicinity of the beach (Fig. 4). In both cases regeneration, i.e. the renewal of former woodland phytocenoses only partially destroyed, is possible. Regression due to anthropogenic factors is common, especially by clearing of tree stands, trampling and local fires. The consequence is a secondary succession which leads to replacement – clearing communities, trampling-resistant communities, sometimes ruderal communities; tree planting is also practised on a large scale. Locally secondary succession may follow natural regression, e.g. where there is slow aeolian accumulation of sand in the forest or where the undergrowth has been buried under sand carried there by storm-waves.

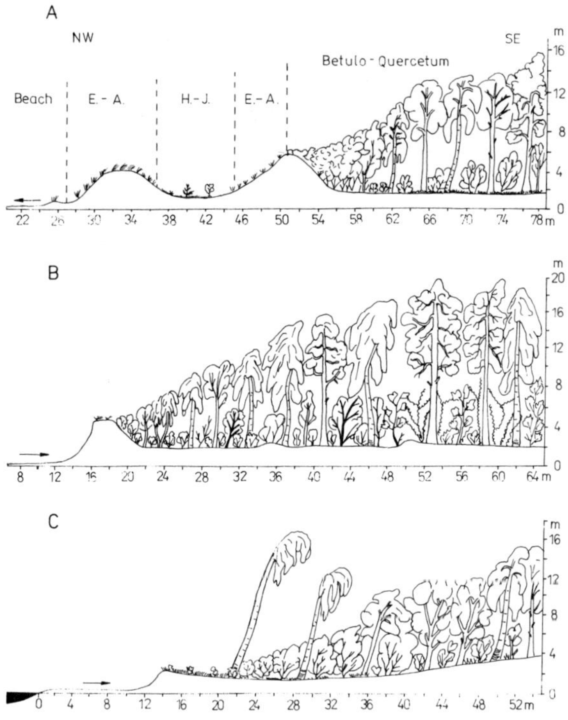

Fig. 4. Full differentiation (A) and regression of coastal communities due to dune abrasion (B, C) and blown-down trees (C). E-A: *Elymo-Ammophiletum*, H-J: *Helichryso-Jasionetum*, Forest: *Betulo-Quercetum*. Transects done on the central Polish coast in 1986 (original).

Over the last 25 years, there has been an intensive renewal of deciduous trees in the older pine plantations of some coastal sections, e.g. on the Vistula Bar. The mixed woodland is trying to return to its former habitats; this process should be considered as secondary succession.

Synanthropization

Synanthropization of the plant cover is part of the directional changes that takes place on the earth as a result of man's activities and is manifested as the replacement of specific components by non-specific ones, of native components by newcomers, and of stenotopic- by eurytopic components (Faliński 1972, 1986; Olaczek 1982; Kornaś 1983).

Because of the extreme habitat conditions on the dunes, synanthropization is less marked here than further inland. None the less, growing pressures from recreation and forestry management have induced considerable changes in the composition of the plant cover; e.g. a comparison of the state of the flora and communities on the dunes on the Vistula Bar today (Piotrowska & Stasiak 1982a, b) with that present at the start of the 20th century (cf. Preuss 1906), shows that: 1. The overall number of vascular plant species has increased by 50%, among the adventive there are many native species atypical of the natural dune flora (e.g. ruderal or meadow species) and anthropophytes. The degree of synanthropization of the flora has increased two and a half times in 80 yr. 2. The *Elymo-Ammophiletum* phytocenoses did not change substantially in composition, but the numerical proportions of the components changed as a result of the planting of psammophytes. 3. The greatest changes occurred in the *Helichryso-Jasionetum* zone, i.e. on the grey dunes. At present there are 350 vascular plant species found in this zone, 20% of which are anthropophytes. The serious synanthropization is due to mechanical damage to the substratum, to its fertilization by peat to encourage the growth of the planted tree- and shrub monocultures, and also to the deposition of rubbish and the accidental introduction of diaspores by holidaymakers. The composition, numerical proportions and structure of most of the *Helichryso-Jasionetum* phytocenoses have changed as a result of past or present planting of *Pinus sylvestria, Rosa rugosa, Salix daphnoides* and others.

Natural spatial changes in the communities

The abrasion of the Pleistocene uplands and the formation of sand-bars has smoothed the shore line of Poland. The period of intensive accumulation on the shore has thus come to an end. In recent decades, abrasion has been exceeding accumulation to an ever increasing extent on ever longer stretches of the coastline. Measurements made on the Polish coast over many years show that the long-shore drift is systematically decreasing, with the result that beaches are getting narrower and dunes lower. The shoreline is receding at a mean rate of 1 m/yr. However, in certain parts losses may go up considerably.

The effects of long-term abrasion, and of the disastrous storm in January 1983 are that along ca 75% of the dune shores of Poland (ca 300 km), the beach is very narrow (15–20 m), and the yellow dune with *Elymo-Ammophiletum* is practically nonexistent; the grey dunes with *Helichryso-Jasionetum* are in a fragmented state if they exist at all. Locally, forests are now right behind the beach (Figs. 4, 5); some of them were swept away by the sea in 1983, the surviving tree-stands are being destroyed by the wind. Most of the sand-bars have become narrower by less than 20 to 50 m, and some of them were broken by storm waves in 1983, e.g. the Hel Peninsula and the Kopań Bar.

Anthropogenic spatial changes in the communities

The effect of the intensification of aeolian processes following the destruction of woods on the dunes, and later human activities, is that the configuration of the dunes and the distribution and composition of communities have changed from the 16th to the 20th centuries. The aeolian accumulation of high mobile dunes has raised the absolute and relative height of the terrain by a factor of 2–3. These dunes have covered the former low rampart dunes and restricted the moist habitats to the deflation hollows. On the other hand, initial psammophytic stages have spread. As the shifting masses of sand were menacing human settlements, stabilization of the dunes was undertaken in the 19th century. First psammophyte grasses were sown, then pine monocultures were tried (cf. Westhoff 1961, 1985). Now the former moving dunes on the Vistula Bar, the Sarbska Bar and on the western part of the Leba Bar are almost completely stabilized and covered by woodland (Piotrowska & Stasiak 1982b; Piotrowska in press).

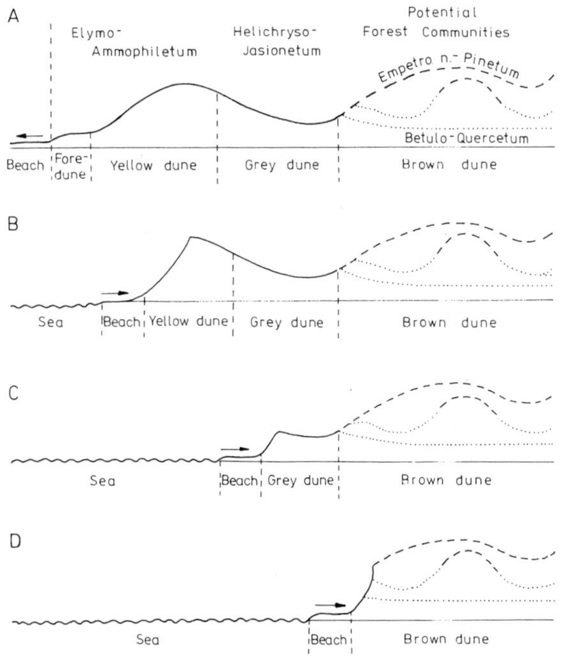

Fig. 5. The zonal distribution of communities on the Polish coastal dunes (A) and the reduction of their area as a result of ongoing abrasion (B, C, D) (original).

During the 19th century, a yellow dune rampart was built on the coastal dunes – which reinforced ongoing aeolian accumulation. *Ammophila arenaria*, sometimes other psammophytes, were planted here, and this was a starting point for many of the later *Elymo-Ammophiletum* phytocenoses. At the same time, pine monocultures were started in many places on the landward side of the coastal dunes. Only a narrow strip (30–60 m wide) of *Helichryso-Jasionetum* was left on the seaward side. In the forests that are managed pines are being planted to form the second or third tree generation. The area of these woods is shrinking on the landward side: recreational facilities are provided a.o. camp sites and a network of paths and roads is becoming denser. Twenty-five years ago, the next step in the elimination of the narrow *Helichryso-Jasionetum* belt was taken: along most of the Polish coastline, pine monocultures were established stretching right up to the yellow dunes. A large part of these plantations is being destroyed by abrasion or aeolian processes, though some, e.g. on the Vistula Bar, are surviving and growing. The heliophytic *Helichryso-Jasionetum* phytocenoses are disappearing from these areas (Piotrowska & Stasiak 1982a, b).

References

Doing, H. 1975. Beobachtungen und historische Tatsachen über die Sukzession von Dünen-Ökosystemen in den Niederlanden. In: Schmidt, W. (ed.), Sukzessionsforschung, pp. 107–122. Cramer, Vaduz.

van Dorp, D., Boot, R. & van der Maarel, E. 1985. Vegetation succession on the dunes near Oostvoorne, The Netherlands, since 1934, interpreted from air photographs and vegetation maps. Vegetatio 58: 123–136.

Ellenberg, H. 1986. Vegetation Mitteleuropas mit den Alpen in Ökologischer Sicht. 4th ed. Ulmer, Stuttgart.

Faliński, J. B. 1972. Synantropizacja szaty roślinnej - próba określenia istoty procesu i głównych kierunków badań (Synanthropisation of the plant cover – an attempt to define the nature of the process and of the main fields of investigations. Phytocoenosis 1(3): 157–170.

Faliński, J. B. 1986. Vegetation dynamics in temperate lowland primeval forests. Geobotany 8: 1–537. Junk, The Hague.

Huiskes, A. H. L. 1979. Biological flora of the British Isles – *Ammophila arenaria* (L.) Link. J. Ecol. 67: 363–382.

Kornaś, J. 1983. Man's impact upon the flora and vegetation in Central Europe. In: Holzner, W., Werger, M. J. A. & Ikusima, I. (eds), Man's impact on vegetation. Geobotany 5: 277–286. Junk, The Hague.

van der Maarel, E. 1978. Experimental succession research in a coastal dune grassland. A preliminary report. Vegetatio 38: 21–28.

van der Maarel, E., Boot, R., van Dorp, D. & Rijntjes, J. 1985. Vegetation succession on the dunes near Oostvoorne, The Netherlands; a comparison of the vegetation in 1959 and 1980. Vegetatio 58: 137–187.

Matuszkiewicz, W. 1981. Przewodnik do oznaczania zbiorowisk roślinnych Polski. PWN, Warszawa.

van der Meulen, F. & Wanders, E. A. J. 1985. Dynamics and management of some coastal dune woodlands near The Hague, The Netherlands. Vegetatio 62: 457–465.

Miszalski, J. 1973. Współczesne procesy eoliczne na Pobrzeżu Słowińskim – studium fotointerpretacyjne (Present-day aeolian processes on the Slovinian coastline – A study of photointerpretation) IG PAN, Warszawa.

Olaczek, R. 1982. Synanthropization of phytocoenoses. Memorabilia Zool. 37: 93–112.

Piotrowska, H. 1984. Szata roślinna (Plant cover). In: Augustowski, B., (ed.), Pobrzeże Pomorskie (Pomeranian Coast). pp. 281–317. Ossolineum, Wrocław.

Piotrowska, H. in press. Development of the vegetation in the active deflation hollows on the Łeba Bar. Fragm. Flor. Geobot.

Piotrowska, H. & Celiński, F. 1965. Zespoły psammofilne wysp Wolina i południowo-wschodniego Uznamu (Psammophilous vegetation of the Wolin Island and that of south-eastern Uznam). Bad. Fizjogr. Pol. zach. 16: 123–170.

Piotrowska, H. & Stasiak, J. 1982a. Naturalne i antropogeniczne zmiany strefowe flory naczyniowej bezleśnych wydm nadmorskich Mierzei Wiślanej (The natural and anthropogenic zonal changes of vascular flora on the non-forest coastal dunes of the Vistula Spit. Fragm. flor. geobot. 28(3): 371–396.

Piotrowska, H., Stasiak, J. 1982b. Zbiorowiska na wydmach Mierzei Wiślanej i ich antropogeniczne przemiany (The plant communities on the dunes of the Vistula Spit and their anthropogenic changes). Fragm. flor. geobot. 28(2): 161–180.

Preuss, H. 1906. Die Vegetationsverhältnisse der Frischen Nehrung (westpreussischen Anteils). Kafeman, Danzig.

Ranwell, D. S. 1960. Newborough Warren, Anglesey. II. Plant associes and succession cycles of the sand dune and dune slack vegetation. J. Ecol. 48: 117–141.

Ranwell, D. S. 1972. Ecology of salt marshes and sand dunes. Chapman and Hall, London.

Tobolski, K. 1980. The fossil soils of the coastal dunes on the Łeba Bar and their paleogeographical interpretation. Quaestiones Geographicae 6: 83–97.

Westhoff, V. 1961. Die Dünenbepflanzung in den Niederlanden. Angew. Pflanzensoz. 17: 14–21.

Westhoff, V. 1985. Nature management in coastal areas of Western Europe. Vegetatio 62: 523–532.

Westhoff, V. & Schouten, M. G. C. 1979. The diversity of European coastal ecosystems. In: Jefferies, R. L. & Davy, A. J., (eds), Ecological processes in coastal environments. pp. 3–21. Blackwell, Oxford.

Wojterski, T. 1964a. Bory sosnowe na wydmach nadmorskich na polskim wybrzeżu (Pine forests on sand dunes at the Polish Baltic coast). PTPN. Prace Komis. Biol. 28(2): 1–217.

Wojterski, T. 1964b. Schematy strefowego układu roślinności nadmorskiej na południowym wybrzeżu Bałtyku (Schemata of the zonal vegetation system on the southern coast of the Baltic Sea). Bad. Fizjogr. Pol. zach. 14: 87–105.

Vegetatio 77: 177–183, 1988
© Kluwer Academic Publishers, Dordrecht – Printed in the Netherlands

Vegetation dynamics of the savannization process on Cuba

Attila Borhidi
Institute of Ecology and Botany, Hungarian Academy of Sciences, H-2163 Vácrátót, Hungary

Accepted 2.9.1988

Keywords: Savanna, Determinant factor, Fire, Grazing, Human Impact, West Indies

Abstract

The main determinants of tropical savannas are shortly discussed, notably climate, insufficient drainage, fire and their interactions. The savannization process on Cuba is basically induced by man as shown for six different habitats. Different savanna types develop from different forests and shrublands through regular burning and grazing. In dry habitats and on oligotrophic soils, savannization is a short term, usually irreversible process, while in humid habitats and on fertile soils it is a long-term reversible process, with spontaneous recolonization of forest elements after savannization has been stopped.

Introduction

Savanna was originally considered as a climatically controlled tropical grassland with a relatively long dry period each year (Schimper 1898; Warming & Graebner 1918; and others). Although this opinion was criticized by Troll (1935), Walter (1962) and others, the seasonal tropical climate is still treated as a savanna climate and its geographical region as a savanna-belt.

Lanjouw (1936), Hardy (1945) and others pointed to the low nutrient level of the extremely leached soils and Waibel (1948) to certain nutrient-poor bedrocks (sandstone, cristalline slate, white sand and serpentine) as the main conditioning factors of savannas.

Later, bad drainage was put forward by pedologists (Bennett & Allison 1928) and ecologists (Beard 1944, 1953; Hills & Randall 1968; Vesey-FitzGerald 1963, etc.). According to them a shallow unpenetrable ironpan or hardpan layer is responsible for an extremely fluctuating water table causing inundations during the wet and desert-like habitat in the dry season.

Busse (1908), Myers (1936) and many others consider fire a dominant factor controlling savanna vegetation and regard savannas as a fire climax. Although in certain regions, e.g. Africa, atmospheric electric activity is high, this factor should not be overestimated (Komarek 1971).

More recent authors (Cole 1960; Eden 1964; Hills 1969; Goodland 1970; Monasterio 1971; Monasterio & Sarmiento 1971; Eiten 1972) point to the interaction of several factors in the evolution and maintenance of the savanna ecosystem. Some may be major conditioning or predisposing factors, others may be additional determinants while still others may be important in the genesis of the ecosystem but not in its maintenance. Since this concept includes a set of simultaneous and/or successive agents, it is called holocenotic by Sarmiento & Monasterio (1975). This holocenotic concept was further developed by Walker (1987) as one of the possible models.

The whole constellation of the interacting factors considered in the model changes fundamentally when man interferes in the system and becomes the determinant of savanna structure controlling and using the different determinant factors for his purpose.

Many ecologists (e.g. Aubréville 1947, 1965;

Budowski 1956; Cole 1960, 1986; Vareschi 1960, 1962) consider that man has the most important role in the formation and maintenance of savannas.

Savannas on Cuba

The bioclimatic map of Cuba (Borhidi 1974; Borhidi & Muñiz 1980, 1984) shows clearly that in the lowlands and hilly regions of Cuba a seasonal tropical climate is found although these regions are generally considered to have a savanna climate. While actually only 8% of the national territory of Cuba is covered by natural forests, it has been considered (Bennett & Allison 1928; Seifriz 1943; Marie-Victorin & León 1942, 1944; León 1946; Beard 1953) that originally 25–50% was covered by natural grasslands. However, according to own studies of the natural vegetation and its correlation to climate and soil types the country must have been basically a forested land with only 5–8% natural grassland (Borhidi 1973; Borhidi & Herrera 1977; Borhidi & Muñiz 1980, 1984).

Almost all the actually existing grasslands of Cuba have been developed under a long-term human impact of combined logging, cutting, burning and grazing.

Case studies

Savannization processes were studied in seven different habitats.

1) A seasonal evergreen lowland forest belt on fertile deep ferrallitic red soils (Matanzas clay). After logging of the valuable timber trees – as *Swietenia, Cedrela, Hibiscus elatus, Calophyllum* – a *Roystonea-Ceiba* woodland develops which is converted into a tall grassland with scattered *Roystonea* and *Ceiba* by repeated burning and into a pasture by intensive grazing combined with burning (Fig. 1A).

2) The savannization process of the swamp forest starts by logging of timber species and fuel woods as *Bucida, Rhizophora* and *Avicennia*, and *Tabebuia* used for charcoal-burning. After logging a *Sabal* palm swamp woodland develops and by burning and grazing it is converted into a *Sabal* wet grassland and lastly into a wet pasture with scattered *Sabal* palms (Fig. 1B).

3) A semideciduous forest belt. A similar process went on with stages of a *Roystonea-Samanea-Spondias* woodland and a *Roystonea-Samanea* tall grassland and ending in a pasture with scattered *Roystonea* palms (Fig. 1C).

Redevelopment patterns of these forest types were also studied. The process starts with an invasion of the colonizing species from the edge of the still existing forest patches and also by scattered colonizing centres developing around the palms and trees left. During this phase N-fixing shrubs as *Cassia, Myrica, Caesalpinia, Belairia, Behaimia* and *Brya*, further fast-growing deciduous soft-wood trees, like *Bursera, Zuelania, Casearia* and *Trema*, and in the hilly regions *Muntingia* and *Cecropia*, play a dominant role.

In the colonization centres evergreen shrubs and deciduous trees distributed by birds, like Myrtaceae (*Eugenia*), Boraginaceae (*Cordia, Bourreria*), *Spondias*-species, palms, and by wind, like Compositae shrubs e.g. *Eupatorium* and *Vernonia* are important. Around the *Ceiba* trees the colonization process usually starts with the establishment of royal palms distributed by both birds and mammals. (See Guevara *et al.* 1986 for the role of remnant forest trees in Mexico, including *Ceiba*).

4) Hardpan soils or mocarrero soils are largely considered as habitats of natural grasslands. In spite of this we found that the original vegetation of these soil types is a thorny scrub-woodland consisting of endemic palms and trees. This vegetation had no economic value and had been converted by burning into *Sabal* palm grasslands already long ago, or in eastern Cuba into *Copernicia* palm grasslands, and intensively used as pasture. Recently these pastures are being invaded by extremely agressive alien shrubs like the African *Dichrostachys cinerea*, originally introduced for its ornamental flowers some decades ago (Fig. 1D).

Four different stages of this degradation process are presented in Table 1, with corresponding life form and phytogeographical spectra.

Savannas are more obviously expected on dry, nutrient-poor soils, e.g. in the serpentine and white sand areas, where several of the determinant factors

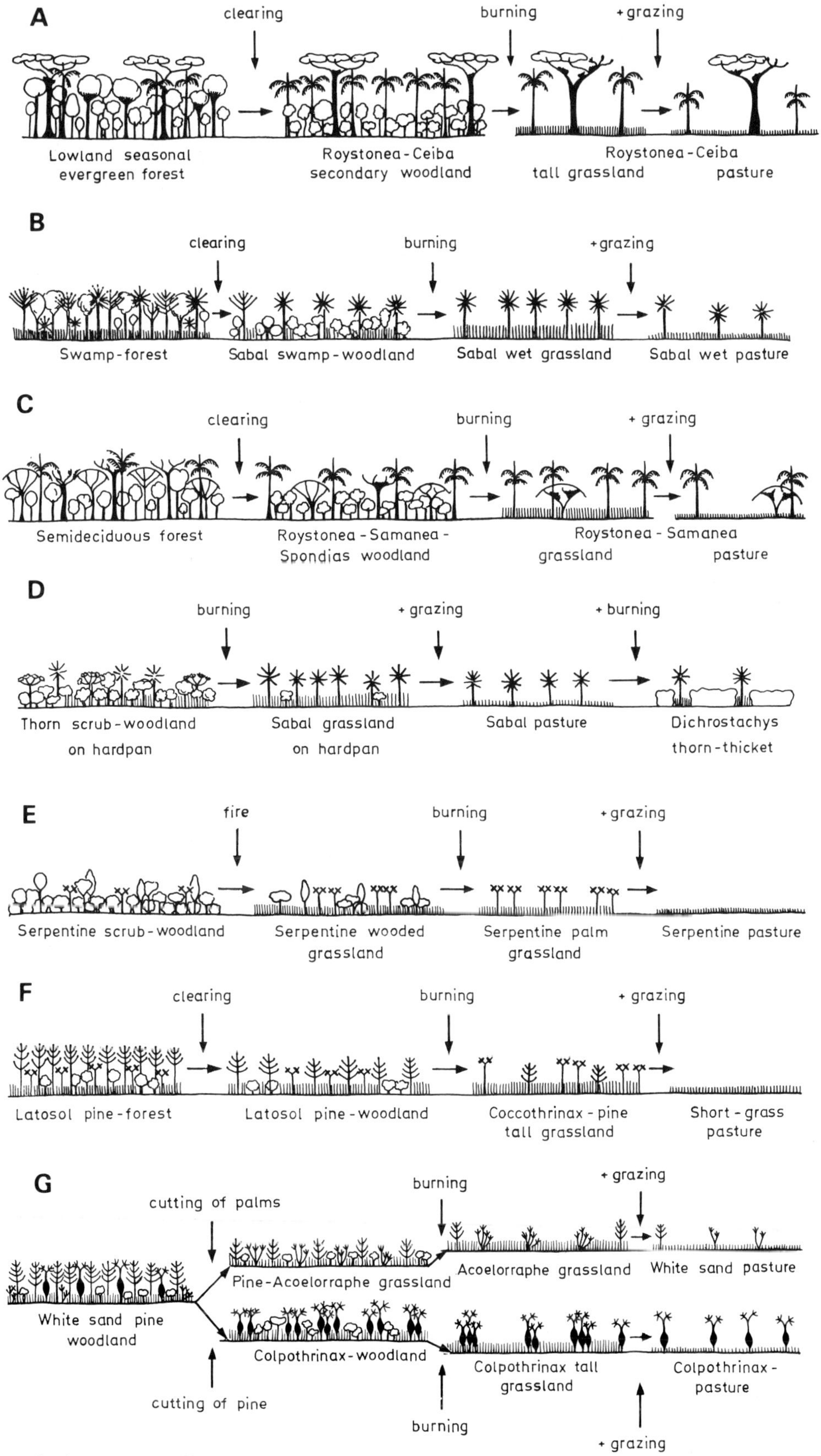

Fig. 1. The savannization process on different soiltypes of Cuba. – A: Red ferrallitic soil on limestone (Matanzas clay); B: Humid tropical black soil; C: Tropical brown soil; D: Yellow ferrallitic soil; E: Reddish-brown fersiallitic serpentine soil; F: Red ferritic serpentine soil; G: White sand.

Table 1. Life form and phytogeographical spectrum of some hyperxerophytic communities on hardpan soil representing different stages of savannization and invasion of aliens.

A: Starting stage: *Belairio-Bucidetum subinermis*, microphyllous semideciduous forest, 5 rel.
B: Intermediate stage: *Camerario retusae-Sabaletum*, palm-savanna woodland, 5 rel.
C: Final stage: *Byrsonimo-Andropogonetum teneris*; savanna community, 7 rel. (Balátová-Tuláčková & Surli 1983)
D: Invaded stage: *Dichrostachys cinerea* thicket 5 rel.

Communities:	A	B	C	D
Tree layer, average cover %	50	10	–	–
Shrub layer	70	20	5	100
Herb layer	20	60	90	–
Number of species	134	80	69	10
Life forms	%	%	%	%
Macro-Phanerophytes	6	2	–	–
Meso-Phanerophytes	9	6	1	–
Micro-Phanerophytes	13	9	8	10
Nano-Phanerophytes	21	13	8	–
Rosulate Meso-Phanerophytes	1	1	–	–
Rosulate Micro-Phanerophytes	2	2	3	10
Woody epiphytes	1	–	–	–
Woody lianes	8	6	1	10
Chamaephytes	5	8	12	–
Hemicryoptophytes	15	33	44	20
Geophytes	4	4	4	–
Therophytes	2	6	13	10
Herb lianes	9	9	6	40
Rosulate epiphytes	3	1	–	–
Other epiphytes	1	–	–	–
Phytogeographical elements				
Endemics, local + regional	9	6	4	–
Endemics, pancuban	27	26	22	10
Endemics, total	36	32	26	10
Greater Antillean	12	10	11	10
Antillean	5	2	1	–
Antillean-Bahaman	4	4	3	–
Cuban-Bahaman	5	5	3	10
North-Caribbean	8	5	4	–
South-Caribbean	1	1	1	–
Pan-Caribbean	5	4	6	–
Neotropical	22	28	29	30
Pantropical	1	5	6	10
Pan-American	–	1	3	–
Cosmopolitan	1	1	4	10
Alien	1	2	3	20

listed above are simultaneously present: long dry period, low nutrient level, eventual fires and occasionally also insufficient drainage. In spite of this, the original natural vegetation of these habitats consists of different woody plant communities and not of grasslands.

5) In the serpentine areas on deep mature ferritic soils the climax vegetation is a dense pine forest of *Pinus caribaea* in West-Cuba and of *Pinus cubensis* in the East. On shallow or gravelly serpentine soils a very dense, thorny evergreen thicket is the original vegetation, which is transformed by burning and grazing into short grass savanna with scattered dwarf palms of the genera *Coccothrinax* and *Copernicia* (Fig. 1E).

With a highly intensive use of palm-savannas by burning and grazing, even palms disappear and savannas without palms develop. Stages and details of this degradation process were studied in the Matanzas province, where original serpentine shrub forest is represented by the *Pseudocarpidio-Bucidetum ophiticolae*, the palm savanna by the *Eugenio-Coccothrinacetum roseocarpae* and the savanna without palms by the *Cassio lineatae-Aristidetun neglectae*. Trends of the degradation process are expressed in Table 2 as changes in life form and phytogeographical spectra.

In contrast to what Seifriz (1943), Beard (1953) and others thought, Cuban palms are not savanna palms. Tall palms – like *Roystonea* and *Sabal* are forest palms. Their young individuals need shade and cannot develop in open savannas. Low and dwarf palms are elements of dry forests and thickets and become savanna palms only secondarily.

6) The degradation of pine forests growing on ferritic soils is described in Fig. 1F, showing examples from the Cajalbana hills and the Nipe Mountains. After a deforestation the originally closed pine forests become at first a woodland vegetation and after burning they turn into a *Pinus-Coccothrinax* wooded grassland and finally by overgrazing into a short grass pasture.

In the white sand areas, on very nutrient poor acid soils the optimum of the vegetation is a pine-woodland with a well-developed herb layer and scattered *Colpothrinax* palms with barrel-shaped trunks in the canopy.

Table 2. Life form and phytogeographical spectrum of some serpentine communities representing different stages of the savannization process

A: Starting stage; *Pseudocarpidio-Bucidetum ophiticolae* serpentine scrub, 5 rel.
B: Intermediate stage: *Eugenio-Coccothrinacetum roseocarpae* serpentine palm-woodland, 5 rel.
C: Final stage: *Cassio lineatae-Aristidetum neglectae* serpentine grassland, 5 rel. (Balátová-Tuláčková & Capote 1985).

Communities:	A	B	C
Tree layer cover in %	20	15	–
Shrub layer	80	20	15
Herb layer	10	70	90
Number of species:	156	99	70
Life forms	%	%	%
Macro-Phanerophytes	1	1	–
Meso-Phanerophytes	6	4	2
Micro-Phanerophytes	9	7	7
Nano Phanerophytes	30	22	13
Rosulate micro-phaneroph.	2	2	2
Woody lianes	6	6	4
Woody succulents	1	1	–
Chamaephytes	8	8	6
Hemicryptophytes	14	29	46
Geophytes	2	2	2
Therophytes	3	5	7
Herb lianes	14	12	11
Rosulate epiphytes	3	1	–
Other epiphytes	1	–	–
Phytogeographical elements			
Endemics, local + regional	9	4	3
Endemics, pancuban	30	22	16
Endemics, total	39	26	19
Greater Antillean	13	15	19
Antillean	6	5	4
Antillean-Bahaman	3	5	7
Cuban-Bahaman	6	8	3
North-Caribbean	7	10	6
South-Caribbean	3	1	1
Pan-Caribbean	5	4	3
Neotropical	13	20	26
Pan-tropical	1	3	6
Pan-American	1	–	–
Cosmopolitan	1	1	1
Alien	1	2	4

7) When palms are logged, which is uncommon in other communities and with other palms, a pine-dwarf palm savanna develops first and by burning and grazing it is converted into an *Acoelorraphe* palm savanna, and in the end, into a white sand pasture with scattered pines and dwarf palms, (Fig. 1G) as explained by Samek (1969).

When pine is logged, a *Colpothrinax* woodland develops which is converted by burning into a *Colpothrinax* tall grassland and by grazing into a *Colpothrinax* pasture (Fig. 1G).

It is emphasized, that in the cases of serpentine and white sand, savannization proceeds in a much shorter and easier way than on the other bedrock types because of the low nutrient level of the habitat. Therefore it is often an irreversible process because of the repeated exploitation of the deficient nutrient budget, which leads to exhaustion. The distribution of these, easily savannizable habitats have been mapped by Borhidi & Herrera 1977; Borhidi & Muñiz 1980, 1984.

Historical and actual grassland pattern

To understand the anthropic process of savannization, not only the actual vegetation but also the historical bibliography and archive material of the economical and land use history of Cuba were studied (Borhidi & Herrera 1977). On the basis of these studies we tried to estimate the size of grassland areas and their extent at the time of the discovery of America, and to compare these estimates with the present status (Table 3). According to our estimation, grasslands covered 15 – 22% of the island at the time of Columbus' first landing. Of these 3 – 5% may have been natural grasslands, whereas the semi-anthropic grasslands accounted for 7 – 10% and the secondarily derived savannas 5 – 10%. Our conclusion is that 95 – 97% of Cuba was originally covered by natural forests and thickets. However, we may assume that in the precolumbian period Indian cultures had brought about changes which cannot be disregarded. At present 87 – 92% of Cuba is not forested. The natural grasslands make up 1 – 2%, the semi-anthropic grasslands 7 – 10%, the anthropic savannas 20 – 25% and cultivated land 50 – 60% (see Table 3).

In conclusion, the form and colourful Cuban savannas are basically secondary formations and – with the exception of flooded riverside moist savan-

Table 3. Savanna types and their distribution in Cuba.

	Flora		Ecosystem		Cover in % of the national territory		
	Fire as control-ling factor	Indigenous + anthropic −	Adequate + non adequate − to the catena	Successional pattern	originally	in 1924	in 1976
1. Natural grasslands	-	+	+	progressive seral	3–5	3–5	1
2. Semianthropic grasslands	±	+	−	regressive seral	2	5–7	7–10
3. Antrophic or derivate grasslands	+	−	−	regressive seral	−	5–10	20–25
4. Artificial grasslands and cultivated belts	+	−	−	non-seral	−	2	50–60
5. Grasslands, total					3–6	15–22	90–92
6. Forests					94–97	78–85	8–10

nas – products of a long-term human impact consisting of combined cutting, burning and grazing activities.

References

Aubréville, A. M. 1947. The disappearance of the tropical forests of Africa. Unasylva 1: 5–11.

Aubréville, A. M. 1965. Les étranges savanes des llanos de l'Orénoque. Adansonia 5: 1–13.

Balátová-Tuláčková, E. & Capote, R. 1985. The *Cassio lineatae-Aristidetum neglectae*, a new association of the serpentine in Cuba. Folia Geobot. Phytotax. 20: 177–183.

Balátová-Tuláčková, E. & Surli, M. 1983. Contribution to the phytosociological characteristics of the Yaguaramas Savanna (Cienfuegos Province, Cuba). Folia Geobot. Phytotax. 18: 1–12.

Beard, J. S. 1944. Climax vegetation in tropical America. Ecology 25: 127–158.

Beard, J. S. 1953. The savanna vegetation of northern Tropical America. Ecol. Monogr. 23: 149–215.

Bennett, H. H. & Allison, R. V. 1928. The soils of Cuba. Washington.

Borhidi, A. 1973. Fundamentals of geobotany of Cuba. Thesis D. Sc. Habil. Budapest.

Borhidi, A. 1974. Kubában térképeztünk. Mapping in Cuba. In Hungarian. Budapest, Gondolat.

Borhidi, A. & Herrera, R. 1977. Génesis, caracteristicas y clasificación de los ecosistemas de sabana en Cuba. Cien. Biol. Acad. Cien. Cuba 1: 115–130.

Borhidi, A. & Muñiz, O. 1980. Die Vegetationskarte von Kuba. Acta Bot. Acad. Sci. Hung. 26: 25–53.

Borhidi, A. & Muñiz, O. 1984. Mapa de Vegetación de Cuba. Imp. Juceplan, La Habana.

Budowski, G. 1956. Tropical savannas, a sequence of forest felling and repeated burnings. Turrialba 15: 40–42.

Busse, W. 1908. Die periodischen Grassbrande im tropischen Africa, ihr Einfluss auf die Vegetation und ihre Bedeutung für die Landeskultur. Mitt. Deutsch. Schutzgeb. 21, 2.

Cole, M. M. 1960. Cerrado, Caatinga and Pantanal: the distribution and origin of the savanna-vegetation of Brazil. Geogr. J. 126: 168–179.

Cole, M. M. 1986. The Savannas. Biogeography and geobotany. Acad. Press. Inc. London.

Eden, M. J. 1964. The savanna ecosystem – northern Rupununi, British Guayana, McGill Univ. Savanna Res. Ser. 1: 1–216.

Eiten, G. 1972. The Cerrado vegetation of Brasil. Bot. Rev. 38: 201–341.

Goodland, R., 1970. The savanna controversy. McGill Univ. Savanna Res. Ser. 15: 1–66.

Guevara, S., Purata, S. E. & van der Maarel, E. 1986. The role of remnant forest trees in tropical secondary succession. Vegetatio 66: 77–84.

Hills, T. L. 1969. The savanna landscape of the Amazon basin. McGill Univ. Savanna Res. Ser. 14: 1–41.

Hills, T. L. & Randall, R. E. (eds) 1968. The ecology of the forest savanna boundary. McGill Univ. Savanna Res. Ser. 13: 1–128.

Komarek Sr, E. V. 1971. Lightning and fire ecology in Africa. Proc. Ann. Tall Timbers Fire Ecol. Conf. 11: 473–511.

Lanjouw, J. 1936. Studies on the vegetation of the Surinam savannahs and swamps. Nederl. Kruidk. Arch. 46: 823–851.

León, H. Sauget 1946. Flora de Cuba. Vol. I. Contr. Ocas. Mus.

Hist. Nat. Col. La Salle, No. 8. 1–441 La Habana.
Marie-Victorin, F. & León, H. Sauget 1942, 1944. Itinéraires botaniques dans l'ile de Cuba. Contr. Inst. Bot. Univ. Montreal 41: 1–496, 50: 1–410.
Medina, E. 1982. Physiological ecology of Neotropical savanna plants. In. Huntley, B. J. & Walker, B. H. (eds), Ecology of tropical savannas, pp. 308–335. Springer, Berlin.
Monasterio, M. 1971. Ecologia de las sabanas de America tropical, I. Análisis macroecológico de los llanos de Calabozo, Venezuel. Cuad. Geogr. 4: 1–126.
Monasterio, M. & Sarmiento, G. 1971. Ecologia de las sabanas de America tropical. II. Caracterización ecológica del clima en los llanos de Calabozo, Venezuel. Rev. Geogr. 21: 5–38.
Myers, J. G. 1936. Savannah and forest vegetation of the interior Guiana Plateau. J. Ecol. 24: 163–184.
Samek, V. 1969. La vegetación de la Isla de Pinos. Acad. Cien. Cuba, Ser. Isla de Pinos 28: 1–28.
Sarmiento, G. 1983. The savannas of tropical America. In: Bowliere, F. (ed.), Ecosystems of the world. vol. 13. Tropical savannas. Elsevier, Amsterdam.
Sarmiento, G. & Monasterio, M. 1975. A critical consideration of the environmental conditions associated with the ocurrence of savanna ecosystems in tropical America. In: Golley, F. B. & Medina, E. (eds), Tropical ecological systems. Ecol. Stud. 11: 223–250.
Schimper, A. F. W. 1898. Pflanzengeographie auf physiologischer Grundlage. Berlin.
Seifriz, W. 1943. The plant life of Cuba. Ecol. Monogr. 13: 375–426.
Troll, C. 1935. Gedanken und Bemerkungen zur ökologischen Pflanzengeographie. Geogr. 14: 1–380.
Vareschi, V. 1960. La estación biológica de los llanos de la Sociedad Venezolana de Ciencias naturales y su tarea. Bol. Soc. Venezol. Cien. Nat. 21, 96: 107–117.
Vareschi, V. 1962. La quema como factor ecológico en los llanos. Bol. Soc. Venezol. Cien. Nat. 23, 101. 9–26.
Vesey-FitzGerald, D. F. 1963. Central African grasslands. J. Ecol. 51: 243–274.
Waibel, L. 1948. Vegetation and land-use in the Planalto Central of Brazil. Geogr. Rev. 38: 529–544.
Walker, B. H. 1987. A general model of savanna structure and function. In: Walker, B. H. (ed.), Determinants of tropical savannas. IUBS Monogr. Ser. No. 3: 1–12.
Walter, H. 1962. Die Vegetation der Erde in ökophysiologischer Betrachtung. I. Die tropischen und subtropischen Zonen. Fischer, Stuttgart-Jena.
Walter, H. 1970. El problema de la sabana. Bol. Soc. Venezol. Cien. Nat. 115–116: 121–144.
Warming, E. & Graebner, P. 1918. Lehrbuch der ökologischen Pflanzengeographie. Berlin.

Vegetatio 77: 185–191, 1988
© Kluwer Academic Publishers, Dordrecht – Printed in the Netherlands

Successional pathways of Mediterranean evergreen vegetation on Sicily

Emilia Poli Marchese, Lidia Di Benedetto & Giuseppe Maugeri
Cattedra di Botanica, Università di Catania, Via Antonino Longo 19 – 95125 Catania, Italia

Accepted 3.9.1988

Keywords: Development stage, Dynamic series, Progressive succession, Regressive succession

Abstract

The Mediterranean evergreen vegetation of Sicily, comprised in the belt of the *Quercetea ilicis,* occupies a large part of the island. Human intervention (cutting, fire, pasture) has brought about a degradation of the natural vegetation. This study is based on our phytosociological research of the *Quercetea ilicis* belt on Sicily.

With the 'habitat comparison' method, the dynamical relations between the different vegetation units have been defined.

We distinguish the following stages, with reference to their vegetation structure:

– a herbaceous stage formed by steppic vegetation, preceded by various types of nitrophilous-ruderal vegetation on abandoned fields;
– a garrigue stage dominated by half-shrubs;
– a macquis stage with various distinct plant communities, four communities being important in regressive successions, and three in progressive ones;
– a woodland and shrub-woodland stage with three different substages: pre-existent forests, present woodlands, and woodlands which tend towards the final, stable stage of vegetation (potential natural vegetation).

The dynamic relationships both in progressive and regressive successions have been synthesized in a scheme. In this scheme we have shown the main stages of the vegetation in their dynamics and we have constructed different series of vegetation types in two altitudinal belts, which are determined by varying environmental conditions of today.

The results also show that in some cases the progressive series follow different pathways than the regressive series, and the final stage of the progressive series is different from the original vegetation.

Introduction

This study deals with the dynamics of Mediterranean evergreen vegetation on Sicily, in both regressive and progressive successions. The belt of this vegetation, belonging to the *Quercetea ilicis*, is widely distributed in Sicily. It extends from the coast up to about 1400 m.

The area is subject to heavy human impact which as in the rest of the Mediterranean region, began a long time ago and has intensified ever since. All this has had disastrous consequences for the vegetation. Instead of the original forests we now find mainly herbaceous and ligneous cultivations, a number of urban centres, and vast areas of abandoned cultivated areas.

Nevertheless, one can still find examples of original woodland, together with development stages of natural vegetation, such as shrub, half-shrub and herb communities.

In the course of our long-term investigations of the *Quercetea ilicis* on Sicily we have described many plant communities corresponding to the different progressive and regressive stages. Their dynamical relationships have been put together in a scheme. A better knowledge of the vegetation dynamics considered, will enable a more precise definition of the *Quercetea ilicis*.

The environment

The territory of Sicily is extremely varied because of its geomorphological, geological and climatic characteristics. This variation is reflected in the vegetation, especially the woodlands, which are fairly diversified even when found on the same altitudinal level. The major elevations of the island are to be found in the north: Mount Etna (3370 m), the Madonie (1979 m), the Nebrodi (1874 m) and the Peloritani mountains (1374 m). Bedrock is varied with a prevalence of limestone and dolomite in the Madonie and in the mountains of Palermo, numidic flysh with quartzarenites and clay siltose in the Nebrodi territory, phyllites, mica-schist, granite and flysch in the Peloritani and vulcanite in the Etna region.

Southern Sicily, dominated in the south-east by the Iblean plateau (914 m), is largely made up by hilly regions and narrow, alluvial plains. Limestone, calcarenites and marne occur in the Iblean territory and vulcanite is present in the northern part. In central and south-western Sicily clay, marne, gypsum, limestone, sand, sandstone and formations of the chalky sulphurous series ('serie gessoso-solfifera') are found.

Quercetea ilicis communities are found in regions with a thermo-mediterranean and meso-mediterranean climate (Emberger *et al.* 1963). Particularly the southern coast of the island is thermo-mediterranean. The meso-mediterranean climate prevails in the northern part of the island, especially at lower altitudes and along the northern and north-eastern coast. The thermo-mediterranean climate is found only here and there, at higher altitudes.

The geographical position and the morphological characteristics of the island influence the distribution of rainfall in a particular manner. Rainfall is higher in the north and north-east of Sicily, usually above 700 mm/yr. In the south of Sicily, rainfall rises above 700 mm only in a few areas of the interior. As regards temperature, the January isotherms range from 12° to 8 °C from the coast towards the interior. The summer (July) isotherms range from 25° to 29 °C.

Method

Vegetation sequences have been studied with the Braun-Blanquet (1964) approach in numerous localities, chosen in such a way that each area includes the different stages of vegetation. By comparison of the plant communities by habitat ('Habitat comparison', Braun-Blanquet 1964) we are able to show the principle dynamical stages both in progressive and regressive successions. This approach is based on estimating development age of vegetation types under identical habitat conditions. Each vegetation type should have a well defined structure, a precise ecological indication and a known dynamic role. They are indicated with the name of the characterising or dominating species. In this way, it has been possible to put together the principal stages of the vegetation dynamics in a clear and concise manner in a dynamic scheme (Fig. 1). Regressive successions range from the pre-existent vegetation, which can not always be reconstructed, to the most degraded stages of real vegetation. Progressive successions start from actual early stages and follow lines of potential vegetation development. It is recognized that in some cases the final stage in the succession cannot be reached any more due to irreversible changes in the soil.

Vegetation dynamics

Dry herbaceous vegetation

The lowest stage of vegetation, comprised in the belt of the *Quercetea ilicis* class is composed of a vast range of herbaceous communities. We can place such communities in two large categories:

- Dry grasslands with steppe characteristics belonging to the *Thero-Brachypodietea* or *Helianthemetea guttati*.

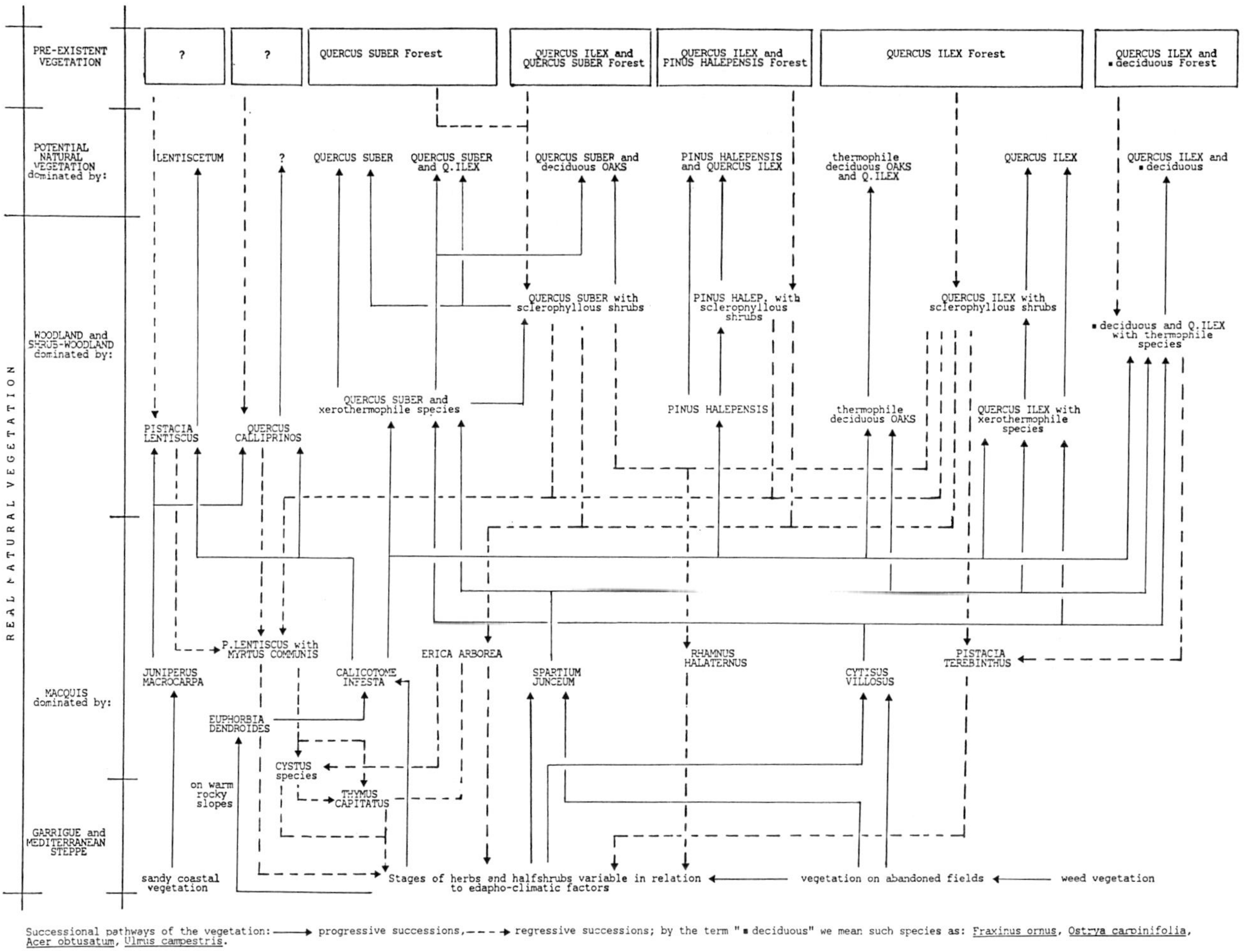

Fig. 1. Successional pathways of Mediterranean evergreen vegetation on Sicily.

- Grassland with nitrophilous and ruderal characteristics of abandoned fields and other areas under major human influence, to be assigned to the *Brometalia rubenti-tectori* (Maugeri 1979).

Among the most widely distributed steppic communities we can mention:

- the *Ampelodesmos mauritanica*-community on steep slopes;
- the *Cymbopogon hirtus*-community on well-drained soils
- the *Asphodelus microcarpus*-community on very poor thin soils
- the *Ononis ramosissima*-community on sandy soils in southern Sicily
- the *Sarcopoterium spinosum*-community in prevalently rocky stations.

Macquis

The shrub stages which establish a macquis are variable due to climatic and soil conditions. The most significant types are *Pistacia lentiscus, Erica arborea, Rhamnus alaternus* and *Pistacia terebinthus* macquis, all belonging to the *Quercetea ilicis. Pistacia lentiscus* macquis, with *Myrtus communis* and *Chamaerops humilis,* is found in the warmest and driest stations. It may have been formed by progressive degradation of woodland dominated by *P. lentiscus, Quercus calliprinos, Pinus halepensis, Quercus suber* or *Quercus ilex.* The *Erica arborea* macquis which often includes *Arbutus unedo,* is found in warm, but less dry stations, preferring permeable and loose soils. As in other Mediterranean areas, such macquis may have been formed by degradation, mainly of *Quercus suber* forests (Loisel 1976) and also *Q. ilex* and *Pinus halepensis* forests. The less developed shrub and half-shrub garrigue communities are mainly connected to these two types of macquis: *Cystus, Thymus capitatus* and *Rosmarinus officinalis*.

The *Rhamnus alaternus* macquis prevails in less warm stations. *Phillyrea latifolia* is frequent or dominant here. The *Pistacia terebinthus* macquis, located in relatively warm but humid areas is characterized by evergreen sclerophyllous species (*Quercus ilex, Rhamnus alaternus, Phillyrea latifolia*) as well as by some deciduous species: *Fraxinus ornus, Crataegus monogyna, Pirus amygdaliformis*. The presence of these deciduous trees differentiates this macquis from that dominated by *Rhamnus alaternus*. From a dynamical point of view the *Pistacia terebinthus* macquis is principally connected to woodland dominated by *Q. ilex* and deciduous trees (*Fraxinus ornus, Ostrya carpinifolia, Ulmus campestris, Acer obtusatum*). Here also soil degradation is the probable reason for the establishment of the deciduous trees.

The ecological differences between these four types of macquis, relative to temperature and humidity can be represented in a scheme as follows:

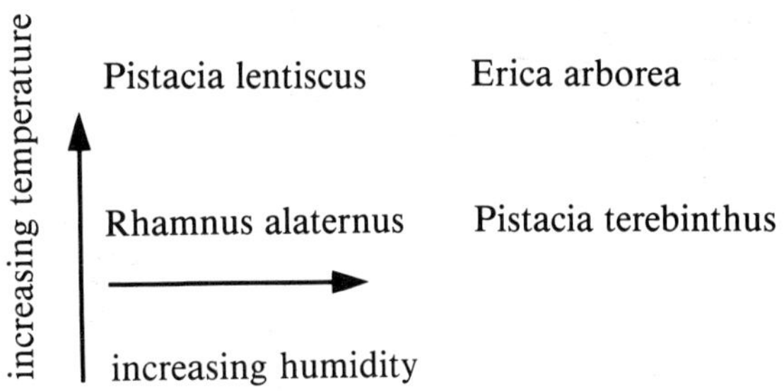

Among the shrub-communities belonging to progressive successions we mention the *Juniperus macrocarpa, Euphorbia dendroides, Calicotome infesta, Spartium junceum*, and *Cytisus villosus* macquis. The *Juniperus macrocarpa* macquis is located in sandy coastal areas; it develops towards *Pistacia lentiscus* (Maugeri & Leonardi 1974) or *Quercus calliprinos* shrubland. In the warmest rocky stations on initial soils macquis characterized by *Euphorbia dendroides* (*Euphorbietum dendroidis*) is located (Maugeri *et al.* 1979). This is often a permanent stage.

In sufficiently warm areas we find a very variable type of macquis of *Calicotome infesta* with more or less xero-thermophilous species like *Prasium majus, Asparagus albus*, and *Daphne gnidium*. Deciduous shrubs may also be associated. This macquis develops from steppic types. Its further development leads to various secondary woodlands.

The two examples of macquis, which become more and more enriched with trees, evolve towards various woodlands. In particular the *Spartium junceum* macquis shows dynamic links with *Quercus suber* woods. The *Cytisus villosus* macquis would develop to more mesophilic woodlands including *Quercus suber, Q. ilex*, and deciduous formations. The types of macquis deriving from the steppic grasslands (*Thero-Brachypodietea* and *Helianthemetea guttati*) or the nitrophile-ruderal communities (*Brometalia rubenti-tectori*) are actually constituted of numerous communities, corresponding to many physiognomical types. There are three different types of environment as follows from the following scheme:

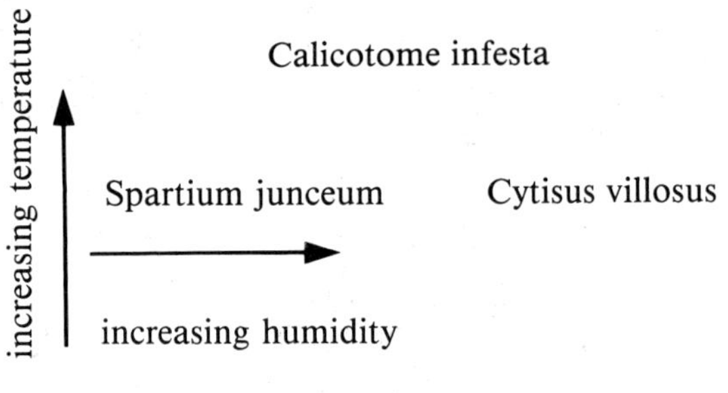

Woodlands and shrub-woodlands

The woody vegetation, the most complex stage of real natural vegetation, is formed by different plant communities of the *Quercetea ilicis*. Some of these are situated in the hottest areas of the island; those which are more mesophilic, situated at higher altitudes, towards the limit of the Mediterranean evergreen vegetation, are enriched by elements of the *Querco-Fagetea*.

Thermophilous coastal shrub-woodlands
First we mention two examples which have now become very rare. They are both shrub-woodlands,

dominated by *Pistacia lentiscus* and *Quercus calliprinos* respectively. The *Pistacia lentiscus* type reaches the maximum level of development in the southernmost part of the island, near the mouth of the River Irminio, behind coastal dunes (Maugeri & Leonardi 1974). *Pistacia lentiscus* attains the dimension of a tree only under the most thermophilic conditions found on our island today. Less evolved stages dominated by *P. lentiscus,* with *Chamaerops humilis* and *Myrtus communis* are present in different areas of the thermo-mediterranean belt of Sicily; they seem to be soil-indifferent. The shrub-woodland dominated by *Quercus calliprinos* can be found today in a few areas on southern Sicily, in the warmest and driest zones (Gentile 1962; Brullo & Marcenò 1984).

These two examples of shrub-woodland on coastal sandy soils show dynamical connections with a macquis dominated by *Juniperus macrocarpa,* to which *Juniperus phoenicea* may be associated.

Pinus halepensis *woodlands*

In southern Sicily, on the surfacing rocks of the calcareous plateau (Lentini 1984) and in warm zones we find formations of *Pinus halepensis* with a pioneer character. Such formations contain many thermophilous elements of the Mediterranean macquis and with soil evolution they are enriched with sclerophyll, e.g. *Q. ilex*. This would be the final stage of potential natural vegetation. According to De Marco & Caneva (1984) and Bartolo *et al.* (1985) the *Pinus halepensis*-community would have a climax character. Our hypothesis is that the original vegetation was established rather by a mixed *Q. ilex-Pinus halepensis* forest, which after degradation became Pine forest with *Q. ilex* and other evergreen sclerophylls.

Quercus suber *woodlands*

The woods dominated by the Cork-oak are located on different kinds of loose soil, incoherent and always permeable. Such more or less open woods generally occupy stations which are warmer and more humid than those preferred by *Q. ilex* (e.g. Giacomini 1958; Loisel 1976). The largest Cork-oak formations are found along the lowest northern-facing slopes of the Nebrodi and Madonie mountains and exposed to humid seawinds, on the vulcanite of the Iblean mountains and on the sandy soil to the west of these mountains.

Concerning these Cork-oak forests, we are able to distinguish:

- pure *Quercus suber* communities, which we consider as secondary woods (Poli Marchese, Maugeri & di Benedetto 1985).
- mixed communities with other sclerophyllous evergreens as *Quercus ilex* and *Phillyrea latifolia* originating by degradation from pre-existent *Quercus suber* forests or *Q. ilex-Q. suber* forests. Progressive vegetation dynamics after human influence would cease, would lead to *Quercus suber* forest again.

The further development of these two types of woodlands would bring about examples of the potential natural vegetation constituted by:

- pure *Q. suber* forests in warm humid zones;
- *Q. suber-Q. ilex* forests in the warm but less humid areas in different zones of southern Sicily;
- *Q. suber* forests with deciduous oaks (including Q. gussonei, Brullo 1984) mainly at higher altitudes in northern Sicily.

Quercus ilex *woodlands*

In the areas once covered by *Q. ilex* forests we now find cultivations, abandoned fields and areas of natural vegetation having a different evolution level (Tomaselli 1976). The distribution area of *Q. ilex* extends over the whole Sicilian Mediterranean belt, irrespective of the nature of the soil. The species shows ecological plasticity by adapting to relatively low temperatures, when the environment is arid, and to relatively high temperatures under humid conditions. Examples are found in deep valleys in southern Sicily and on the mountains of North Sicily, in the climatic belt of the deciduous broadleaved trees, and in rocky and particularly dry stations (Hofmann 1960).

In particular circumstances (Gentile 1969a, b), *Q. ilex* woodland is capable of growing upwards until it comes into direct contact with *Fagus* woodland as can be observed on the north-western slopes of Mount Etna (Poli *et al.* 1983) and on the northern slopes of the Nebrodi mountains.

Basically we can distinguish two types among the

present *Q. ilex* formations. One is dominated almost exclusively by *Q. ilex* and the other is a mixed type of *Q. ilex*, and other evergreen sclerophylls, such as *Rhamnus alaternus* and *Phillyrea latifolia*. The first type is considered as secondary vegetation, originating from earlier stages. The second type, characterized by many species is considered as a stage of degradation of the pre-existing forests. In the *Q. ilex* series, the dynamic stages of a lower level corresponding to macquis are more numerous as compared with other series.

The stages of degradation of the *Q. ilex* woodland, situated in the warm and arid stations, analogous to the *Quercus suber* formations are often established by communities which are near to those of the series of *Pinus halepensis, Q. calliprinos* and *P. lentiscus*, with *Pistacia lentiscus, Myrtus communis* and *Chamaerops humilis*. At their upper limits the *Q. ilex* woodlands meet with deciduous oak formations. As stated by Brullo & Ronsisvalle (1984), the deciduous oaks of Sicily indicated as *Q. pubescens* s.l. would also include *Q. virgiliana, Q. amplifolia, Q. congesta, Q. dalechampii, Q. leptobalana, Q. gussonei* and *Q. fontanesii*. In this contact zone we find other *Q. ilex* formations, belonging to the *Quercetea ilicis* class, but with a considerable presence of *Querco-Fagetea* elements.

In those areas which are sufficiently warm and with a relatively high rainfall, formations of *Quercetea ilicis*, are found at lower altitudes characterized by various deciduous trees such as *Fraxinus ornus* and *Ostrya carpinifolia*, and also *Q. ilex*. These formations have a certain similarity to the *Orno-Quercetum ilicis* of the northern Adriatic and Dalmatian regions (Horvatic 1934; Lausi & Poldini 1962; Wraber 1967). The Sicilian formations are different in that they have more deciduous species. Such species are particularly represented in secondary formations. In original but degraded vegetation *Q. ilex* is always well represented, as we have observed in the Dalmation regions. In warm areas, where intense and prolonged human impact causes a certain soil aridity a woodland dominated by thermophilous deciduous oaks such as *Quercus virgiliana* and *Q. amplifolia* is formed. During further succession such woodland tends towards a mixed community dominated by thermophyllous deciduous oaks.

References

Austin, M. P., Williams, O. B. & Belbin, L. 1981. Grassland dynamics under sheep grazing in an Australian Mediterranean type climate. Vegetatio 47: 201–211.

Ballatore, G. P. & Fierotti, G. 1967. Carta dei suoli della Sicilia. Geografica, Firenze.

Bartolo, G., Brullo, S., Minissale, P. & Spampinato, G. 1985. Osservazioni fitosociologiche sulle pinete a *Pinus halepensis* Miller del bacino del fiume Tellaro (Sicilia sud-orientale). Boll. Acc. Gioenia Sc. Nat. 18, 325: 255–270.

Bridgewater, P. B. & Backshall, D. J. 1981. Dynamics of some western Australian ligneous formations with special reference to the invasion of exotic species. Vegetatio 46: 141–148.

Brullo, S. 1983. Contributo alla conoscenza della vegetazione delle Madonie (Sicilia settentrionale). Boll. Acc. Gioenia Sc. Nat. 16, 322: 351–420.

Brullo, S. & Marcenò, C. 1984. Contributo alla conoscenza della classe *Quercetea ilicis* in Sicilia. Not. Fitosoc. 19: 183–227.

De Marco, G. & Caneva, G. 1984. Analisi sintassonomica e fitogeografica comparata di alcune significative cenosi a *Pinus halepensis* Miller in Italia. Not. Fitosoc. 19: 155–176.

Emberger, L., Gaussen, H., Kassas & de Philippis, A. 1963. Carte bioclimatique de la région méditerranéenne. Notice explicative. UNESCO-FAO, Paris.

Floret, C. 1981. The effects of protection on steppic vegetation in the Mediterranean arid zone of southern Tunisia. Vegetatio 46: 117–128.

Forgeard, F. & Touffet, J. 1980. La récolonisation des landes et des pelouses dans la région de Paimpont. Evolution de la végétation au cours des trois années suivant l'incendie. Bull. Ecol. 11: 349–358.

Gentile, S. 1962. Frammenti di macchia particolarmente con *Quercus calliprinos* Web. nel territorio di Ragusa (Sicilia meridionale-orientale). Delpinoa 4: 125–142.

Gentile, S. 1968. Memoria illustrativa della carta della vegetazione naturale potenziale della Sicilia. Ist. Bot. Lab. Critt. Univ. Pavia 40: 1–114.

Gentile, S. 1969. Remarques sur les chênaies d'Yeuse de l'Apennin méridional et de la Sicilie. Vegetatio 17: 214–231.

Giacomini, V., in: Giacomini, V. & Fenaroli, L. 1958. La flora. T.C.I., Milano.

Gloaguen, J. C. & Gautier, N. 1981. Pattern development of the vegetation during colonization of a burnt heathland in Brittany (France). Vegetatio 46: 167–176.

Hofmann, A. 1960. Il faggio in Sicilia. Flora et Vegetatio italica, 2, Sondrio.

Horvat, I. 1958. Laubwerfende Eichenzonen Südosteuropas in pflanzensoziologischen, klimatischen und bodenkundlichen Betrachtungen. Angew. Pflanzensoz. 15: 50–62.

Horvatic, S. 1934. Flora und Vegetation der nordadriatischen Inseln. Pag. Prir. istraz. Jug. Akad. Zagreb 19: 116–372.

Lausi, D. & Poldini, L. 1962. Il paesaggio vegetale della costiera triestina. Boll. Soc. Adriatica Sc. 52: 3–63.

Lentini, F. 1984. Carta geologica della Sicilia sud-orientale. S.EL. CA., Firenze.

Loisel, R. 1976. La végétation de l'étage méditerranéen dans le sud-est continental français. Thesis, Université D'Aix-Marseille III.

Maugeri, G. & Leonardi, S. 1974. Esempio di macchia a ginepro e lentisco nella Sicilia meridionale. Arch. Bot. Biogeog. Ital. 50, 4, 20 (I, II): 51–60.

Ozenda, P. 1964. Biogéographie végetale. Doin Deren, Paris.

Peet, R. K. 1981. Forest vegetation of the Colorado Front Range. Vegetatio 45: 3–75.

Poli, E. & Maugeri, G. 1974. I boschi di leccio del versante nord-occidentale dell'Etna. Boll. Acc. Gioenia Sc. Nat. 12 (5–6): 729–739.

Poli, E., Maugeri, G. & Ronsisvalle, G. 1981. Note illustrative della carta della vegetazione dell'Etna. C.N.R. AQ/1/131: 1–29, Roma.

Poli Marchese, E., Maugeri, G. & di Benedetto, L. 1985. Analisi della copertura vegetale in tre territori rappresentativi della Sicilia per interventi in sistemi marginali. Atti Convegno 'Stato della ricerca sulle terre marginali della Sicilia': 49–99, Palermo.

Quézel, P. 1976. Les forêts du pourtour meditérranéen. Note technique du MAB, 2: 9–33. UNESCO, Paris.

Rivas-Martinez, S., Costa, M. & Izco, J. 1984. Sintaxonomia de la clase *Quercetea ilicis* en el Mediterraneo occidental. Not. Fitosoc. 19: 71–98.

Ronsisvalle, G. A., Giorgi, G. & Silvetti, E. 1984. Indagine tassonomica in querce caducifoglie siciliane. Pubbl. Ist. Bot. Univ. Catania: 1–15.

Tomaselli, R. 1976. La dégradation du maquis méditerranéen. Note technique du MAB, 2: 35–76. UNESCO, Paris.

Trinajstic, I. 1984. Sulla sintassonomia della vegetazione sempreverde della classe *Quercetea ilicis* Br.-Bl. del litorale adriatico jugoslavo. Not. Fitosoc. 19: 77–98.

Wraber, M. 1967. Oekologische und pflanzensoziologische Characteristik der Vegetation des Slowenischen küstenländischen Karstgebietes. Mitt. Ostalp.- Dinarische Pflanzensoz. Arbgm. 7: 3–31.

Vegetatio 77: 193–200, 1988
© Kluwer Academic Publishers, Dordrecht – Printed in the Netherlands

Species diversity of mainland- and island forests in the Pacific area

Syuzo Itow
Plant Ecology Laboratory, Faculty of Liberal Arts, Nagasaki University, Nagasaki 852, Japan

Accepted 25.5.1988

Keywords: Bonin, Galápagos, Isolation, Micronesia, Ryukyu, Species richness, Warmth index

Abstract

Alpha diversity, or species richness, of East Asian mainland evergreen broadleaved forests, expressed by indices of Fisher's alpha (α) and $S_{(100)}$, a new index showing species number in a 100-individual sample, is significantly correlated with the climatic favorableness, expressed by Kira's warmth index. On the contrary, diversity values of insular forests studied on Kyushu satellites of Japan, the Bonins, the Eastern Carolines of Micronesia, and the Galápagos in the eastern Pacific, are below those expected from the climate of respective oceanic islands. Species-individual curves, comparing mainland- and insular communities, also support clearly the above conclusion of species poverty in the insular communities studied.

Abbreviations: WI = Kira's (1977) warmth index.

Nomenclature: I. L. Wiggins & D. M. Porter 1971. Flora of the Galápagos Islands. Palo Alto; R. Kanehira 1972. Flora Micronesica. Tokyo.

Introduction

Biotic saturation of oceanic islands is one of the both old and modern concerns of biologists studying insular ecosystems. In island studies we have to distinguish the species richness of a community from the floristic magnitude of an island. A plant community usually consists of species unique to the community and of species shared with other communities. The vegetation of an island comprises various plant community types and the flora of the island will include both sets of community-specific and common plant species.

The concept of alpha diversity denotes the species richness within a community, whereas gamma diversity means the richness in a region or on an island (Whittaker 1977). The present paper only deals with the alpha diversity of mainland- and insular communities.

Alpha diversity is a structural feature of the plant community, just like stratification and physiognomy. It depends on various factors such as climatic conditions, continentality, soil moisture and fertility, topography, and disturbance (Whittaker 1972, 1977; Glenn-Lewin 1977; Peet 1978; Peet *et al.* 1983). From the dynamic point of view alpha diversity also depends on the successional status of the community, and it generally increases towards the equilibrium level at the climax stage (Auclair & Goff 1971; Peet *et al.* 1983). In my study (Itow 1985) of mainland communities of East Asian warm-temperate and subtropical climates, the alpha diversity of evergreen broadleaved forests is correlated with climatic favourableness, but that of insular communities is not.

The present study will continue along these lines while including data from tropical mainland and insular communities (Itow 1986) and of nine equatorial South American communities. The communities studied are all climatic climaxes, considered to represent maximum alpha diversity at equilibrium level, or nearly so.

Study areas and their climate

Mainland and insular communities will be compared between and within climatic regions. Three pairs of mainland areas and oceanic islands in different climatic regions will be compared. The communities actually studied were all natural- and climax forests, being free of disturbance by man and alien plants.

Warm-temperate region

The Japanese Archipelago is an island chain, which was connected to the Asian continent in the glacial period, some 20000 years ago. The northern temperate region of the archipelago supports a summergreen forest, while an evergreen broadleaved forest is found in the southern warmer part. The present paper deals only with the evergreen broadleaved climax forests found in Kyushu, southwestern Japan (31°N – 34°N). The forests are dominated by evergreen *Castanopsis* and *Quercus* (*Cyclobalanopsis*) species (Miyawaki & Sasaki 1985).

Three satellite islands of Kyushu are used as counterpart of the Kyushu main island. They are Meshima of the Danjo Islands, Biryo-jima of the Goto Islands and Okinoshima. The Fagaceae species are absent on all three islands. This may be because the diaspores (acorns) did not cross the sea when the species were expanding their distribution range towards the north in the period of the rise of the sea level during the post-glacial age. Itow (1984) gives a detailed description of Kyushu satellite communities and the measurement of their diversity.

Subtropical region

The Ryukyu Islands are the southernmost part of the Japanese Archipelago (24°N – 27°N). They are continental islands that were connected with the Asian continent in the early Pleistocene. The climax forests are also dominated by evergreen *Castanopsis* or *Quercus* (*Cyclobalanopsis*) species (Miyawaki & Sasaki 1985). The islands Ishigaki, Iriomote and Yonaguni at 24°30′N were included with data from Itow *et al.* (1984).

The Bonin (Ogasawara) Islands are oceanic islands, located about 1000 km east of the Ryukyus and 1000 km south of Tokyo, between 26°30′N and 27°30′N. The geological structure of the islands is complicated, and their origin is probably late Pliocene. Apparently the islands have never been connected to any continent or large land mass. Acorn-bearing trees are absent on the islands. Canopy species are mainly Lauraceae, Elaeocarpaceae, Myrsinaceae and Aquifoliaceae with seeds carried by birds, either internally or externally (Ono & Sugawara 1981; Shimizu 1983; Okutomi *et al.* 1985). The diversity studies were made on the two islands Chichi-jima and Haha-jima (Itow *et al.* 1984).

Tropical region

The present study uses data on Thailand forests by Ogawa *et al.* (1965: loc. cit. Tables 6 and 7) as continental communities. The forest is tropical rain forest, and very rich in species (Itow & Miyata 1977; Itow 1986).

The Eastern Carolines of Micronesia are oceanic islands of volcanic origin, of which Ponape and Kosrae were visited. The inland forest on both islands is well preserved. The common canopy species are *Campnosperma brevipetiolata* (Anacardiaceae) on Ponape and *Horsfieldia nunu* (Myristicaceae) on Kosrae (Itow 1986; Nakano *et al.* 1986). Field work was conducted in 1985 for tree populations (dbh ≥ 10 cm).

Equatorial South America

The Galápagos Islands are the main target of the present study. For comparison Amazonian equatorial tropical rain forests between 77°W and 78°W (Grubb *et al.* 1963: loc. cit. Table 6) were included. Two communities were lowland and one montane (Table 1).

The Galápagos Islands are located on the equator in the eastern Pacific, about 1000 km west of the South American coast. The islands are of volcanic origin, and at least 3 million yr old (Bailey 1976). Isla Santa Cruz is centrally located in the archipelago, between 0°30′S and 0°50′S. The island supports

Table 1. Tree population sizes (>20 ft high) adapted to a geometric series for three equatorial Amazonian forests of Ecuador reported by Grubb *et al.* (1963). Only samples with more than 50 trees are used.

Locality	Lowland		Montane
	Armenia Vieja	Araki	Borja
No. of individuals	101	52	52
No. of species	42	30	27
Species occurring once	24	22	18
Species 1	8	5	6
2	7	5	5
3	7	4	4
4	6	4	4
5	6	3	4
6	5	3	3
7	5	3	3
8	4	3	3
9	4		3
10	4		
11	3		
12	3		
13	3		
14	3		
15	3		
16	2		
17	2		
18	2		

four vegetation zones: Coastal Arid, Transition, Moist, and Highland Zones (Itow 1971; Wiggins & Porter 1971; Hamann 1979). Although the archipelago as a whole has a dry climate, the Moist Zone receives more than 1500 mm rainfall (Hamann 1979) and supports a dense forest of *Scalesia pedunculata* (Compositae). This tree is 15 m high and 20 cm across at maximum growth (See also Lawesson 1988). The field work was conducted in 1978 and 1981 at three stands of the *Scalesia* forest (Table 2), at one in the Transition Zone and two in the Arid Zone (Table 3).

Climate of the study sites

The study sites, except the Galápagos forests, are all located in humid climates, with a rainfall between 1500 mm and 4500 mm, supporting evergreen broadleaved forest, or rain forest.

The mean annual temperatures range from 14 °C in the warm-temperate climate to 27 °C in the tropical. The thermal favourableness is expressed by

Table 2. Species and number of trees (dbh ≥ 5 cm) in three stands of *Scalesia pedunculata* forest in the Moist Zone on Isla Santa Cruz, Galápagos Islands.

Stand	G-1	G-2	G-3
Altitude (m)	590	640	530
Scalesia pedunculata	82	98	85
Psidium galapageium	14	–	8
Zanthoxylum fagara	2	1	7
Pisonia floribunda	5	–	–
Chiococca alba	1	2	–
Cordia anderssonii	–	3	–
Tournefortia rufo-sericea	–	–	3
Acnistus ellipticus	–	–	2
No. of species	5	4	5
No. of individuals	104	104	105

Table 3. Species and number of trees (dbh ≥ 5 cm) in three stands of *Bursera graveolens* forest in the Arid and Transition Zones on Isla Santa Cruz, Galápagos Islands.

Stand	G-4	G-5	G-6
Alt. (m)	350	210	150
Bursera graveolens	50	65	53
Piscidia carthagensis	18	7	18
Zanthoxylum fagara	13	8	1
Waltheria ovata	1	2	3
Chiococca alba	2	1	5
Tournefortia pubesens	5	2	2
Castela galapagensis	–	6	3
Croton scouleri	–	2	7
Scalesia crockerii	–	4	1
Acacia macracantha	–	1	6
Pisonia floribunda	14	–	–
Scutia pauciflora		1	4
Erythrina velutina	1	–	–
Tournefortia psylostachya	–	1	–
Vallesia glabra	–	1	–
Opuntia echios	–	1	–
No. of species	8	14	11
No. of individuals	104	102	103

Kira's (1977) Warmth Index (hereafter abbreviated as WI). This index represents a ratio of the total amount of heat available for plant life, and is given as

$$WI = \Sigma\,(t - 5), \quad (1)$$

where t is the mean monthly temperature exceeding 5 °C. WI values from 85–180 relate to warm-temperate climates, WI = 180–240 to subtropical, and WI > 240 to tropical ones (Kira 1977). The WI value for each of the forest stands studied was estimated using a lapse rate of 0.57 °C/100 m altitude from the temperature records made at the nearby weather stations. The WI values for the three Amazonian communities could not be calculated.

Field methods, diversity indices and *S-N* curves

The field methods I adopted are the quadrat method, the quarter method (Cottam & Curtis 1956) and minimum sampling program (Itow 1984, 1985). The trees sampled are ≥ 3 cm dbh unless otherwise noted. With the quadrat method, five or ten 10 m × 10 m quadrats were laid out and more than 200 trees were sampled. With the quarter method and the minimum sampling program, a little more than 100 trees were sampled. The minimum program is a method of diversity measurement applicable in the field when time and assistance are limited. Preferably the sampling includes more than 150 individuals and the measurement record of their dbh.

Diversity indices adopted in the present study are Fisher's alpha (α) (Williams 1947),

$$S = \alpha \log e\,(1 + N/\alpha), \quad (2)$$

and Itow's (1984) $S_{(100)}$ calculated by Hurlbert's (1971) equation, as given below,

$$S_{(100)} = \Sigma \left[1 - \binom{N - N_i}{100} \Big/ \binom{N}{100} \right] \quad (3)$$

where S is the number of species, N is the number of individuals, and N_i is the number of individuals of the i-th species found in a community. The $S_{(100)}$ index means the estimated number of species in a sample of 100 individuals, but it does not aim to find the collective number of species in the community at large. $S_{(100)}$ is an index of species diversity, which is useful and easy to understand (Itow 1984, 1985, 1986; Itow *et al.* 1984).

Since Hurlbert's (1971) equation gives the number of species for n individuals in the community, species-individuals (*S-N*) curves for the community can be drawn. The curve is another way to express diversity characteristics (Itow 1984, 1985). It is applicable to any data, even if the sample area is not given. The rate of increase from $S_{(50)}$ to $S_{(100)}$, which is called b in the present paper,

$$b = \frac{S_{(100)} - S_{(50)}}{\log 100 - \log 50} \quad (4)$$

is also useful as an index of species diversity (Itow 1984, 1985, 1986). For the stands from which fewer than 100 trees were sampled, b was calculated with $S_{(50)}$ and $S_{(N)}$ or with $S_{(40)}$ and $S_{(50)}$, and in this case $S_{(100)}$ was obtained by projection using b.

The indices of species diversity used in the present study express the characteristics of community structure as follows. Fisher's α gives an estimation of the number of species represented by a single individual in the community in which the population distribution fits the logarithmic series (Williams 1947). $S_{(100)}$ denotes the species richness found in a sample of 100 individuals. The parameter b represents the population evenness in the community. The *S-N* curve is a graphic presentation of the species richness and the population evenness.

In addition to the above-mentioned indices, Shannon's H' and the complement of Simpson's index, d', were also calculated for comparison with other literature (See Itow 1984, 1985).

For the Amazonian communities, only the number of individuals (> 20 ft high) and the number of species occurring once were given. Here the number of individuals for the species occurring more than once were estimated, on the basis of a geometric series (Table 1). The values of diversity indices for the three Amazonian communities were thus calculated

Table 4. Diversity measurements of tree populations in natural forests in mainland Ecuador (first three stands calculated from Grubb *et al.* 1963) and on the Galápagos Islands (G-1 to G-6).

Stand	Armenia Vieja	Araki	Borja	G-1	G-2	G-3	G-4	G-5	G-6
Alt. (m)	300	300	1710	590	640	530	350	210	150
WI				200	208	205	215	221	222
S	42	30	27	5	4	5	8	14	11
S_1*	24	22	18	1	1	0	2	6	2
N	101	52	52	104	104	105	104	102	103
H'	4.96	4.58	4.41	1.04	0.40	1.05	2.16	2.12	2.37
d'	0.96	0.95	0.94	0.36	0.11	0.33	0.70	0.58	0.69
α	27.4	29.6	22.6	1.1	0.83	1.1	2.0	4.4	3.1
$S_{(50)}$	28.2	29.2	26.3	4.2	3.1	4.6	6.7	10.1	9.3
$S_{(100)}$	41.8	42.6	37.2	5.0	4.0	5.0	7.9	13.9	10.9
b	45.02	44.37	36.12	2.60	2.93	1.41	4.19	12.63	5.38

* Number of species occurring once.

based on that estimated distribution of populations (First 3 stands in Table 4). The estimated numbers of species were used only in drawing the S-N curve.

Results and discussion

Figures 1 and 2 give the regression lines for Fisher's

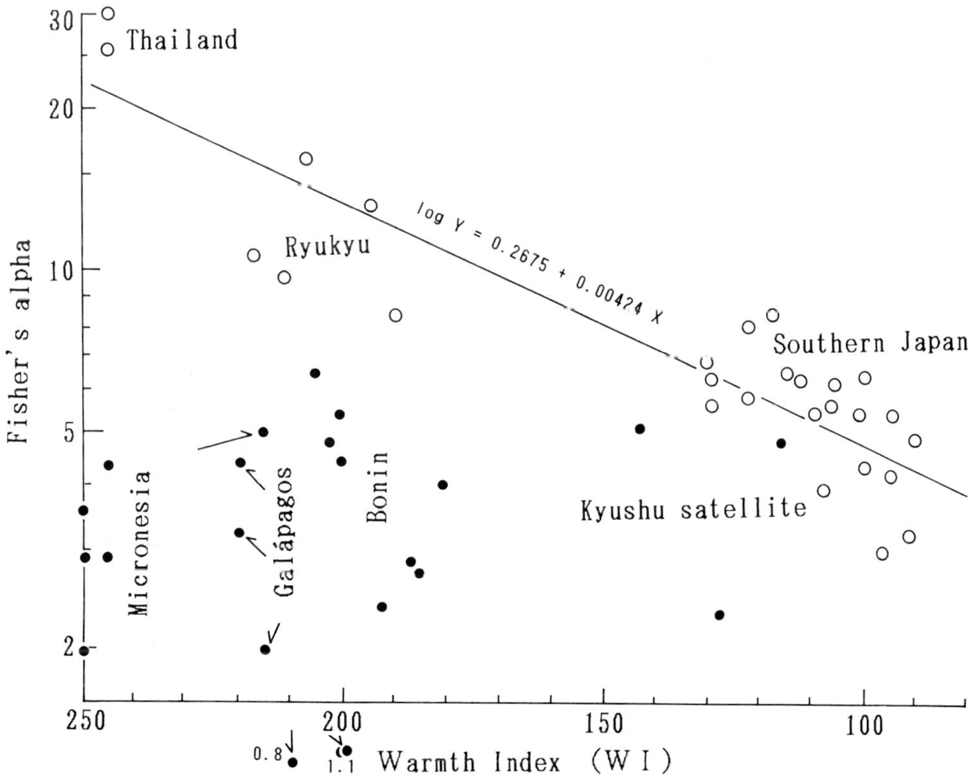

Fig. 1. Alpha-WI correlation in mainland communities (open circles) of humid climate in East Asia. Closed circles show insular communities. Note their positions plotted far below the regression line of the mainland communities.

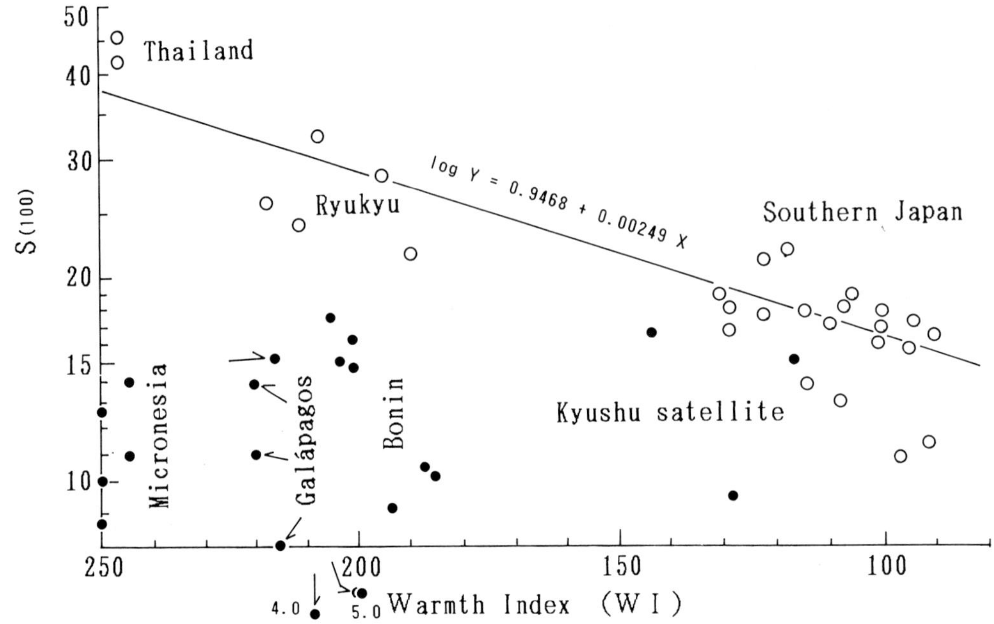

Fig. 2. $S_{(100)}$-WI correlation in mainland communities (open circles). Closed circles show the insular communities.

α on WI and of $S_{(100)}$ on WI respectively. The regression equations are $\log Y = 0.2675 + 0.00424 X$ for the α relation and $\log Y = 0.9468 + 0.00249 X$ for the $S_{(100)}$ relation, where Y is α or $S_{(100)}$, respectively, and X is the WI value. Both regressions are positively significant ($p<0.05$). Thus, alpha diversity of mainland climax communities of humid climates is predictable from the WI index reflecting the thermal favourableness for plant life. Glenn-Lewin (1977) recognized a similar relation between diversity and elevation and latitude in North American temperate forests.

To compare the insular communities with the mainland ones, I plotted the index values of insular communities of Kyushu satellites (Itow 1984), the Bonins (Itow *et al.* 1984), Micronesia (Itow 1986) and the Galápagos communities (Stands G-1 to G-6 in Table 4) in the same graphs (Figs 1 and 2). I used all these Galápagos communities as given in Tables 2 and 3, in spite of their wide range of climatic conditions from dry to humid. It is clear that insular communities under all climatic conditions included have far lower diversities than comparable mainland communities and that no diversity trends can be seen within the set of insular communities (Figs 1 and 2). Clearly, alpha diversity of insular communities depends largely on factors other than the climate. One important factor is isolation.

Table 5 presents a summary of all diversity values. It shows once more that the insular communities are poor in species as compared with the mainland ones. The significant correlation between diversity and climatic favourableness in mainland communities can be interpreted so that the mainland communities are more or less saturated with plant species. The fact that the mainland communities studied had not been invaded by alien plants supports this interpretation. In contrast, the insular communities are not saturated up to and far below the richness level expected from the climatic favourableness.

The *S-N* curve is a graphic presentation of alpha diversity (Itow 1984, 1985, 1986; Itow *et al.* 1984). In Fig. 3 the curves of mainland and insular communities are compared. The level of $S_{(100)}$ shows community species richness and the slope of the curve (b in the present paper) shows the evenness in population size within the community. Apparently, the curves for the mainland communities are much steeper than those for the insular ones (Fig. 3).

The results of the present study do not elucidate

Table 5. Comparison of diversity values as averages and ranges between natural forest communities of mainlands and satellite islands in the warm-temperate, subtropical and tropical climates of East Asia and pacific South America.

	Mainlands		Satellites	
	Warm-Temperate			
	SW Japan		Kyushu satellites	
WI	90–130		116–143	
H'	3.34	(2.35– 3.94)	3.29	(3.27– 3.30)
d'	0.85	(0.71– 0.92)	0.85	(0.84– 0.85)
α	5.7	(3.0 – 8.4)	5.0	(4.9 – 5.1)
$S_{(100)}$	16.9	(10.7 –21.6)	15.8	(15.2 –16.4)
b	11.9	(5.8 –17.8)	7.3	(3.3 – 9.3)
	Subtropical			
	Ryukyu		Bonins	
WI	190–217		182–205	
H'	3.95	(3.51– 4.23)	2.58	(1.63– 3.43)
d'	0.89	(0.80– 0.94)	0.71	(0.49– 0.89)
α	11.6	(8.1 –16.4)	4.15	(2.39– 6.45)
$S_{(100)}$	26.0	(21.0 –32.2)	13.2	(9.0 –17.3)
b	22.5	(16.1 –34.7)	8.7	(5.7 –11.8)
	Tropical			
	Thailand		Micronesia	
WI	244		210–260	
H'	5.26	(5.20– 5.31)	2.46	(0.95– 3.30)
d'	0.96	(0.96– 0.96)	0.72	(0.29– 0.87)
α	28.0	(26.2 –30.0)	3.4	(1.8 – 4.9)
$S_{(100)}$	43.0	(41.5 –44.4)	11.7	(8.5 –15.0)
b	46.5	(44.4 –48.6)	7.4	(2.9 –10.5)
	Equatorial			
	South America			
	Amazonian upstreams		Galápagos	
WI			200–208	
H'	4.65	(4.41– 4.96)	1.52	(0.40 2.37)
d'	0.95	(0.94– 0.96)	0.46	(0.11– 0.70)
α	26.5	(22.6 –42.6)	2.1	(0.8 – 4.4)
$S_{(100)}$	40.5	(37.2 –42.6)	7.8	(4.0 –13.9)
b	41.8	(36.1 –45.0)	4.9	(1.4 –12.9)

possible causes for species poverty of insular communities. Here a special study is needed of the islands' geology and history and of the ecological consequences of species poverty, such as habitat expansion of plant species, invasion by aliens, and vulnerability of island ecosystems as a whole.

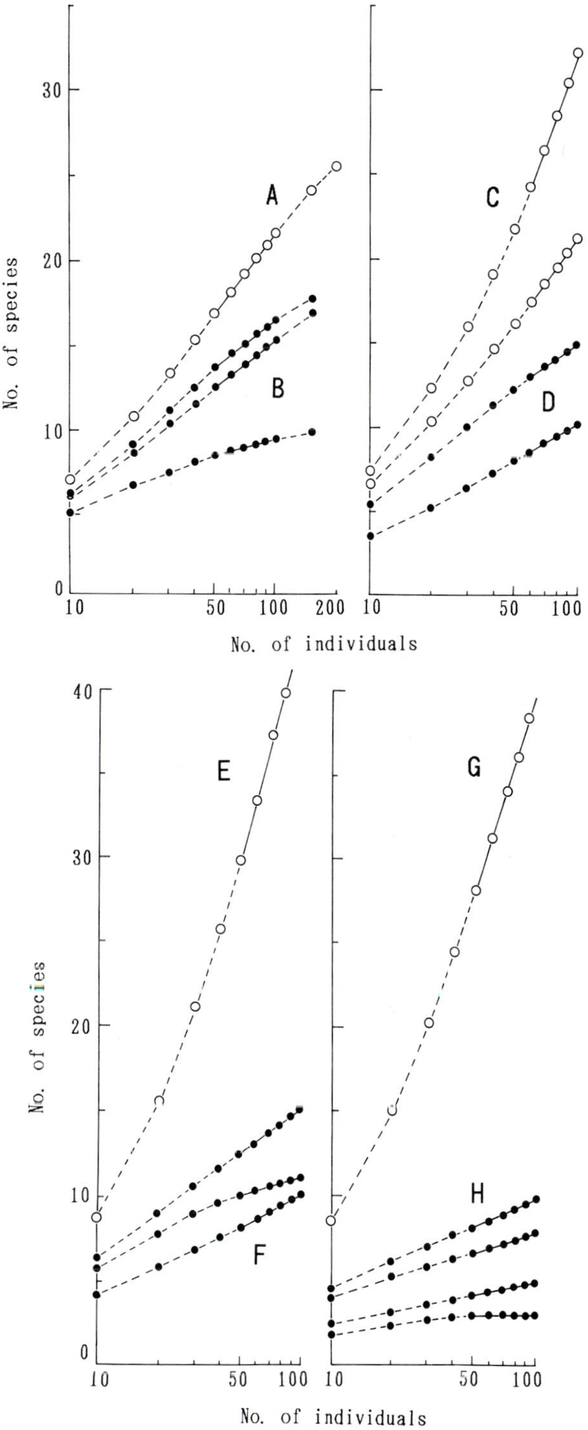

Fig. 3. *S-N* curves, comparing mainland (open circle) and insular (closed circle) communities. a. A–D: warm-temperate (A: Kyushu, B: Kyushu satellite islands) and subtropical climates (C: Ryukyu, D: Bonin); b. E–H: tropical climates (E: Thailand, F: Micronesia), and in equatorial South America (G: Amazon in Mainland Ecuador, H: Galápagos). The curves of the Thailand and Amazonian communities were based on calculations from Ogawa *et al.* (1965) and Grubb *et al.* (1963), respectively.

Acknowledgements

I am grateful to Hendrik Hoeck and Friedman Köster, directors of Charles Darwin Research Station, Galápagos, who kindly arranged my stay in the Station and field work on Isla Santa Cruz in 1978 and 1981, respectively.

References

Auclair, A. N. & Goff, F. G. 1971. Diversity relations of upland forests in the western Great Lakes area. Amer. Nat. 105: 499–528.

Bailey, K. 1976. Potassium-Argon ages from the Galápagos Islands. Science 192: 465–467.

Cottam, G. & Curtis, J. T. 1956. The use of distance measurements in phytosociological sampling. Ecology 37: 451–460.

Glenn-Lewin, D. C. 1977. Species diversity in North American temperate forests. Vegetatio 33: 153–162.

Grubb, P. J., Lloyd, J. R., Pennington, T. D. & Whitmore, T. C. 1963. A comparison of montane and lowland rain forest in Ecuador. I. The forest structure, physiognomy, and floristics. J. Ecol. 51: 567–601.

Hamann, O. 1979. On climatic conditions, vegetation types and leaf size in the Galápagos Islands. Biotropica 11: 101–122.

Hurlbert, S. H. 1971. The nonconcept of species diversity: A critique and alternative parameters. Ecology 52: 577–586.

Itow, S. 1971. A study of vegetation in Isla Santa Cruz, Galápagos Islands. Not. Galápagos 17: 10–13.

Itow, S. 1984. Species diversity of Fagaceae-absent evergreen broadleaf forests on three NW-Kyushu satellite islands. Jap. J. Ecol. 34: 225–228.

Itow, S. 1985. Species diversity-climate-isolation relationships in warm-temperate and subtropical evergreen broadleaved forests on Kyushu, the Ryukyus and the Bonins. In: Hara, H. (ed.), Origin and evolution of diversity in plants and plant communities. Academia, Tokyo.

Itow, S. 1986. Species diversity of equatorial insular forests on Ponape and Kosrae, Micronesia. Ecol. Res. 1: 223–227.

Itow, S. & Miyata, I. 1977. Species diversity of plant communities. In: Itow, S. (ed.), Species composition and structure of plant communities. Asakura, Tokyo. (In Japanese).

Itow, S., Ono, M. & Seki, T. 1984. Species diversity of sub-tropical evergreen broadleaf forests on the Ryukyu and the Bonin Islands. Jap. J. Ecol. 34: 467–472.

Kanehira, R. 1972. Flora Micronesica. Inoue-Shoten, Tokyo.

Kira, T. 1977. A climatological interpretation of Japanese vegetation zones. In: Miyawaki, A. & Tüxen, R. (ed.), Vegetation science and environmental protection. Maruzen, Tokyo.

Lawesson, J. E. 1988. The stand-level dieback and regeneration of forests in the Galapagos islands. Vegetatio.

Miyawaki, A. & Sasaki, Y. 1985. Floristic changes in the *Castanopsis cuspidata* var. *sieboldii*-forest communities along the Pacific Ocean coast of the Japanese Islands. Vegetatio 59: 225–234.

Nakano, K., Itow, S., Une, S. & Cornelius, G. 1986. Overview of habitat-types and horticultural land-use in Kosrae, Pohnpei and Truk Islands. Prompt Report of 4th Sci. Survey, South Pacific, 1–7. Kagoshima Univ., Kagoshima.

Ogawa, H. *et al.* 1965. Comparative ecological study on three main types of forest vegetation in Thailand. I. Structure and floristic composition. Nature and Life in SE Asia 4: 13–48.

Okutomi, K. *et al.* 1985. Vegetation of the Bonin Islands. In: Ono, M. & Okutomi, K. (eds), Endemic plants and vegetation of the Bonin Islands. Aboc-sha, Kamakura. (Japanese with English summary.)

Ono, M. & Sugawara, T. 1981. An analysis of the flowering plant flora of the Ogasawara (Bonin) Islands with their mode of dispersal. Ogasawara Res. 5: 25–40.

Peet, R. K. 1978. Forest vegetation of the Colorado front range: Patterns of species diversity. Vegetatio 37: 65–78.

Peet, R. K., Glenn-Lewin, D. C. & Walker Wolf, J. 1983. Prediction of man's impact on plant species diversity – A challenge for vegetation science. In: Holzner, W., Werger, M. J. A. & Ikushima, I. (eds), Man's impact on vegetation, pp. 41–54. Dr W. Junk, Dordrecht.

Shimizu, Y. 1983. Phenological studies of the subtropical broad-leaved evergreen forests at Chichijima Island in the Bonin (Ogasawara) Islands. Jap. J. Ecol. 33: 135–147.

Whittaker, R. H. 1972. Evolution and measurement of species diversity. Taxon 21: 213–251.

Whittaker, R. H. 1977. Evolution of species diversity in land communities. Evol. Biol. 10: 1–67.

Wiggins, I. L. & Porter, D. M. 1971. Flora of the Galápagos Islands. Stanford Univ. Press, Palo Alto.

Williams, C. B. 1947. The logarithmic series and its application to biological problems. J. Ecol. 34: 253–272.